Native Crops in Latin America

Food Biotechnology and Engineering
Series Editor:
Octavio Paredes-López

Bioenhancement and Fortification of Foods for a Healthy Diet
Edited by Octavio Paredes-López, Oleksandr Shevchenko, Viktor Stabnikov, and Volodymyr Ivanov

Volatile Compounds Formation in Specialty Beverages
Edited by Felipe Richter Reis and Caroline Mongruel Eleutério dos Santos

Native Crops in Latin America: Biochemical, Processing, and Nutraceutical Aspects
Edited by Ritva Repo-Carrasco-Valencia and Mabel C. Tomás

Native Crops in Latin America

Biochemical, Processing, and Nutraceutical Aspects

Edited by
Prof. Dr. Ritva Repo-Carrasco-Valencia
Principal Professor | Specialist in Andean Crops
Centro de Investigación e Innovación en Productos Derivados de Cultivos Andinos (CIINCA), Facultad de Industrias Alimentarias, Universidad Nacional Agraria La Molina (UNALM)

Prof. Dr. Mabel Cristina Tomás
Centro de Investigación y Desarrollo en Criotecnología de Alimentos (CIDCA)—Facultad de Ciencias Exactas (UNLP)-CONICET-CIC (Argentina)

CRC Press is an imprint of the
Taylor & Francis Group, an **informa** business

Cover description: Upper line: Quinoa (*Chenopodium quinoa*) Murtilla (*Ugni molinae* Turcz) Sacha inchi (*Plukenetia spp.*) Andean Maize Capia Garrapata

Bottom line: Lupin (*Lupinus spp.*) Amaranth (*Amaranthus spp.*) Chili Pepper (*Capsicum chinense*) Chia (*Salvia hispanica* L.) and seeds

First edition published 2022
by CRC Press
6000 Broken Sound Parkway NW, Suite 300, Boca Raton, FL 33487–2742

and by CRC Press
2 Park Square, Milton Park, Abingdon, Oxon, OX14 4RN

CRC Press is an imprint of Taylor & Francis Group, LLC

Library of Congress Cataloging-in-Publication Data
Names: Repo-Carrasco-Valencia, Ritva, editor. | Tomás, Mabel Cristina, editor.
Title: Native crops in Latin America : biochemical, processing and nutraceutical aspects / edited by Ritva Repo-Carrasco-Valencia, Principal Professor, Specialist in Andean crops, Centro de Investigación e Innovación en Productos Derivados de Cultivos, Andinos (CIINCA), Facultad de Industrias Alimentarias, Universidad Nacional Agraria La Molina (UNALM), Mabel Cristina Tomás, Centro de Investigación y Desarrollo en Criotecnología de Alimentos (CIDCA), Facultad de Ciencias Exactas (UNLP)-CONICET-CIC (Argentina).
Description: First edition. | Boca Raton : CRC Press, 2022. | Series: Food biotechnology and engineering series | Includes bibliographical references and index.
Identifiers: LCCN 2021037188 (print) | LCCN 2021037189 (ebook) | ISBN 9780367531409 (hardcover) | ISBN 9781032187778 (paperback) | ISBN 9781003087618 (ebook)
Subjects: LCSH: Crops—Latin America. | Crops—Nutrition—Latin America. | Crops—Health aspects—Latin America.
Classification: LCC SB99.L29 N38 2022 (print) | LCC SB99.L29 (ebook) | DDC 572/.429—dc23
LC record available at https://lccn.loc.gov/2021037188
LC ebook record available at https://lccn.loc.gov/2021037189

ISBN: 978-0-367-53140-9 (hbk)
ISBN: 978-1-032-18777-8 (pbk)
ISBN: 978-1-003-08761-8 (ebk)

DOI: 10.1201/9781003087618

Typeset in Kepler Std
by Apex CoVantage, LLC

Contents

Series Preface

BIOTECHNOLOGY—OUTSTANDING FACTS

The beginning of agriculture started about 12,000 years ago, and it has played a key role in food production ever since. We look to the farmers to provide the food we need but at the same time, now more than ever, to farm in a manner compatible with the preservation of the essential natural resources of the earth. Additionally, besides the remarkable positive aspects that farming has had throughout history, several undesirable consequences have been generated. The diversity of plants and animal species that inhabit the earth is decreasing. Intensified crop production has had undesirable effects on the environment (e.g., chemical contamination of groundwater, soil erosion, exhaustion of water reserves). If we do not improve the efficiency of crop production in the short term, we are likely to destroy the very resource base on which this production relies. Thus, the role of so-called sustainable agriculture in the developed and underdeveloped world, where farming practices are to be modified so that food production takes place in stable ecosystems, is expected to be of strategic importance in the future—but the future has already arrived.

The biotechnology of plants is a key player in these scenarios of the 21st century. Nowadays, especially molecular biotechnology is receiving increasing attention because it has the tools of innovation for the agriculture, food, chemical, and pharmaceutical industries. It provides the means to translate an understanding of, and ability to modify, plant development and reproduction into enhanced productivity of traditional and new products. Plant products from seeds, fruits, components, and extracts are being produced with better functional properties and longer shelf life; they need to be assimilated into commercial agriculture to offer new options to small (and more than small) industries and finally to consumers. Within these strategies it is imperative to select crops with larger proportions of edible parts as well, thus generating less waste; it is also imperative to consider the selection and development of a more environmentally friendly agriculture.

The development of research innovations for products is progressing, but the constraints of relatively long times to reach the marketplace, intellectual property rights, uncertain profitability of the products, consumer acceptance, and even caution and fear with which the public may view biotechnology are tempering the momentum of all but the most determined efforts. Nevertheless, it appears uncontestable that food biotechnology of plants and microbials is and will emerge as a strategic component to provide food crops and other products required for human well-being.

FOOD BIOTECHNOLOGY AND ENGINEERING SERIES/ OCTAVIO PAREDES-LOPEZ, PH.D., SERIES EDITOR

The Food Biotechnology and Engineering Series aims at addressing a range of topics around the edge of food biotechnology and the food microbial world. In the case of foods, it includes molecular biology, genetic engineering and metabolic aspects, science, chemistry, nutrition, medical foods and health ingredients, and processing and engineering with traditional- and innovative-based approaches. Environmental aspects to produce green foods cannot be left aside. At the world level, there are foods and beverages produced by different types of microbial technologies to which this series will give attention. It will also consider genetic modifications of microbial cells to produce nutraceutical ingredients, and advances of investigations on the human microbiome as related to diets.

NATIVE CROPS IN LATIN AMERICA: BIOCHEMICAL, PROCESSING, AND NUTRACEUTICAL ASPECTS EDITORS: RITVA REPO-CARRASCO-VALENCIA AND MABEL CRISTINA TOMÁS

As the editors of this book indicate, the supply of foods at the world level is mostly based in three crops—wheat, rice, and maize—with the risks in food production and with the nutritional restrictions involved. One of these major crops originated in Latin America; but this region has given birth to, or has acted as center of distribution of, a high number of additional food crop treasures (or so-called superfoods) that may play a key role in the nutrition of mankind at the world level, and at the same time, as rich sources of nutraceutical and medical compounds. This book aims to be of help in the opening of food strategies with much wider and intelligent options as compared to the past and present strategies.

This multiauthor book is meant to be a comprehensive set of reviews of our present knowledge of biotechnology, food science and technology, and nutrition, including traditional knowledge of Latin American societies, associated with the production and consumption of plant food crops. The advantage is that each subject is explored by an expert or experts in that field.

Mirroring the diversity of crops described in this book, including their outstanding potentials for food, nutrition, nutraceuticals, and medicine, the authors come from diverse institutions and organizations in Argentina, Chile, Finland, Peru, Spain, and the United States. This book will enrich the knowledge of the academicians, technical people, students, and communities involved, in different roles, with the key crops able to improve well-being among the people of the 21st century.

Thanks are due to the editors and to all of the authors for their excellent contributions to this book, and to the editorial staff of CRC Press, especially to Mr. Stephen Zollo and to Ms. Laura Piedrahita.

Preface

The global food supply is dominated by only three crops: wheat, rice, and maize. These grains contribute to nearly 60% of calories obtained by humans from plants. Plant genetic diversity has been abandoned as farmers worldwide have changed their local crops and cultivars for genetically uniform, high-yielding varieties; consequently, countries are dependent on the same staple crops today. Reliance on few grains presents a risk of crop failure and, from a nutritional point of view, it seems to be detrimental to long-term health. Nutritional issues, such as hidden hunger derived from micronutrient deficiencies, food allergies, and chronic diseases such as obesity and cardiovascular and degenerative diseases, represent an enormous challenge for the future.

Latin America is the region where several food crops originated, and which are nowadays cultivated worldwide, such as maize, common beans, potato, peanut, and tomato. However, there are many other crops that have not received much attention beyond their own place. These crops could be regarded as true natural treasures, or superfoods, for consumers due to their high nutritional and nutraceutical values. Andean lupin and maize, quinoa, kañiwa, amaranth, chia, sacha inchi, native cocoa, chili peppers, and native berries are all excellent sources of macro- and micronutrients, as well as bioactive compounds. These crops are traditionally consumed by local inhabitants directly, without further processing and transformation. Indigenous communities have valuable knowledge about the traditional use of these crops as food and medicinal sources; present-day scientific research seeks to recover this ancestral know-how.

Researchers from different countries have oriented their studies toward the potential use of these native crops as sources of bioactive and health-promoting compounds (proteins, hydrolysates, peptides, antioxidants, essential lipids, dietary fiber, pre- and probiotics) as ingredients and extracts in functional foods.

This book focuses on recent findings related to the impact of the research in diverse aspects: cardiovascular and gastrointestinal systems, gut microbiota, delivery systems, product development, and gastronomy. It may be a useful resource for food scientists, food technologists, nutritionists, ingredient manufacturers, technical people at industrial organizations, and health care professionals. It should be relevant knowledge for food science and nutrition departments in universities as well. We hope that this book will contribute to increasing the knowledge on Latin American crops, which may significantly influence the nutritional status, health, and well-being of consumers all over the world.

Ritva Repo-Carrasco-Valencia
Mabel Cristina Tomás

Acknowledgments

The editors would like to express their gratitude to CYTED (Spain) for its kind support through Project 119RT0567, Universidad Nacional Agraria La Molina (UNALM, Peru), Facultad de Ciencias Exactas de la Universidad Nacional de La Plata (UNLP), and CONICET (Argentina), and to Prof. Octavio Paredes-Lopez for his valuable advice. The editors' kind acknowledgment to Julio Mauricio Vidaurre Ruiz for designing the book cover and Luciana Magdalena Julio for her collaboration in the index creation.

Editor Biography

Ritva Repo-Carrasco-Valencia (PhD Food Chemistry, University of Turku; MSc. Cereal Chemistry and Technology, University of Helsinki) has been passionate about Andean native grains ever since she first learnt about them in the 1980s while living in Cusco, Peru, the capital of the old Inca Empire. Through her research, she has raised awareness of the numerous nutritional benefits of quinoa, amaranth, kañiwa and tarwi (Andean lupin). Although consumed by indigenous peoples for centuries, these grains had been shunned by the wider Peruvian population in favour of a Western diet for most of the past century. Today, their high nutritional value is better understood and quinoa is now ubiquitous in health stores worldwide.

Dr. Repo-Carrasco-Valencia is currently a professor and research scientist at the National Agrarian University La Molina (UNALM), in Lima, Peru. She is director of the Center of Innovation for Andean Grains at UNALM and leads various international research projects. Among these is the European Union-funded Protein2Food (P2F) project, which aims at developing innovative, cost-effective and resource-efficient plant proteins. In 2017, Dr Repo was awarded the Order of the Lion of Finland in recognition of her work.

Mabel C. Tomás, PhD, is a Principal Research Scientist of Consejo Nacional de Investigaciones Científicas y Técnicas (CONICET), and conducts her research activities at Centro de Investigación y Desarrollo at Criotecnología de Alimentos (CIDCA- CONICET-CIC-UNLP). (Argentina). She earned her degree in Chemistry at Facultad de Ciencias Exactas y Naturales of the Universidad de Buenos Aires (UBA (1981) and received her PhD in Chemical Sciences from the Universidad Nacional de La Plata (UNLP) (Argentina) (1988). She worked by post-doctoral research stages at Instituto del Frío and Instituto of Fermentaciones Industriales (CSIC, Spain) and Universidad Autónoma de Yucatán (Mexico).
She is a Professor of Food Toxicology at the Facultad de Ciencias Exactas (UNLP) and in the post-graduate Master in Food Technology and Hygiene with a great contribution in human resources formation (PhD, Master, and undergraduate Thesis Supervision) at (UNLP).

Her research interests are related to the development, characterization, and application of delivery systems of different bioactive compounds in foods. The development of functional foods enriched with omega-3 fatty acids, mainly from chia seeds, and their nutraceutical aspects represents for her activities an important challenge. The major area of her research is on functional lipids, (omega-3 fatty acids, phospholipids, lecithin), phenolic bioactive, and other natural antioxidants of plants and preservation technologies to retard lipid oxidation contributing with benefits on human health.

M. C. Tomás has published more than 70 research papers, edited 2 books, more than 50 technical reports for the oil and food industry. She has participated as a speaker in many scientific conferences with over 170 presentations and leads different research projects. She received awards like best papers awards at AOCS, the Willem van Nieuwenhuyzen Award of the International Lecithin and Phospholipid Society (ILPS), the National Academy of Pharmacy and Biochemistry of Argentina, the Argentine Forum of Nutrition and Health, and four honorable mentions from the Argentine Chemical Association related to the high level of several PhD. Dissertations under her supervision.

She has participated as a member of scientific societies serving as Chair and Vice Chair of the Phospholipid Division of AOCS, President of the International Society for Fat Research (ISF) (2013–2015), and President of the Scientific Committee for the World Congress of Oils and Fats and 31st ISF Lectureship Series in Rosario, Argentina. At the 2015 ISF, she organized Short Courses, chaired sessions, gave presentations, and contribute to young researchers who were able to attend the meeting. Also, she organized Short Courses, Session Chair and speaker at the XI and XIII Latin American Congress of Fats and Oils (ASAGA - AOCS) in Rosario and Buenos Aires in 2005 and 2009. She is a founding member of the International Chia i-link Network since 2015 and has participates in the CYTED la ValSe-Food or Iberoamerican Valuable Seeds Group since 2019 maintaining academic collaboration with more than 20 research groups from 16 countries.

Contributors

Alarcón Rivera, Rafael
Universidad de Lima, Peru

Añón, María Cristina
CIDCA-CONICET-UNLP
Buenos Aires, Argentina

Arancibia, Marcelo
CIESAL, Facultad de Medicina
Universidad de Valparaíso
Valparaíso, Chile

Cabanillas, Billy
Universidad Peruana Cayetano
Heredia, Peru

Chasquibol Silva, Nancy
Universidad de Lima
Lima, Peru

Cisneros, Fausto H.
USIL
Lima, Peru

Cisneros-Zevallos, Luis
Texas A&M University
College Station, Texas

Cotabarren, Juliana
CIPROVE, Fac. Cs. Exactas UNLP
Buenos Aires, Argentina

García Fillería, Susan F.
CIDCA-CONICET-UNLP
Buenos Aires, Argentina

Gimenez, Alejandra
UNJU
Jujuy, Argentina

Gómez-Coca, Raquel B.
Instituto de la Grasa CSIC Seville, Spain

Ixtaina, Vanesa Yanet
CIDCA-CONICET-UNLP
Buenos Aires, Argentina

Julio, Luciana Magdalena
CIDCA-CONICET-UNLP
Buenos Aires, Argentina

Lobo, Manuel
UNJU
Jujuy, Argentina

Lutz, Mariane
CIESAL, Facultad de Medicina
Universidad de Valparaíso
Valparaíso, Chile

Morales-Soriano, Eduardo
UNALM
Lima, Peru

Moreda, Wenceslao
Instituto de la Grasa CSIC Seville, Spain

Muñoz, Loreto A.
LabCial, Escuela de Ingeniería
Universidad Central de Chile
Santiago, Chile

Nardo, Agustina E.
CIDCA-CONICET-UNLP
Buenos Aires, Argentina

Obregón, Walter David
CIPROVE, Fac. Cs. Exactas UNLP
Buenos Aires, Argentina

Parisi, Mónica Graciela
UNLu
Buenos Aires, Argentina

Pérez-Camino, M. Carmen
Instituto de la Grasa CSIC
Seville, Spain

Portales, Rosario
Centro de Innovación del Cacao
Universidad Peruana Cayetano Heredia
Lima, Peru

Quiroga, Alejandra V.
CIDCA-CONICET-UNLP
Buenos Aires, Argentina

Repo-Carrasco-Valencia, Ritva
UNALM
Lima, Peru

Rojas, Rosario
Centro de Innovación del Cacao
Universidad Peruana Cayetano Heredia
Lima, Peru

Rosso, Adriana Mabel
UNLu
Buenos Aires, Argentina

Ruiz, Candy
Centro de Innovación del Cacao
Universidad Peruana Cayetano Heredia
Lima, Peru

Sabbione, Ana C.
CIDCA-CONICET-UNLP
Buenos Aires, Argentina

Sammán, Norma
UNJU
Jujuy, Argentina

Scilingo, Adriana A.
CIDCA-CONICET-UNLP
Buenos Aires, Argentina

Segundo, Cristina
UNJU
Jujuy, Argentina

Suárez, Santiago E.
CIDCA-CONICET-UNLP
Buenos Aires, Argentina

Suomela, Jukka-Pekka
University of Turku
Turku, Finland

Talavera, Martin J.
Kansas State University
Manhattan, Kansas

Tironi, Valeria A.
CIDCA-CONICET-UNLP
Buenos Aires, Argentina

Tomás, Mabel Cristina
CIDCA-CONICET-UNLP
Buenos Aires, Argentina

Ugás, Roberto
UNALM
Lima, Peru

Vidaurre-Ruiz, Julio Mauricio
UNALM
Lima, Peru

Chapter 1

Andean Native Grains, Quinoa, and Lupin as Sources of Bioactive Components

Jukka-Pekka Suomela,
Ritva Repo-Carrasco-Valencia, and Mariane Lutz

CONTENTS

1.1 INTRODUCTION

Quinoa (*Chenopodium quinoa*; Figure 1.1) is an indigenous crop from the Andean mountains (ranging from Ecuador to southern Chile). Quinoa together with canihua (*Chenopodium pallidicaule*) and kiwicha (*Amaranthus caudatus*) belong to the Amaranthaceae family of plants. Quinoa seeds contain ca. 60% starch and 6% lipids. The typical proximate composition of quinoa is presented in Table 1.1. The seeds contain a relatively high amount of protein (13%–17%) with a balanced essential amino acid composition compared with common cereals, where lysine and tryptophan are typically limiting amino acids. From the point of view of digestibility, bioavailability, available lysine, and net utilization, proteins of pseudocereals like quinoa are often a better option compared

DOI: 10.1201/9781003087618-1

Figure 1.1 Quinoa plant (*Chenopodium quinoa*).

Source: Photo taken by Ritva Repo-Carrasco-Valencia.

TABLE 1.1 PROXIMATE COMPOSITION OF QUINOA AND ANDEAN LUPIN (G/100 G DRY WEIGHT)

Grain	Protein	Fat	Ash	Fiber	Carbohydrates
Quinoa[a]	11.3–16.1	5.7–7.6	2.6–3.5	10.1	55.7–65.8
Lupin[b]	40.7–47.4	13.9–18.0	2.9–4.8	18.8	31.7–45.0

[a] IICA (2015), Alarcón et al. (2020).
[b] Córdova-Ramos et al. (2020), Johnson et al. (2017), Alarcón et al. (2020).

to cereal proteins, especially for people following a vegetarian diet. Quinoa seeds also contain a high amount of calcium, magnesium, iron, copper, and zinc (Ando et al. 2002; Repo-Carrasco et al. 2003; Covarrubias et al. 2020). Quinoa does not contain gluten and is rich in phenolic compounds that may have various physiological functions in the human body. Major phenolic compounds in quinoa include vanillic acid, ferulic acid (together with its derivatives), and flavonoids kaempferol and quercetin, as glycosides (Hemalatha et al. 2016; Tang et al. 2015).

Quinoa has been consumed for hundreds of years as cooked and roasted whole grains or flours in traditional foods such as cooked dishes, desserts, and salads. Nowadays there is a large selection of quinoa-based products available on the market. These include beverages, breakfast cereals, pastries, snacks, chocolates, gluten-free products, and baby foods (Pellegrini et al. 2018). The world trade of quinoa has grown fast in recent years; the United States and China are the major markets, but also in the EU countries there is a high demand for this grain. The growth has been due to a rise in health awareness and increased middle-class purchase power in Asia. Current export of quinoa from the Andean region is mostly in the form of unprocessed seeds. Quinoa seeds and their fractions exhibit radical scavenging activity evaluated by different antioxidant activity assays, which is at least partially due to phenolic compounds (Hemalatha et al. 2016). Boiling in water, a typical treatment of quinoa seeds in food preparation, may cause a significant loss of antioxidant capacity in water (Dini et al. 2010). In recent times, several quinoa varieties have been introduced to various "non-traditional" quinoa regions; hence, it is important to study the variability in the profiles and contents of secondary metabolites in different varieties and between different growth locations. Quinoa hulls make up to 12% of the kernel. Hulls are also the site where triterpenoid saponins, responsible for an unpleasant bitter taste, are mainly concentrated. Removal of saponins in food use is a necessary processing step. Saponins have antitumor, antioxidant, and antimicrobial activity and can potentially be used in chemical industry for food additives and detergent production (Woldemichael and Wink 2001; Kuljanabhagavad and Wink 2009).

The genus *Lupinus* is a member of the legume family Fabaceae. Oceania and Eurasia contribute over 90% of the 1.3 Mt annual world lupin production, followed by the countries of Africa (5%–7%) and South, North, and Central America (3%–5%; Lucas et al. 2015; Vargas-Guerrero et al. 2014). Lupin seeds are gluten-free, and when consumed they exert very low glycemic impact (Mattila et al. 2018). Lupin can be consumed after cooking as whole seeds but also as flour in the production of bread, gluten-free cakes, and dairy products, and in fermented form in tofu and tempeh. Lupin has been recognized as a highly nutritious grain providing relatively high quantity of proteins compared to traditional legumes, as well as a high content of essential fatty acids and dietary fiber (Carvajal-Larenas et al. 2016). The nutritive value of the seeds is dependent on species, genotype and location. An Andean indigenous lupine species, *L. mutabilis* (Figure 1.2), has the highest levels of seed oil of all domesticated lupin species (14%–24%; Hatzold et al. 1983; Santos et al. 1997). Typical proximate

Figure 1.2 Andean lupin plant (*Lupinus mutabilis*).

Source: Photo taken by Ritva Repo-Carrasco-Valencia.

composition of the seeds is presented in Table 1.1. All the other common lupin species (*L. angustifolius, L. albus*, and L. *luteus*) contain less oil: *L. albus* typically 10%–12% and *L. angustifolius* and *L. luteus* 5%–7% (Rybinski et al. 2018). Carvajal-Larenas et al. (2016) reported an average seed protein content (g/100 g dw) of 38.2 for *L. albus* (white lupin), 33.9 for *L. angustifolius* (narrow-leafed lupin), 42.2 for *L. luteus* (yellow lupin), and 43.3 for *L. mutabilis*. *L. mutabilis* is grown at altitudes between 2000 and 3850 m above sea level at the high Andean areas (mainly Ecuador, Bolivia, and Peru), and also in Colombia and Argentina. The average fiber content of lupin seeds varies from 8.2 g/100 g dw in *L. mutabilis* to 16.0 g/100 g dw in *L. angustifolius* (Carvajal-Larenas et al. 2016). Of the total amount of fiber, 75%–80% is soluble and 18%–25% insoluble (Bähr et al. 2014). Raffinose family (raffinose, stachyose, verbascose) oligosaccharides are a major dietary fiber fraction in lupin seeds, which could cause flatulence (Karnpanit et al. 2016). Dehulling decreases the fiber content. In-detail studies on the potentially bioactive components (e.g., polyphenols) in lupins are still to date rather scarce.

A limiting factor for wider use of some lupins, especially *L. mutabilis*, has been the high content of alkaloids that confer bitter taste to food products and may have acute anticholinergic toxicity, characterized by symptoms such as headache, nervousness, nausea, and relaxation of the nictitating membrane of the eye (Khan et al. 2015). Preliminary studies on their toxicity suggest the acute-lethal dose as present in lupin seeds is 10 mg/kg body weight (bw) for infants and 25 mg/kg bw for adults (Carvajal-Larenas et al. 2016). Unlike quinoa, in lupin the antinutrients are not concentrated only in the outer layers of the seeds. More than 170 quinozidine alkaloids have been identified in different *Lupinus* species, some of which are neurotoxic (Schrenk et al. 2019). Based on this, some countries have set a limit of 200 mg/kg of total lupine alkaloids in foods (Luckett 2010). Lupin alkaloids are used in Chinese and Japanese traditional medicine due to their pharmacological activity. New variants of *L. angustifolius*, also known as Australian sweet lupine, grown in largest quantities in Australia, Germany, and Poland, have only traces of alkaloids and therefore lack bitter taste and require no special treatment like soaking in salt water.

Germination and kilning processes have been studied in cereals such as wheat, sorghum, rice, and barley. However, there are only a few studies focused on evaluating the effect of the processes on phenolic and nutritional components in pseudocereals such as quinoa. Also, most of the studies have not identified the phenolic compounds generated during the germination and subsequent kilning process. The same applies to the studies on the effects of germination of lupin seeds on their bioactive compounds. The published data is reviewed in this chapter, most of which is very recent.

1.2 BIOACTIVE COMPONENTS IN QUINOA

Abderrahim et al. (2015) studied physical features, bioactive compounds, and total antioxidant capacity of colored quinoa varieties from Peruvian highlands (Altiplano). Total phenolic (1.23–3.24 mg gallic acid equivalents/g) and flavonol contents (0.47–2.55 mg quercetin equivalents/g) had a high correlation ($r = 0.910$). Betalain content (0.15–6.10 mg/100 g) positively correlated with darkness of the seeds (measured as L parameter), total phenolics, and flavonol content, measured spectrophotometrically. High total antioxidant capacity values (119.8–335.9 mmol Trolox equivalents/kg), measured by a modified direct QUENCHER-CUPRAC procedure, positively correlated with darkness, betalains ($r = 0.730$), as well as with free ($r = 0.639$), bound ($r = 0.558$), and total phenolic compounds ($r = 0.676$). Pereira et al. 2019 studied several types of black, red, and white quinoa; higher contents of tocopherols were found in black and red quinoa compared with white grains.

Repo-Carrasco-Valencia et al. (2010) studied phenolic compounds of various Peruvian quinoa samples. The phenolic acid content was low compared with common cereals like wheat and rye but was similar to levels found in oat, barley, corn, and rice. The flavonoid content was rather high, varying from 36.2 to 72.6 mg/100 g. In a study by Hirose et al. (2010), four flavonol glycosides were identified and quantified in Japanese quinoa seeds: quercetin and kaempferol 3-O-(2′,6′-di-O-alpha-rhamnopyranosyl)-beta-galactopyranosides, quercetin 3-O-(2′,6′-di-O-alpha-rhamnopyranosyl)-beta-glucopyranoside, and quercetin 3-O-(2′-O-beta-apiofuranosyl-6′-O-alpha-rhamnopyranosyl)-beta-galactopy ranoside. In a study by Hemalatha et al. (2016), milling of whole-grain white quinoa resulted in an approximate 30% loss of total phenolic content in milled grain. However, dehulled and milled grain fractions showed significantly higher metal chelating activity than other fractions. In addition to other phenolic compounds, the researchers detected daidzein, an isoflavone, in the whole grains but not in the milled grains.

In a study by Balakrishnan and Schneider (2020), 7 out of 11 flavonoid compounds (derivatives of the flavonols quercetin and kaempferol) identified by LC-electrospray ionization-MS^3 were found intact after *in vitro* gastrointestinal digestion. Their concentrations were significantly increased, likely as a result of being released from other substances in the seed material. Antioxidant capacity, measured as oxygen radical absorbance capacity (ORAC) assay and 2,2(1), diphenyl-1-picrylhydrazyl (DPPH) scavenging assay, increased approximately twofold as a result of the treatment. Pellegrini et al. (2018) studied white, red, and black varieties of quinoa by LC-MS/MS. In all samples analyzed after consecutive extraction with 80% methanol and 70% acetone, the main phenolic compound was 4-hydroxybenzoic acid, except for white Spanish quinoa, which also contained neohesperidin, a flavanone glycoside, as a major component. In the study, kaempferol was more abundant than quercetin. Total phenolic content (as measured by Folin-Ciocalteu reagent) as well as total flavonoid content (colorimetric measurement) were the smallest in white quinoa, but the differences did not reach significance.

Chen et al. (2019) studied 28 quinoa varieties cultivated in Washington State, and unlike most of the previous research, specifically focused on the oil fraction. The researchers found variability in phenolic and carotenoid contents, which was reflected in antioxidant capacity measurements (ferric reducing antioxidant capacity [FRAP] and DPPH-based assays). Interestingly, high amounts of phytosterols β-sitosterol, stigmasterol, and campesterol (range of averages of varieties 19.6–46.6, 8.7–20.1, and 0.4–2.0 mg/g oil, respectively) as well as squalene (range of averages of varieties 27.3–67.8 mg/g oil) were found in the samples.

Non-extractable polyphenols (NEPP), not accessible to organic solvents, are common for quinoa and various other seeds. NEPP are commonly associated with cell wall materials and bound to cellulose and hemicellulose; thus the amount of phenolic compounds can potentially be underestimated (Multari et al. 2018a; Chen et al. 2019). Therefore, caution is required when making conclusions based on results on phenolic composition reported in scientific articles. Bound phenolic compounds may be beneficial from a nutritional standpoint, as they are released in the colon by the action of enzymes or microbiota. Multari et al. (2018a) studied the effects of different drying temperatures (i.e., room temperature and the range from 40 °C to 70 °C at 10 °C intervals) on the phenolic and carotenoid contents of quinoa seeds. Drying of the seeds is a necessary step in their post-harvest processing. Both the extractable and non-extractable components were analyzed in the study. Drying at 70 °C allowed the greatest recovery of total cumulative phenolic compounds, 994 ± 28 mg/kg; ferulic acid and quercetin (deglycosylated due to hydrolysis) were the main phenolics identified. Drying at 60 °C allowed the greatest recovery of cumulative carotenoids, 2.39 ± 0.05 mg/kg. The carotenoids identified were xanthophylls (e.g., lutein, zeaxanthin, and neochrome). In a study by Miranda et al. (2010), thermal degradation, especially at 60 °C, 70 °C, and 80 °C, resulted in a notable reduction in total phenolic content. However, vitamin E showed an important increase at 70 °C and 80 °C. It needs to be taken into consideration that in both studies, the duration of heat treatment was shorter when the temperature was higher.

Liu et al. (2020) analyzed the composition of free and bound phenolic extracts from white, red, and black quinoa. Phenolic compounds were mainly in bound form in red and black quinoa and in free form in white quinoa. Red quinoa showed the highest total phenolic content and total flavonoid content, and black quinoa had the highest phenolic acids concentration. Red quinoa also showed the best radical scavenging activities measured by 2,2′-azino-bis(3-ethylbenzot hiazoline-6-sulfonate) or ABTS radical scavenging assay and DPPH assay. Based on the results of Laus et al. (2012) on bitter variety "Real" quinoa seeds may be a promising source of natural antioxidant compounds originating in particular from the free-soluble antioxidant fraction. Vega-Galvez et al. (2018) determined the contents of dietary fiber, polyphenols, flavonoids, and isoflavones as well as antioxidant and antimicrobial activity of six ecotypes of quinoa cultivated in three different zones of Chile. The results indicated that all Chilean quinoa ecotypes could be considered as good sources of dietary fiber (12.23 g/100 g dw) and polyphenols (161.32 mg gallic acid equivalents [GAE]/100 g dw). Also, the isoflavones daidzein and genistein were detected, the north and central ecotypes exhibiting the highest concentrations. In an earlier study of the research group (Lutz et al. 2013), a northern ecotype contained more isoflavones than central and southern ecotypes. In all samples, the concentration of daidzein (avg. 0.70–2.05 mg/100 g) was higher than that of genistein (avg. 0.04–0.41 mg/100 g). Results of Tang et al. (2015) on white, red, and black quinoa demonstrated the presence of at least 23 phenolic compounds found in either free or conjugated forms (liberated by alkaline and/or acid hydrolysis). The majority of these compounds were phenolic acids, mainly vanillic acid and ferulic acid and their derivatives as well as the flavonoids quercetin and kaempferol and their glycosides. Dark quinoa seeds had highest phenolic concentration and antioxidant activity (measured by FRAP and DPPH assays). In another study by Tang et al. (2016), the concentration of bound phenolics was highest in black quinoa followed by red and white quinoa, regardless of the hydrolysis method (acid, alkaline, or enzymatic). The authors conclude that the carbohydrases, which effectively liberated bound phenolics in the study, are known to be secreted by colonic bacteria and may thus exert antioxidant and anti-inflammatory effects in the large intestine during colonic fermentation.

Based on their results on white, red, and black quinoa, Tang et al. (2015) suggested that the pigments of red and black quinoa are betacyanins. Escribano et al. (2017) studied betalains in quinoa varieties of various colors, and identified betanin, isobetanin, amaranthin, and isoamaranthin in red-violet varieties. The presence of betalains correlated with the results of various antioxidant capacity assays (FRAP, ABTS, and ORAC). Dihydroxylated betaxanthins were detected both in red-violet and in yellow-orange grains. These molecules are of special interest due to their high antioxidant activities. In the Peruvian quinoa samples of various colors studied by Repo-Carrasco-Valencia et al. (2010), betalains were not detected.

In addition to flavonoid glycosides and other more familiar secondary bioactive metabolites, quinoa contains phytoecdosteroids, a class of triterpenoids. In some studies, they have demonstrated insulin-sensitizing, fat-reducing, and fitness-enhancing activities in mammals without inducing androgenic or estrogenic effects (Graf et al. 2016). In their study, Graf et al. (2016) investigated differences in phytoecdosteroid and flavonoid glycoside contents among 17 distinct quinoa sources and studied their correlations to genotypic and physicochemical characteristics. Phytoecdosteroid and flavonoid glycoside concentrations exhibited over fourfold differences across quinoa sources. Phytoecdosteroid content was positively and consistently correlated with oil content. Also, quinoa polysaccharides are potentially bioactive components, but systematic studies focusing on them have been scarce. Hu et al. (2017) optimized an ultrasound-assisted extraction process for quinoa polysaccharides and obtained a high content of a fraction with low molecular weight consisting of galacturonic acid and glucose monosaccharides, which exhibited significant antioxidant effect against DPPH and ABTS radicals.

Table 1.2 contains a summary of bioactive components identified in quinoa.

TABLE 1.2 BIOACTIVE COMPONENTS IDENTIFIED IN QUINOA AND/OR LUPIN

Bioactive component	Quinoa references	Lupin references
Betalains		
betaxanthins, betacyanins	Abderrahim et al. (2015), Escribano et al. (2017),[a] Tang et al. (2015)[b]	
Steroids		
β-sitosterol	Chen et al. (2019)	Kalogeropoulos et al. (2010),[jk] Rumiyati et al. (2013)[l]
stigmasterol	Chen et al. (2019)	Kalogeropoulos et al. (2010),[jk] Rumiyati et al. (2013)[l]
campesterol	Chen et al. (2019)	Kalogeropoulos et al. (2010),[jk] Rumiyati et al. (2013)[l]
Δ5-avenasterol		Kalogeropoulos et al. (2010)[jk]
20-hydroxyecdysone and structural analogs	Graf et al. (2016)	
Squalene	Chen et al. (2019)	Kalogeropoulos et al. (2010)[jk]
Carotenoids	Chen et al. (2019),[c] Multari et al. (2018a)[d]	Multari et al. (2018b)[m]
Tocopherols	Miranda et al. (2010),[e] Perreira et al. (2019)[e]	Lampart-Szczapa et al. (2003a)[en]

(*Continued*)

TABLE 1.2 (CONTINUED)

Bioactive component	Quinoa references	Lupin references
Phenolic acids		
caffeic acid	Repo-Carrasco-Valencia et al. (2010), Tang et al. (2015, 2016)	Siger et al. (2012)[o]
dicaffeoyl quinic acid		Zhong et al. (2019)[l]
chlorogenic acid (or derivatives)	Hemalatha et al. (2016), Tang et al. (2016)	Lampart-Szczapa et al. (2003b)[p]
cinnamic acid derivative		Zhong et al. (2019)[l]
ferulic acid (aglycone or glucoside)	Hemalatha et al. (2016), Liu et al. (2020), Multari et al. (2018a), Repo-Carrasco-Valencia et al. (2010), Tang et al. (2015, 2016)	Krol et al. (2018),[l] Lampart-Szczapa et al. (2003b),[n] Zhong et al. (2019),[l] Sosulski and Dabrowski (1984)[jq]
8,5′-diferulic acid	Tang et al. (2016)	
isoferulic acid	Liu et al. (2020), Tang et al. (2015)	
gallic acid	Hemalatha et al. (2016), Multari et al. (2018a)	Siger et al. (2012)[o]
p-coumaric acid (aglycone or glucoside)	Hemalatha et al. (2016), Liu et al. (2020), Repo-Carrasco-Valencia et al. (2010), Tang et al. (2015, 2016)	Krol et al. (2018),[l] Lampart-Szczapa et al. (2003b),[j] Siger et al. (2012),[o] Sosulski and Dabrowski (1984),[jr] Zhong et al. (2019)[l]
o-coumaric acid	Tang et al. (2016)	
p-hydroxy benzoic acid	Liu et al. (2020), Multari et al. (2018a), Repo-Carrasco-Valencia et al. (2010), Tang et al. (2015, 2016)	Lampart-Szczapa et al. (2003b),[o] Siger et al. (2012),[o] Sosulski and Dabrowski (1984)[js]
dihydroxy benzoic acids	Hemalatha et al. (2016), Tang et al. (2015, 2016)	
protocatechuic acid	Liu et al. (2020), Tang et al. (2016)	Magalhaes et al. (2017),[n] Lampart-Szczapa et al. (2003b),[o] Siger et al. (2012),[o] Zhong et al. (2019)[l]
quinic acid derivative		Zhong et al. (2019)[l]
rosmarinic acid	Tang et al. (2016)	

Bioactive component	Quinoa references	Lupin references
sinapinic acid	Hemalatha et al. (2016), Liu et al. (2020), Tang et al. (2016)	Krol et al. (2018)[l]
syringic acid	Liu et al. (2020), Tang et al. (2016)	Sosulski and Dabrowski (1984)[js]
vanillic acid (aglycone, glucoside, or other derivatives)	Hemalatha et al. (2016), Liu et al. (2020), Multari et al. (2018a), Pilco-Quesada et al. (2020), Repo-Carrasco-Valencia et al. (2010), Tang et al. (2015, 2016)	
Phenolic aldehydes		
syringaldehyde	Multari et al. (2018a)	
vanillin (or derivatives)	Tang et al. (2015, 2016)	Lampart-Szczapa et al. (2003b)[o]
Flavonols		
quercetin (as aglycones or glycosides)	Balakrishnan and Schneider (2020),[f] Hemalatha et al. (2016), Hirose et al. (2010),[g] Pilco-Quesada et al. (2020), Repo-Carrasco-Valencia et al. (2010),[h] Tang et al. (2015, 2016)	
kaempferol (as aglycones or glycosides)	Balakrishnan and Schneider (2020),[f] Hemalatha et al. (2016), Hirose et al. (2010),[g] Pilco-Quesada et al. (2020), Repo-Carrasco-Valencia et al. (2010),[h] Tang et al. (2015, 2016)	
myricetin	Hemalatha et al. (2016)	
Flavan-3-ols		
catechin	Hemalatha et al. (2016); Tang et al. (2016)	
epicatechin	Tang et al. (2015)	
epigallogatechin	Tang et al. (2015)	
Isoflavones		

(*Continued*)

TABLE 1.2 (CONTINUED)

Bioactive component	**Quinoa references**	**Lupin references**
daidzein	Hemalatha et al. (2016), Lutz et al. (2013),[i] Vega-Galvez et al. (2018)[i]	
genistein	Lutz et al. (2013), Vega-Galvez et al. (2018)	Cortés-Avendaño et al. (unpubl.),[t] Ranilla et al. (2009),[tu] Zhong et al. (2019)[l]
prunetin/biochanin A (derivatives)	Tang et al. (2015, 2016)	
puerarin	Tang et al. (2015)	
Flavones		
luteolin	Hemalatha et al. (2016)	
apigenin/apigenin-methylether/questin	Hemalatha et al. (2016), Pilco-Quesada et al. (2020), Tang et al. (2015)	Cortés-Avendaño et al. (unpubl.),[tv] Magalhaes et al. (2017),[ov] Siger et al. (2012),[ov] Zhong et al. (2019)[lv]
vicenin 2		Zhong et al. (2019)[l]
Flavanones		
naringenin/naringin	Hemalatha et al. (2016); Tang et al. (2015)	
Flavanonols		
aromadendrin glycoside		Zhong et al. (2019)[l]
Raffinose family oligosaccharides (raffinose, stachyose, verbascose)		Martinez-Villaluenga et al. (2005)[o]

[a] betacyanins (betanin, isobetanin, amaranthin, and isoamaranthin) detected in red-violet quinoa; [b] betacyanins (betanin and isobetanin) detected in red and black quinoa; [c] carotenoid species not specified; [d] zeaxanthin, lutein, neochrome A, and neochrome B detected; [e] gamma-tocopherol as the major species; [f] detected as various glycosides; [g] detected as triosides; [h]analyzed as aglycones; [i] among isoflavones, daidzein always more abundant than genistein; [j] in *L. albus*; [k] cooked lupin studied; [l] in *L. angustifolius*; [m] isomers of luteoxanthin, violaxanthin, lutein, zeaxanthin, β-carotene, and neoxanthine as well as epoxy-β-carotene; [n] in *L. luteus* and *L. albus*; [o] in *L. luteus*, *L. albus*, and *L. angustifolius*; [p] in *L. albus* and *L. angustifolius*; [q] *trans*-ferulic acid ester; [r] as trans-*p*-coumaric acid ester; [s] as ester; [t] in *L. mutabilis*; [u] also derivative; [v] as glycosides.

1.3 BIOACTIVE COMPONENTS IN LUPIN

Although the high nutritional quality of lupin has been recognized for decades, not much is known about the content of phytochemicals in this legume. Many of these compounds are bioactive and may exert beneficial effects on health. They include polyphenols, isoflavones, carotenoids, phytosterols, tocopherols, polysaccharides, and peptides with antioxidant, antimicrobial, anticarcinogenic, and anti-inflammatory activities, among others (Duenas et al. 2009; Campos-Vega et al. 2010; Khan et al. 2015; Zhu et al. 2018). Similarly to other constituents of lupin seeds, the relative amounts of phenolic compounds differ among genotypes, locations (Oomah et al. 2006), and agricultural conditions. Magalhaes et al. (2017) analyzed 29 European lupin seeds and other legumes and observed that *L. luteus* and *L. angustifolius* had the highest total phenolic content. In lupin, flavonoids were predominant over phenolic acids, whereas the reverse was the case for other species. Lampart-Szczapa et al. (2003a) analyzed the content of antioxidants (phenolic compounds and tocopherols) in flour obtained from Polish varieties of *L. luteus, L. albus*, and *L. angustifolius* seeds, produced under industrial conditions, to study the impact of milling and temperature. The researchers described the antibacterial activity of lupin seeds extracts on both gram-positive and gram-negative strains and observed that it is directly correlated with the amount of total polyphenols (Lampart-Szczapa et al. 2003b). As in other oil plants, dominance of γ-tocopherol over other tocopherols was observed; also, its proportion was higher in the yellow variety (88.5%) than in the white variety (86.1%). Pastor-Cavada et al. (2010) compared the antioxidant activity of four wild lupin species (*L. micranthus, L. hispanicus, L. angustifolius*, and *L. luteus*) and reported the highest value in *L. hispanicus* and the lowest value in *L. angustifolius*. Oomah et al. (2006) analyzed genotypes of *L. angustifolius* cultivated in Canada and found no association between parameters of antioxidant activity (antioxidant capacity, lag time, and antioxidant index) and phenolic content. On the contrary, Krol et al. (2018) observed that the phenolic compounds correlated well with the antioxidant capacity of the seeds of *L. angustifolium*. Also, the bitter cultivars contained more phenolic compounds than the sweet cultivars analyzed. However, it is difficult to compare the values reported by different authors, since the methodologies used to measure total polyphenols and antioxidant capacity vary.

Results by Sosulski and Dabrowski (1984) indicated that *L. albus* contains soluble esters of *trans*-ferulic, *p*-hydroxybenzoic, syringic, and trans-*p*-coumaric acids, while the hulls (12.6% of the whole-seed phenolics) consist mainly of *trans*-ferulic and *p*-hydroxybenzoic acids. Later, phenolic acids such as protocatechuic, *p*-hydroxybenzoic, chlorogenic, vanillic, *p*-coumaric, and ferulic acids were identified, and they have been shown to be mainly present in the external parts of the seed (Lampart-Szczapa et al. 2003a). Krol et al. (2018) described ferulic, sinapinic, and *p*-coumaric acids in seeds of *L. angustifolius*. Siger et al. (2012) studied the content of total phenolic compounds in *L. albus*, *L. luteus*, and *L. angustifolius* seeds and reported the highest content in *L. luteus* and the lowest in *L. albus*. The results demonstrate that the most abundant phenolic acids were protocatechuic in *L. luteus* and *p*-hydroxybenzoic in *L. angustifolius*. *L. luteus* also exhibited the highest content of apigenin glycosides. Zhong et al. (2019) identified three flavones (apigenin-7-O-β-apiofuranosyl-6,8-di-C-β-glucopyranoside, vicenin 2, and apigenin-7-O-β-glucopyranoside), isoflavone genistein, and a dihydroflavonol derivative (aromadendrin-6-C-β-D-glucopyranosyl-7-O-[β-D-apiofuranosyl-(1→2)]-O-β-D-glucopyranoside) along with several hydroxybenzoic and hydroxycinnamic acid derivatives in Australian sweet lupin (*L. angustifolius*) seed coats. The predominant phenolic compound in the seed coats analyzed was apigenin-7-O-β-apiofuranosyl-6,8-di-C-β-glucopyranoside.

Another type of antioxidant phenolic compounds in lupin seeds are isoflavones, which are categorized as phytoestrogens due to their structural similarity with estradiol (17-β-estradiol, E2). The main isoflavones in legumes are genistein (7,4′-dihydroxy-6-methoxyisoflavone), daidzein (7,4′-dihydroxyisoflavone), glycitein (7,4′-dihydroxy-6-methoxyisoflavone), biochanin A (5,7-dihydroxy-4′-methoxyisoflavone), and formononetin (7-h ydroxy-4′-methoxyisoflavone; Krizova et al. 2019). The intake of the major isoflavones (i.e., genistein and daidzein) may reduce risk factors of breast cancer and osteoporosis, the symptoms of menopause, and exert a hypocholesterolemic effect (Khan et al. 2015). Barcelo and Muñoz (1989) identified genistein, 20-hydroxygenistein, luteone, and wighteone in sprouted hypocotyls of *L. albus* cv multolupa, which may be related to the lignification of the cell wall. Sirtori et al. (2004) reported low levels of these compounds in *L. albus* seeds compared to the amount in soybeans. Ranilla et al. (2009) reported the presence of isoflavones only in *L. mutabilis*, while they were absent in *L. albus* and *L. angustifolius* seeds. These authors reported a total isoflavone content in *L. mutabilis* ranging from (mg genistein/100 g fw) 9.8 to 87 in seed coat, 16.1 to 30.8 in cotyledon, and 1.3 to 6.1 in hypocotyl fractions.

Peptides that may be related to hypotensive and lipid-lowering activities have been described in lupin protein hydrolysates (Bettzieche et al. 2009). Hypertension is a very common non-communicable disease that is treated with drugs to lower blood pressure, such as inhibitors of the angiotensin I converting enzyme (ACE). In addition to drugs, a variety of peptides obtained from plant food proteins act as ACE inhibitors; sources include soybean, quinoa, and lupin seeds (Boschin et al. 2014). A major characteristic of lupin proteins is the presence of conglutin-γ protein fraction, which is of interest in controlling insulin resistance and diabetes (Terruzzi et al. 2011).

Phytosterols constitute another type of bioactive phytochemicals in lupin beans. Ryan et al. (2007) reported the phytosterols content in various legumes. In most seeds, the main phytosterol is β-sitosterol that is accompanied by small quantities of campesterol, Δ5 avenasterol, and stigmasterol (Kalogeropoulos et al. 2010; Rumiyati et al. 2013). Lupin seeds also contain oligosaccharides of the raffinose family (raffinose, stachyose, and verbascose). The oligosaccharides are indigestible carbohydrates (α-galactosides) that are fermented in the colon and selectively stimulate the establishment and development of beneficial bacteria such as *Bifidobacterium*, which may have several beneficial effects on the health of the host (Gibson and Roberfroid 2008). In most cultivars of *L. albus, L. luteus*, and *L. angustifolius*, the quantity of stachyose was higher compared with other raffinose family oligochosaccharides and sucrose (Martinez-Villaluenga et al. 2005; Muzquiz et al. 2012).

Table 1.2 contains a summary of bioactive components identified in lupin.

1.4 EFFECT OF PROCESSING ON THE BIOACTIVE COMPONENTS OF QUINOA

Nickel et al. (2016) studied the effects of washing, washing followed by hydration, cooking (with or without pressure), and toasting on the phenolic content and antioxidant capacity (DPPH and FRAP) of quinoa grains. The highest content of phenolic compounds was obtained after cooking under pressure after washing; however, these components also increased with washing only. Toasting caused the greatest loss of phenolics. The antioxidant capacity of the grains was similarly increased by cooking treatments. Alvarez-Jubete et al. (2010) examined the polyphenol composition and antioxidant properties of methanolic extracts from quinoa and other grains and evaluated how these properties were affected by sprouting. Total phenolic content and antioxidant activity (DPPH and FRAP) increased in most samples as a result of sprouting. Laus et al. (2017) studied the effects of germination for 4 days at 20 °C and 70% humidity on phenolic content as well as on Trolox equivalent antioxidant capacity (TEAC) and ORAC to measure reducing activity and peroxyl radical scavenging capacity, respectively. Both TEAC and ORAC assays showed a significantly higher (about 2- and 2.8-fold, respectively) antioxidant capacity of sprouts compared to untreated seeds, while phenolic content of sprouts was about 2.6 times higher compared to untreated seeds.

In a study by Paucar-Menacho et al. (2018), the optimum conditions to maximize the content of total phenolic content as well as antioxidant activity

(ORAC assay) in quinoa were sprouting at 20 °C for 42 hours. Sprouts exhibited increase of 80% and 30% in total phenolic content and antioxidant activity, respectively, compared to non-germinated seeds. Kaempferol-O-dirhamnosyl-galactopyranose and quercetin-O-glucuronide experienced the most noticeable increase in quinoa after germination. In a study by Ujiroghene et al. (2019), yogurt was produced from germinated and ungerminated Chinese quinoa seeds of two different cultivars ('Mengli 1' and 'Mengli 2'). Total phenolic and total flavonoid contents were the highest in germinated quinoa yogurt of both cultivars. Germinated quinoa yogurt demonstrated high antioxidant capacities with DPPH, FRAP, ABTS+ radical scavenging, and ORAC assays.

Pilco-Quesada et al. (2020) studied the effect of germination and malting (germination with subsequent kilning) on the phenolic compounds and proximate composition of selected Peruvian varieties of quinoa ('Chullpi') and kiwicha ('Oscar Blanco'). Both germination and malting increased the protein content of the samples but decreased lipid content. Germination and kilning clearly increased the concentration of total phenolic compounds in both quinoa and kiwicha. Germination for 72 hours, either with or without the kilning process, resulted in a significant increase in the total content of phenolics compared to untreated materials, which was especially due to increase in coumaric acid and a kaempferol tri-glycoside in quinoa. Pachari et al. (2019) compared the tocopherol content of four Peruvian quinoa cultivars ('Amarilla de Maranganf', 'Blanca de Juli', INIA 415 'Roja Pasankalla', INIA 420 'Negra Collana') during germination. The quinoa cultivars from Puno and Cusco were germinated in the dark at 25 °C and analyzed for tocopherols by HPLC with a fluorescence detector during 72 hours of germination. Only the 'Blanca de Juli' white seed cultivar showed a decrease in total tocopherol content at the end of this treatment. Aguilar et al. (2019) studied the effects of malting on three varieties of quinoa: 'Negra Collana', 'Pasankalla Roja', and INIA 'Salcedo'. Malting consisted of hydration (4 hours at 25 °C), germination (48 hours at 25 °C), drying (24 hours at 55 °C), and deculming (radicle removing) steps. The process significantly affected the nutritional properties by increasing phenolic compounds, flavonoids, antioxidant capacity (DPPH-based assay), and ascorbic acid, and reducing sugars in all varieties. However, ash, protein, and fat contents were reduced due to germination, except in variety 'Negra Collana'.

Carciochi et al. (2016a) investigated the effect of malting process of quinoa seeds on the antioxidant compounds and antioxidant capacity. The researchers observed maximum increase in phenolic compounds, Maillard reaction products, and antioxidant activity (DPPH and reducing power) in samples germinated for 27 hours and subsequently roasted at 145 °C for 30 minutes. Treatment at a higher temperature (190 °C) decreased the levels of all evaluated variables.

In another study (Carciochi et al. 2016b), ascorbic acid and total tocopherols were significantly increased after 72 hours of germination process in comparison with raw quinoa seeds, whereas fermentation caused a decrease in the quantity of the compounds. Phenolic compounds and antioxidant capacity increased using both bioprocesses, but this effect was more noticeable for germination process (101% of increase after germination of 72 hours). Based on the results by Castro-Alba et al. (2019), fermentation may be effective for degrading phytate in pseudocereal flours and improving mineral accessibility. Motta et al. (2017) analyzed the contents of total folates (five folate isomers) by LC-MS/MS in quinoa, amaranth, and buckwheat seeds before and after cooking or malting. Boiling and steaming increased the total folate content in quinoa by 10%–15%, but decreased it in amaranth. Malting significantly increased total folate content in amaranth by 21% and buckwheat by 27%, but no significant change was observed in quinoa.

1.5 EFFECT OF PROCESSING ON THE BIOACTIVE COMPONENTS OF LUPIN

Fabaceae seeds are consumed both for their nutritional value as well as for their health benefits, and their processing may change these properties. Germination is a common and effective process for improving the quality of legumes, and germinated legumes are widely consumed all around the world (Gulewicz et al. 2014). Germination and fermentation are inexpensive means of improving the nutritional profile of lupin flour (Taraseviciene et al. 2019). Processing can reduce the levels of antinutritional and antidigestible factors and increase antioxidant capacity (Aguilera et al. 2013). In a study by Chilomer et al. (2013), the content of alkaloids and raffinose family oligosaccharides decreased in seeds of yellow and blue lupin as a result of germination.

During germination, various nutrients (vitamins and trace elements) and bioactive compounds such as γ-aminobutyric acid and polyphenols accumulate in seeds and sprouts (Gan et al. 2017). Germination increases the fiber content but decreases the lipid content of the seeds (Rumiyati et al. 2013). Andor et al. (2016) evaluated the presence of cinnamic acid derivatives and genistein in germinated and ungerminated seed extracts of *L. albus* and *L. angustifolius*, observing that germination can be employed for the augmentation of some bioactive secondary metabolites, such as isoflavones and cinnamic acid derivatives (ferulic, caffeic, rosmarinic, and coumaric acids) that may exert beneficial effects on health. Similarly, Danciu et al. (2017) studied the antiproliferative effects of the extracts obtained from the same germinated and ungerminated lupin seeds against A375 human melanoma cells and observed that germination increased the amount of bioactive compounds in the seeds of the two studied species of lupin. The analysis of the samples revealed the presence of genistein and cinnamic acids derivatives; however, the extracts showed only weak antiproliferative potential.

Duenas et al. (2009) studied the effect of germination of lupin seeds (*L. angustifolius*, cv Zapaton) on bioactive phenolic compounds and on the antioxidant activity. Phenolic compounds were analyzed by HPLC coupled with diode array detection and with electrospray ionization-MS. Germination produced changes in the profiles of flavonoids and non-flavonoid phenolic compounds with a significant increase in flavonoids. In the analyzed samples, isoflavones, flavones, and dihydroflavonols in free and conjugated forms were identified. An increase in the antioxidant activity (DPPH) was observed as a consequence of the process. Also, in a study by Fernandez-Orozco et al. (2006), germination of lupin seeds (*L. angustifolius*, cv Zapaton) enhanced their antioxidant capacity measured by superoxide dismutase-like activity (SOD-like activity), peroxyl radical-trapping capacity (PRTC), TEAC, and inhibition of lipid peroxidation in unilamellar liposomes of egg yolk phosphatidylcholine. The amount of glutathione decreased after 6 and 9 days of germination. Germinated seeds provided more vitamin C, vitamin E, and polyphenols than untreated seeds, of which the largest amounts were found after 6 days of germination.

Rumiyati et al. (2013) studied the effect of germination of *L. angustifolius* seeds for 9 days on the concentration of bioactive compounds and the radical scavenging activity in the resulting flour. In the methanolic extracts of germinated flour, phenolic contents and the antioxidant activity (DPPH) were increased following germination (700% and 1400%, respectively). In the oil extract, the concentration of phytosterols and the antioxidant activity were also increased compared to ungerminated lupin flour (300% and 800%, respectively). The relative proportions of β-sitosterol, stigmasterol, and campesterol were 60%, 30%, and 10%, respectively. Based on their results, Khan et al. (2018) concluded that germination of *L. angustifolius* seeds for a short period can be used as a pretreatment to obtain higher quantities of bioavailable phytochemicals such as phenolic compounds in fermented lupin tempeh. Mohammed et al. (2017) compared the effects of germination, boiling, and fermentation on alkaloids, phytate, total phenolics, and antioxidant activity (DPPH) of a newly developed lupin cultivar of *L. albus*. Regardless of the processing method and compared to untreated seeds, phytate level was significantly decreased while total phenolics significantly increased. Antiradical activity significantly increased for germinated seeds and decreased for fermented ones.

Table 1.3 contains a summary of the effects of processing on key components of quinoa and lupin.

TABLE 1.3 EFFECT OF PROCESSING ON KEY COMPONENTS OF QUINOA AND LUPIN

Grain/process	Effect	References
Quinoa/washing	Increase of phenolic compounds content	Nickel et al. (2016)
Quinoa/pearling	Decrease of phenolic compounds content	Gomez-Caravaca et al. (2014)
Quinoa/abrasive milling	Minimum loss of proteins, fat, and ash	D'Amico et al. (2019)
Quinoa/extrusion	Decrease of total dietary fiber, increase of soluble dietary fiber	Repo-Carrasco-Valencia and Astuahuman Cerna (2009)
Quinoa/roasting and boiling	No effect on mineral dialyzability	Repo-Carrasco-Valencia et al. (2010)
Quinoa/cooking	Increase of phenolic compounds content	Nickel et al. (2016)
Quinoa/ germinating and kilning	Increase of proteins and total phenolic compounds	Pilco-Quesada et al. (2020)
Quinoa/sprouting and germination	Increase of phenolic compounds content	Alvarez Jubete et al. (2010) Laus et al. (2017) Paucar et al. (2018)
Quinoa/malting	Increase of amino acid content	Motta et al. (2019)
Lupin/aqueous debittering	Decrease of alkaloids content	Carvajal-Larenas et al. (2014) Cortes-Avendaño et al. (2020)
Lupin/sprouting and germination	Increase of phenolic compounds content	Fernández Orozco et al. (2006) Duenas et al. (2009) Aguilera et al. (2013) Rumiyati et al. (2013) Andor et al. (2016) (Isoflavones) Carciochi et al. (2016a) Gan et al. (2017) Mohammed et al. (2017) Aguilar et al. (2019)
Lupin/germination	Decrease in phytic acid	Mohammed et al. (2017)
Lupin/germination	Increase of protein content; decrease of fats, carbohydrates, and alkaloids content	Dagnia et al. (1992)

(*Continued*)

TABLE 1.3 (CONTINUED)

Grain/process	Effect	References
Lupin/ germinating, cooking, soaking, and fermentation	Decrease of alkaloids content	Erbas, Mustafa (2010) Chilomer et al. (2013)
Lupin/debittering, extrusion, and spray-drying	Enhanced nutritional value and digestibility, without inducing relevant heat damage	Córdova et al. (2020)

1.6 ANTINUTRITIONAL FACTORS IN QUINOA

Amaranthaceae family plants are characterized by the presence of betalains, flavonoids, phenolic acids, essential oils, sesquiterpenes, diterpenes, and triterpenes. The species also contain high amounts of triterpenoid saponins. Saponins can be classified into two groups based on chemical structure of their aglycone skeleton (Vincken et al. 2007) in steroidal and triterpenoid saponins. Triterpenoid saponins include monodesmosidic molecules with one carbohydrate chain, bidesmosidic saponins with two carbohydrate chains, and tridesmosidic saponins with three carbohydrate chains. Quinoa saponins carry one or more carbohydrate chains, which basically consist of arabinose, glucose, galactose, glucuronic acid, xylose, and rhamnose glycosidically linked to a hydrophobic aglycone, mainly oleanane- and phytolaccagenane-type (Kuljanabhagavad and Wink 2009). Quinoa is the richest species in saponins within the Amaranthaceae family; it contains at least 26 different molecules, although some researchers estimate the presence of a minimum of 80 triterpenic compounds (Madl et al. 2006; Mroczek 2015).

The saponins of quinoa are located in pericarp. They cause a bitter taste and tend to foam in aqueous solutions. Saponins are reported to be toxic for cold-blooded animals and they have been used as fish poison by the inhabitants of South America (Zhu et al. 2002). Saponins exert some adverse physiological effects, since they are membranolytic on cells of the small intestine and possess hemolytic activity. Monodesmoside saponins exhibit higher hemolytic activity than bidesmoside saponins (Woldemichael and Wink 2001). Saponins form complexes with iron and zinc and may reduce the absorption of these minerals. Besides the negative effects, they also may exert positive effects by reducing serum cholesterol levels and by exerting anti-inflammatory and antioxidant activity as well as by having insecticidal, immunomodulatory, cytotoxic, antitumor, antimutagenic, antihepatotoxic, antidiabetic, hemolytic, antiviral, antibacterial, and fungicidal properties (Kuljanabhagavad and Wink 2009; Woldemichael and Wink 2001). Consequently, in recent years interest on the characterization of saponins in plants and the investigation of their biological properties has increased (Mroczek 2015). However, studies on the potential bioactivities of saponins have been conducted mainly on other species than quinoa. Estrada et al. (1998) indicated that quinoa saponins have a potential to act as immunological adjuvants in vaccines. Crude saponin fraction and individual saponins obtained from quinoa seeds were examined for antifungal activity against *Candida albicans* (Woldemichael and Wink 2001), and it was observed that the total saponin fraction of quinoa inhibited the growth of *C. albicans* at 50 lg/ml.

The quantity of saponins is variable between different quinoa types. Quinoa genotypes can be classified as bitter or sweet according to their saponin content. The bitter genotypes contain 4.7–11.3 g/kg of sapogenins (saponin aglycones), whereas sweet genotypes contain 0.2–0.4 g/kg sapogenins (Mastebroek et al. 2000). Before the consumption of quinoa seeds, saponins must be removed, which is traditionally done by washing the grains in running water. The main problem with this method is the high consumption of water and pollution of the rivers. To avoid these problems, saponins can be removed by physical abrading (pearling) of the seeds, the method used by most quinoa processing companies. However, it does not completely eliminate the bitter taste, and pearled quinoa must be washed in households before consumption. Gomez-Caravaca et al. (2014) studied the effect of pearling of quinoa on the residual saponin content in the grain. The saponin content after pearling was 50 mg/100 g, which is below the limit of bitter-tasting quinoa (110 mg/100 g; Koziol 1991). Various chromatographic approaches have been utilized in saponin analysis in order to develop analytical strategies allowing a rapid detection and saponin content determination (Mastebroek et al. 2000; Maria Gomez-Caravaca et al. 2011). These techniques could be particularly suitable for monitoring the content of bioactive saponins in plants, and also for quality control and screening of extracts designated for pharmaceutical, agricultural, and industrial applications (Mroczek 2015).

Like most grains, quinoa contains phytic acid. In cereals, phytic acid is located in the germ. In quinoa seeds, phytic acid is found in the external layers as well as in the endosperm. Phytate forms complexes with multivalent metal ions such as iron, calcium, magnesium, and zinc, consequently reducing their bioavailability. According to Ruales and Nair (1993), the content of phytic acid in quinoa seeds is about 1% of the dry matter. Scrubbing and washing reduce the phytic acid content of the seeds by about 30%. The authors detected neither protease inhibitors nor tannins in quinoa grains. In a study by Valencia-Chamorro (2003), phytic acid concentration in five quinoa varieties was in average 1.18 g/100 g, and trypsin inhibitor concentration in eight varieties of quinoa (range 1.36–5.04 TIU/mg) was lower than in soybean (24.5 TIU/mg).

1.7 ANTINUTRITIONAL FACTORS IN LUPIN

Wild and some cultivated varieties of lupin contain antinutrients, including bitter, toxic quinolizidine alkaloids, such as lupinine, lupanine, and α-isolupanine (Resta et al. 2008). Apart from being toxic, alkaloids have been described as immunosuppressive, anti-arrhythmic, and hypocholesterolemic agents that can exert teratogenic and anticholinergic effects (Carvajal-Larenas et al. 2016; Otterbach et al. 2019). Lupin seeds can accumulate up to 5% of their dry weight in the form of these alkaloids (Frick et al. 2017). The undesirable bitter compounds are a family of numerous secondary metabolites, and wild lupin species may contain over 10,000 mg/kg alkaloids. In 2019, the EFSA Panel on Contaminants in the Food Chain (CONTAM) considered that the impact of the uncertainties on the risk assessment of quinolizidine alkaloids (QA) in lupin seeds and lupin-derived products is substantial due to the limited data on toxicity, occurrence, and consumption (Schrenk et al. 2019), and recommended the generation of more data related to the toxicokinetics, toxicity, and occurrence of QA as well as to the consumption of lupin seeds and lupin-based foods in order to refine the risk assessment. The content of alkaloids can be reduced to very low amounts by breeding to obtain sweet varieties of lupin (Duranti and Gius 1997) as well as by post-harvest debittering treatments (e.g., germinating, cooking, soaking, fermentation, extraction) that improve the quality of the product as a food ingredient (Erbas 2010). The traditional treatment of lupin involves a process where the grain is boiled and subsequently subjected to extensive leaching in water or brine, which may also remove a large proportion of soluble proteins, minerals, flavonoids, monosaccharides, and sucrose from the seeds (Erbas 2010). According to Prusinski (2017), the total alkaloid content in sweet white lupin cultivars has been significantly reduced in the process of domestication and breeding. Cortés-Avendaño et al. (2020) described alkaloids from ten ecotypes from different areas of Peru before and after aqueous debittering process. Seeds were soaked with seed-to-water ratio 1:6 (w/v) for 12 hours at room temperature. Then, the seeds were cooked for 1 hour at atmospheric pressure (at 241 meters above sea level). During cooking, the water was changed once after 30 minutes. Following cooking, the seeds were washed with running water for 5 days. After washing, the seeds were dried at 50 °C for 18 hours. The results showed that from eight alkaloids identified before the debittering process, only very small amounts of lupanine (avg. 0.0012 g/100 g DM) and sparteine (avg. 0.0014 g/100 g dw) remained in the seeds after the debittering process, and no other alkaloids were detected. The aqueous debittering process reduced the content of alkaloids to levels far below the maximal level allowed by international regulations (≤0.2 g/kg dw). An aqueous debittering process of 5.7 days tested by Carvajal-Larenas et al. (2014) removed 94.9% of the total alkaloids, consumed water almost 62 times the weight of the untreated dry and bitter lupine, and caused a 22% loss of total solids—mostly fat, minerals, and carbohydrates. The alkaloids remain in the aqueous extract of lupins after debittering process. In their study, Santana et al. (2002) removed alkaloids from *L. albus* extract by bacterial treatment. For this purpose, two gram-negative bacterial strains capable of using lupanine as the only carbon source were isolated from soil in which *L. albus* and *L. luteus* had been grown. The results suggest these isolates can potentially remove lupanine and other quinolizidine alkaloids from the effluent and hence be used in wastewater treatment.

Lupin beans contain low amounts of lectins, protease inhibitors, and indigestible oligosaccharides. Lectins are glycoproteins that reduce the absorption of nutrients, and their content in lupin seeds is very low (3×10^{-5} hemagglutination activity units) compared to common beans (840×10^{-5} hemagglutination activity units; Nachbar and Oppenheim 1980), or are undetectable (Petterson and Fairbrother 1996). Bitter saponins are not present in white lupin (Petterson and Fairbrother 1996), while yellow and blue lupin contain 55.0–68.3 mg/kg and 270.1–469.5 mg/kg saponins, respectively (Muzquiz et al. 1993), which is far below the content of soybean seeds, which can reach 3500 mg/kg. Other antinutritional factors, such as trypsin inhibitors, are very limited in lupin (Erbas et al. 2005).

1.8 CONCLUSION

Scientific data on the bioactive components of various quinoa cultivars and different lupin species and varieties have started to accumulate in recent years. However, published literature focusing on the profile and quantities of various types of bioactive components is still very limited, particularly in case of lupin. New information on the effects of varietal differences and growing sites on the components is needed in order for the scientific community and product developers to make more educated choices to where to direct the research and development efforts. The crops have a great potential as part of a balanced diet, not only in the form of seeds or flour but also as value-added products that are ideally produced at the farm level in addition to factories, thus providing more income to the growers who often live in rural, low-income areas.

REFERENCES

Abderrahim, Fatima, Elizabeth Huanatico, Roger Segura, Silvia Arribas, M. Carmen Gonzalez, and Luis Condezo-Hoyos. 2015. "Physical Features, Phenolic Compounds, Betalains and Total Antioxidant Capacity of Coloured Quinoa Seeds (*Chenopodium quinoa* Willd.) from Peruvian Altiplano." *Food Chemistry* 183: 83–90. https://doi.org/10.1016/J.Foodchem.2015.03.029.

Aguilar, Julio, Alberto Claudio Miano, Jesus Obregon, Jose Soriano-Colchado, and Gabriela Barraza-Jauregui. 2019. "Malting Process as an Alternative to Obtain High Nutritional Quality Quinoa Flour." *Journal of Cereal Science* 90: 102858. https://doi.org/10.1016/J.Jcs.2019.102858.

Aguilera, Yolanda, Maria Felicia Diaz, Tania Jimenez, Vanesa Benitez, Teresa Herrera, Carmen Cuadrado, Mercedes Martin-Pedrosa, and Maria A. Martin-Cabrejas. 2013. "Changes in Nonnutritional Factors and Antioxidant Activity During Germination of Nonconventional Legumes." *Journal of Agricultural and Food Chemistry* 61 (34): 8120–8125. https://doi.org/10.1021/Jf4022652.

Alarcón, Marcelo, Michelle Bustos, Diego Méndez, Eduardo Fuentes, Iván Palomo and Mariane Lutz. 2020. "*In Vitro* Assay of Quinoa (*Chenopodium quinoa* Willd.) and Lupin (*Lupinus* spp.) Extracts on Human Platelet Aggregation." *Plant Foods for Human Nutrition* 75: 215–222. https://doi.org/10.1007/s11130-019-00786-y.

Alvarez-Jubete, Laura, Hilde Wijngaard, Elke. K. Arendt, and Eimear Gallagher. 2010. "Polyphenol Composition and *in Vitro* Antioxidant Activity of Amaranth, Quinoa, Buckwheat and Wheat as Affected by Sprouting and Baking." *Food Chemistry* 119 (2): 770–778. https://doi.org/10.1016/J.Foodchem.2009.07.032.

Ando, Hitomi, Yi Chun Chen, Hanjun Tang, Mayumi Shimizu, Katsumi Watanabe, and Toshio Mitsunaga. 2002. "Food Components in Fractions of Quinoa Seed." *Food Science and Technology Research* 8 (1): 80–84. https://doi.org/10.3136/Fstr.8.80.

Andor, Bogdan, Corina Danciu, Ersilia Alexa, Istvan Zupko, Elena Hogea, Andreea Cioca, Dorina Coricovac, et al. 2016. "Germinated and Ungerminated Seeds Extract from Two Lupinus Species: Biological Compounds Characterization and *in Vitro* and *in Vivo* Evaluations." *Evidence-Based Complementary And Alternative Medicine*: 7638542. https://doi.org/10.1155/2016/7638542.

Bähr, Melanie, Anita Fechner, Katrin Hasenkopf, Stephanie Mittermaier, and Gerhard Jahreis. 2014. "Chemical Composition of Dehulled Seeds of Selected Lupin Cultivars in Comparison to Pea and Soya Bean." *LWT-Food Science and Technology* 59 (1): 587–590. https://doi.org/10.1016/J.Lwt.2014.05.026.

Balakrishnan, Gayathri, and Renee Goodrich Schneider. 2020. "Quinoa Flavonoids and Their Bioaccessibility During in Vitro Gastrointestinal Digestion." *Journal of Cereal Science* 95 (9): 103070. https://doi.org/Rg/10.1016/J.Jcs.2020.103070.

Barcelo, Antonia R., and Romualdo Munoz. 1989. "Epigenetic Control of a Cell-Wall Scopoletin Peroxidase by Lupisoflavone in Lupinus." *Phytochemistry* 28 (5): 1331–1333. https://doi.org/10.1016/S0031-9422(00)97740-9.

Bettzieche, Anja, Corinna Brandsch, Klaus Eder, and Gabriele I. Stangl. 2009. "Lupin Protein Acts Hypocholesterolemic and Increases Milk Fat Content in Lactating Rats by Influencing the Expression of Genes Involved in Cholesterol Homeostasis and Triglyceride Synthesis." *Molecular Nutrition & Food Research* 53 (9): 1134–1142. https://doi.org/10.1002/Mnfr.200800393.

Boschin, Giovanna, Graziana Maria Scigliuolo, Donatella Resta, and Anna Arnoldi. 2014. "Ace-Inhibitory Activity of Enzymatic Protein Hydrolysates from Lupin and Other Legumes." *Food Chemistry* 145: 34–40. https://doi.org/10.1016/J.Foodchem.2013.07.076.

Campos-Vega, Rocio, Guadalupe Loarca-Pina, and B. Dave Oomah. 2010. "Minor Components of Pulses and Their Potential Impact on Human Health." *Food Research International* 43 (2): 461–482. https://doi.org/10.1016/J.Foodres.2009.09.004.

Carciochi, Ramiro, Krasimir Dimitrov, and Leandro Galvan Dalessandro. 2016. "Effect of Malting Conditions on Phenolic Content, Maillard Reaction Products Formation, and Antioxidant Activity of Quinoa Seeds." *Journal of Food Science and Technology-Mysore* 53 (11): 3978–3985. https://doi.org/10.1007/S13197-016-2393-7.

Carciochi, Ramiro Ariel, Leandro Galvan-D'Alessandro, Pierre Vandendriessche, and Sylvie Chollet. 2016b. "Effect of Germination and Fermentation Process on the Antioxidant Compounds of Quinoa Seeds." *Plant Foods for Human Nutrition* 71 (4): 361–367. https://doi.org/10.1007/S11130-016-0567-0.

Carvajal-Larenas, Francisco, Anita Linnemann, Martinus Nout, Michael Koziol, and Tiny Van Boekel. 2016. "*Lupinus mutabilis*: Composition, Uses, Toxicology, and Debittering." *Critical Reviews in Food Science and Nutrition* 56 (9): 1454–1487. https://doi.org/10.1080/10408398.2013.772089.

Carvajal-Larenas, Francisco, Tiny Van Boekel, Michael Koziol, Martinus Nout, and Anita Linnemann. 2014. "Effect of Processing on the Diffusion of Alkaloids and Quality of *Lupinus Mutabilis* Sweet." *Journal of Food Processing and Preservation* 38 (4): 1461–1471. https://doi.org/10.1111/Jfpp.12105.

Castro-Alba, Vanesa, Claudia E. Lazarte, Daysi Perez-Rea, Nils-Gunnar Carlsson, Annette Almgren, Bjorn Bergenståhl, and Yvonne Granfeldt. 2019. "Fermentation of Pseudocereals Quinoa, Canihua, and Amaranth to Improve Mineral Accessibility Through Degradation of Phytate." *Journal of the Science of Food and Agriculture* 99 (11): 5239–5248. https://doi.org/10.1002/Jsfa.9793.

Chen, Yu-Shuo, Nicole A. Aluwi, Steven R. Saunders, Girish M. Ganjyal, and Ilce G. Medina-Meza. 2019. "Metabolic Fingerprinting Unveils Quinoa Oil as a Source of Bioactive Phytochemicals." *Food Chemistry* 286: 592–599. https://doi.org/10.1016/J.Foodchem.2019.02.016.

Chilomer, Katarzyna, Malgorzata Kasprowicz-Potocka, Piotr Gulewicz, and Andrzej Frankiewicz. 2013. "The Influence of Lupin Seed Germination on the Chemical Composition and Standardized Ileal Digestibility of Protein and Amino Acids in Pigs." *Journal of Animal Physiology and Animal Nutrition* 97 (4): 639–646. https://doi.org/10.1111/J.1439-0396.2012.01304.X.

Córdova-Ramos, Javier, Patricia Glorio-Paulet, Félix Camarena, Andrea Brandolini and Alyssa Hidalgo. 2020. "Andean Lupin (*Lupinus mutabilis* Sweet): Processing Effects on Chemical Composition, Heat Damage and *in vitro* Protein Digestibility." *Cereal Chemistry* 97: 827–835. https://doi.org/10.1002/CCHE.10303.

Cortes-Avendaño, Paola, Marko Tarvainen, Jukka-Pekka Suomela, Patricia Glorio-Paulet, Baoru Yang, and Ritva Repo-Carrasco-Valencia. 2020. "Profile and Content of Residual Alkaloids in Ten Ecotypes of *Lupinus mutabilis* Sweet After Aqueous Debittering Process." *Plant Foods for Human Nutrition* 75 (2): 184–191. https://doi.org/10.1007/S11130-020-00799-Y.

Covarrubias, Nazar, Soraya Sandoval, Javier Vera, Claudia Núñez, Christian Alfaro, and Mariane Lutz. 2020. "Moisture, Protein and Mineral Content of Ten Varieties of Chilean Quinoa grown in Different Geographic Zones." *Revista Chilena de Nutricion* 47 (5): 730–737. https://doi.org/10.4067/S0717-75182020000500730.

Dagnia, Suzanne G., David S. Petterson, Roma R. Bell, and Frank V. Flanagan. 1992. "Germination Alters the Chemical-Composition and Protein-Quality of Lupin Seeds." *Journal of the Science of Food and Agriculture* 60 (4): 419–423. doi:10.1002/jsfa.2740600403.

D'Amico, Stefano, Sarah Jungkunz, Gabor Balasz, Maike Foeste, Mario Jekle, Sandor Tömösköszi, and Regine Schoenlechner. 2019. "Abrasive Milling of Quinoa: Study on the Distribution of Selected Nutrients and Proteins Within the Quinoa Seed Kernel." *Journal of Cereal Science* 86: 132–138. https://doi.org/10.1016/j.jcs.2019.01.007.

Danciu, Corina, Ioana Zinuca Pavel, Roxana Babuta, Alexa Ersilia, Oana Suciu, Georgeta Pop, Codruta Soica, Cristina Dehelean, and Isidora Radulov. 2017. "Total Phenolic Content, FTIR Analysis, and Antiproliferative Evaluation of Lupin Seeds Harvest from Western Romania." *Annals of Agricultural And Environmental Medicine* 24 (4): 726–731. https://doi.org/10.26444/Aaem/80795.

Dini, Irene, Gian Carlo Tenore, and Antonio Dini. 2010. "Antioxidant Compound Contents and Antioxidant Activity Before and After Cooking in Sweet and Bitter *Chenopodium Quinoa* Seeds." *LWT-Food Science and Technology* 43 (3): 447–451. https://doi.org/10.1016/J.Lwt.2009.09.010.

Duenas, Monserrat, Teresa Hernandez, Isabel Estrella, and David Fernandez. 2009. "Germination as a Process to Increase the Polyphenol Content and Antioxidant Activity of Lupin Seeds (*Lupinus angustifolius* L.)." *Food Chemistry* 117 (4): 599–607. https://doi.org/10.1016/J.Foodchem.2009.04.051.

Duranti, Matteo, and Cristina Gius. 1997. "Legume Seeds: Protein Content and Nutritional Value." *Field Crops Research* 53 (1–3): 31–45. https://doi.org/10.1016/S0378-4290(97)00021-X.

Erbas, Mustafa. 2010. "The Effects of Different Debittering Methods on the Production of Lupin Bean Snack from Bitter *Lupinus albus* L. Seeds." *Journal of Food Quality* 33 (6): 742–757. https://doi.org/10.1111/J.1745-4557.2010.00347.X.

Erbas, Mustafa, Mustafa Certel, and Mustafa Uslu. 2005. "Some Chemical Properties of White Lupin Seeds (*Lupinus albus* L.)." *Food Chemistry* 89 (3): 341–345. https://doi.org/10.1016/J.Foodchem.2004.02.040.

Escribano, Josefa, Juana Cabanes, Mercedes Jimenez-Atienzar, Martha Ibanez-Tremolada, Luz Rayda Gomez-Pando, Francisco Garcia-Carmona, and Fernando Gandia-Herrero. 2017. "Characterization of Betalains, Saponins and Antioxidant Power in Differently Colored Quinoa (*Chenopodium quinoa*) Varieties." *Food Chemistry* 234: 285–294. https://doi.org/10.1016/J.Foodchem.2017.04.187.

Estrada, Alberto, Bing Li, and Bernard Laarveld. 1998. "Adjuvant Action of *Chenopodium quinoa* Saponins on the Induction of Antibody Responses to Intragastric and Intranasal Administered Antigens in Mice." *Comparative Immunology Microbiology and Infectious Diseases* 21 (3): 225–236. https://doi.org/10.1016/S0147-9571(97)00030-1.

Fernandez-Orozco, Rebeca, Mariusz K. Piskula, Henryk Zielinski, Halina Kozlowska, Juana Frias, and Concepcion Vidal-Valverde. 2006. "Germination as a Process to Improve the Antioxidant Capacity of *Lupinus angustifolius* L. Var. Zapaton." *European Food Research and Technology* 223 (4): 495–502. https://doi.org/10.1007/S00217-005-0229-1.

Frick, Karen M., Lars G. Kamphuis, Kadambot H. M. Siddique, Karam B. Singh, and Rhonda C. Foley. 2017. "Quinolizidine Alkaloid Biosynthesis in Lupins and Prospects for Grain Quality Improvement." *Frontiers in Plant Science* 8: 87. https://doi.org/10.3389/Fpls.2017.00087.

Gan, Ren-You, Wing-Yee Lui, Kao Wu, Chak-Lun Chan, Shu-Hong Dai, Zhong-Quan Sui, and Harold Corke. 2017. "Bioactive Compounds and Bioactivities of Germinated

Edible Seeds and Sprouts: An Updated Review." *Trends in Food Science & Technology* 59: 1–14. https://doi.org/10.1016/J.Tifs.2016.11.010.

Gibson, Glean R., and Marcle B. Roberfroid. 2008. *Handbook of Prebiotics*. Boca Raton, FL: CRC Press.

Gomez-Caravaca, Ana, Giovanna Iafelice, Vito Verardo, Emanuele Marconi, and Maria Fiorenza Caboni. 2014. "Influence of Pearling Process on Phenolic and Saponin Content in Quinoa (*Chenopodium quinoa* Willd)." *Food Chemistry* 157: 174–178. https://doi.org/10.1016/J.Foodchem.2014.02.023.

Gomez-Caravaca, Ana, Antonio Segura-Carretero, Alberto Fernandez-Gutierrez, and Maria Fiorenza Caboni. 2011. "Simultaneous Determination of Phenolic Compounds and Saponins in Quinoa (*Chenopodium quinoa* Willd) by a Liquid Chromatography-Diode Array Detection-Electrospray Ionization-Time-of-Flight Mass Spectrometry Methodology." *Journal of Agricultural and Food Chemistry* 59 (20): 10815–10825. https://doi.org/10.1021/Jf202224j.

Graf, Brittany L., Leonel E. Rojo, Jose Delatorre-Herrera, Alexander Poulev, Camila Calfio, and Ilya Raskin. 2016. "Phytoecdysteroids and Flavonoid Glycosides Among Chilean and Commercial Sources of *Chenopodium quinoa*: Variation and Correlation to Physico-Chemical Characteristics." *Journal of the Science of Food and Agriculture* 96 (2): 633–643. https://doi.org/10.1002/Jsfa.7134.

Gulewicz, Piotr, Cristina Martinez-Villaluenga, Malgorzata Kasprowicz-Potocka, and Juana Frias. 2014. "Non-Nutritive Compounds in Fabaceae Family Seeds and the Improvement of Their Nutritional Quality by Traditional Processing—A Review." *Polish Journal of Food And Nutrition Sciences* 64 (2): 75–89. https://doi.org/10.2478/V10222-012-0098-9.

Hatzold, Thomas, Ibrahim Elmadfa, and Rainer Gross. 1983. "Edible Oil and Protein-Concentrate from *Lupinus mutabilis*." *Qualitas Plantarum-Plant Foods for Human Nutrition* 32 (2): 125–132. https://doi.org/10.1007/Bf01091333.

Hemalatha, P., Dikki Pedenla Bomzan, B. V. Sathyendra Rao, and Yadahally N. Sreerama. 2016. "Distribution of Phenolic Antioxidants in Whole and Milled Fractions of Quinoa and Their Inhibitory Effects on Alpha-Amylase And Alpha-Glucosidase Activities." *Food Chemistry* 199: 330–338. https://doi.org/10.1016/J.Foodchem.2015.12.025.

Hirose, Yuko, Tomoyuki Fujita, Toshiyuki Ishii, and Naoya Ueno. 2010. "Antioxidative Properties and Flavonoid Composition of *Chenopodium quinoa* Seeds Cultivated in Japan." *Food Chemistry* 119 (4): 1300–1306. https://doi.org/10.1016/J.Foodchem.2009.09.008.

Hu, Yichen, Jinming Zhang, Liang Zou, Chaomei Fu, Peng Li, and Gang Zhao. 2017. "Chemical Characterization, Antioxidant, Immune-Regulating and Anticancer Activities of a Novel Bioactive Polysaccharide from *Chenopodium quinoa* Seeds." *International Journal of Biological Macromolecules* 99: 622–629. https://doi.org/10.1016/J.Ijbiomac.2017.03.019.

IICA, Instituto Interamericano de Cooperación para la Agricultura. 2015. *El mercado y producción de quinua en el Perú*. Lima, Perú: IICA.

Johnson Stuart K., Jonathan Clements, Casiana B. J. Villarino, and Rarnil Coorey. 2017. "Lupins: Their Unique Nutritional and Health-Promoting Attributes." In John R. N. Taylor and Joseph M. Awika (Eds.), *Gluten-Free Ancient grains*. Cambridge: Woodhead Publishing Series in Food Science Technology and Nutrition, 179–221.

Kalogeropoulos, Nick, Antonia Chiou, Maria Ioannou, Vaios T. Karathanos, Maria Hassapidou, and Nikolaos K. Andrikopoulos. 2010. "Nutritional Evaluation and Bioactive Microconstituents (Phytosterols, Tocopherols, Polyphenols, Triterpenic Acids) in Cooked Dry Legumes Usually Consumed in the Mediterranean Countries." *Food Chemistry* 121 (3): 682–690. https://doi.org/10.1016/J.Foodchem.2010.01.005.

Karnpanit, Weeraya, Ranil Coorey, Jon Clements, Syed M. Nasar-Abbas, Muhammad K. Khan, and Vijay Jayasena. 2016. "Effect of Cultivar, Cultivation Year and Dehulling on Raffinose Family Oligosaccharides in Australian Sweet Lupin (*Lupinus angustifolius* L.)." *International Journal of Food Science And Technology* 51 (6): 1386–1392. https://doi.org/10.1111/Ijfs.13094.

Khan, Muhammad Kamran, Weeraya Karnpanit, Syed M. Nasar-Abbas, Zill-E-Huma, and Vijay Jayasena. 2015. "Phytochemical Composition and Bioactivities of Lupin: A Review." *International Journal of Food Science and Technology* 50 (9): 2004–2012. https://doi.org/10.1111/Ijfs.12796.

Khan, Muhammad Kamran, Weeraya Karnpanit, Syed M. Nasar-Abbas, Zill-E Huma, and Vijay Jayasena. 2018. "Development of a Fermented Product with Higher Phenolic Compounds and Lower Anti-Nutritional Factors from Germinated Lupin (*Lupinus Angustifolius* L.)." *Journal of Food Processing and Preservation* 42 (12): E13843. https://doi.org/10.1111/Jfpp.13843.

Koziol, Michael. 1991. "Afrosimetric Estimation of Threshold Saponin Concentration for Bitterness in Quinoa (*Chenopodium Quinoa* Willd)." *Journal of the Science of Food and Agriculture* 54 (2): 211–219. https://doi.org/10.1002/Jsfa.2740540206.

Krizova, Ludmila, Katerina Dadakova, Jitka Kasparovska, and Tomas Kasparovsky. 2019. "Isoflavones." *Molecules* 24 (6): 1076. https://doi.org/10.3390/Molecules24061076.

Krol, Angelika, Ryszard Amarowicz, and Stanislaw Weidner. 2018. "Content of Phenolic Compounds and Antioxidant Properties in Seeds of Sweet and Bitter Cultivars of Lupine (*Lupinus angustifolius*)." *Natural Product Communications* 13 (10): 1341–1344.

Kuljanabhagavad, Tiwatt, and Michael Wink. 2009. "Biological Activities and Chemistry of Saponins From *Chenopodium Quinoa* Willd." *Phytochemistry Reviews* 8 (2): 473–490. https://doi.org/10.1007/S11101-009-9121-0.

Lampart-Szczapa, E., J. Korczak, M. Nogala-Kalucka, and R. Zawirska-Wojtasiak. 2003a. "Antioxidant Properties of Lupin Seed Products." *Food Chemistry* 83 (2): 279–285. https://doi.org/10.1016/S0308-8146(03)00091-8.

Lampart-Szczapa, E., A. Siger, K. Trojanowska, M. Nogala-Kalucka, M. Malecka, and B. Pacholek. 2003b. "Chemical Composition and Antibacterial Activities of Lupin Seeds Extracts." *Nahrung-Food* 47 (5): 286–290. https://doi.org/10.1002/Food.200390068.

Laus, Maura N., Mariagrazia P. Cataldi, Carlo Robbe, Tiziana D'Ambrosio, Maria L. Amodio, Giancarlo Colelli, Giuditta De Santis, Zina Flagella, and Donato Pastore. 2017. "Antioxidant Capacity, Phenolic and Vitamin C Contents of Quinoa (*Chenopodium quinoa* Wild.) as Affected by Sprouting." *Italian Journal of Agronomy* 12 (1): 63–68. https://doi.org/10.4081/Ija.2017.816.

Laus, Maura N., Anna Gagliardi, Mario Soccio, Zina Flagella, and Donato Pastore. 2012. "Antioxidant Activity of Free and Bound Compounds in Quinoa (*Chenopodium quinoa* Willd.) Seeds in Comparison with Durum Wheat and Emmer." *Journal of Food Science* 77 (11): C1150–C1155. https://doi.org/10.1111/J.1750-3841.2012.02923.X.

Liu, Mengjie, Kaili Zhu, Yang Yao, Yinhuan Chen, Huimin Guo, Guixing Ren, Xiushi Yang, and Jincai Li. 2020. "Antioxidant, Anti-Inflammatory, and Antitumor Activities of Phenolic Compounds from White, Red, and Black *Chenopodium quinoa* Seed." *Cereal Chemistry* 97 (3): 703–713. https://doi.org/10.1002/Cche.10286.

Lucas, M. Mercedes, Frederick L. Stoddard, Paolo Annicchiarico, Juana Frias, Cristina Martinez-Villaluenga, Daniela Sussmann, Marcello Duranti, Alice Seger, Peter M. Zander, and Jose J. Pueyo. 2015. "The Future of Lupin as a Protein Crop in Europe." *Frontiers in Plant Science* 6: 705. https://doi.org/10.3389/Fpls.2015.00705.

Luckett, David. 2010. "Lupini Bean—A Bitter Contamination Risk for Sweet Albus Lupins. Primefact 682." *NSW Government, Industry & Investment*. https://www.dpi.nsw.gov.au/__data/assets/pdf_file/0003/186672/Lupini-bean-a-bitter-contamination-risk-for-lupins.pdf.

Lutz, Mariane, Angelica Martinez, and Enrique A. Martinez. 2013. "Daidzein And Genistein Contents in Seeds of Quinoa (*Chenopodium quinoa* Willd.) from Local Ecotypes Grown in Arid Chile." *Industrial Crops and Products* 49: 117–121. https://doi.org/10.1016/J.Indcrop.2013.04.023.

Madl, Tobias, Heinz Sterk, and Martin Mittelbach. 2006. "Tandem Mass Spectrometric Analysis of a Complex Triterpene Saponin Mixture of *Chenopodium quinoa*." *Journal of the American Society for Mass Spectrometry* 17 (6): 795–806. https://doi.org/10.1016/J.Jasms.2006.02.013.

Magalhaes, Sara C. Q., Marcos Taveira, Ana R. J. Cabrita, Antonio J. M. Fonseca, Patricia Valentão, and Paula B. Andrade. 2017. "European Marketable Grain Legume Seeds: Further Insight into Phenolic Compounds Profiles." *Food Chemistry* 215: 177–184. https://doi.org/10.1016/J.Foodchem.2016.07.152.

Martinez-Villaluenga, Cristina, Juana Frias, and C. Vidal Valverde. 2005. "Raffinose Family Oligosaccharides and Sucrose Contents in 13 Spanish Lupin Cultivars." *Food Chemistry* 91 (4): 645–649. https://doi.org/10.1016/J.Foodchem.2004.06.034.

Mastebroek, Dick, Harry Limburg, Tijs Gilles, and Hans J. P. Marvin. 2000. "Occurrence of Sapogenins in Leaves and Seeds of Quinoa (*Chenopodium quinoa* Willd)." *Journal of The Science of Food and Agriculture* 80 (1): 152–156. https://doi.org/10.1002/(Sici)1097-0010(20000101)80:13.3.Co;2-G.

Mattila, Pirjo H., Juha-Matti Pihlava, Jarkko Hellström, Markus Nurmi, Merja Eurola, Sari Makinen, Taina Jalava, and Anne Pihlanto. 2018. "Contents of Phytochemicals and Antinutritional Factors in Commercial Protein-Rich Plant Products." *Food Quality and Safety* 2 (4): 213–219. https://doi.org/10.1093/Fqsafe/Fyy021.

Miranda, Margarita, Antonio Vega-Galvez, Jessica Lopez, Gloria Parada, Mariela Sanders, Mario Aranda, Elsa Uribe, and Karina Di Scala. 2010. "Impact of Air-Drying Temperature on Nutritional Properties, Total Phenolic Content and Antioxidant Capacity of Quinoa Seeds (*Chenopodium quinoa* Willd.)." *Industrial Crops and Products* 32 (3): 258–263. https://doi.org/10.1016/J.Indcrop.2010.04.019.

Mohammed, Mohammed A., Elshazali Ahmed Mohamed, Abu Elgasim A. Yagoub, Awad R. Mohamed, and Elfadil E. Babiker. 2017. "Effect of Processing Methods on Alkaloids, Phytate, Phenolics, Antioxidants Activity and Minerals of Newly Developed Lupin (*Lupinus Albus* L.) Cultivar." *Journal of Food Processing and Preservation* 41 (1): E12960. https://doi.org/10.1111/Jfpp.12960.

Motta, Carla, Ines Delgado, Ana Sofia Matos, Gerard Bryan Gonzales, Duarte Torres, Mariana Santos, Maria V. Chandra-Hioe, Jayashree Arcot, and Isabel Castanheira. 2017. "Folates in Quinoa (*Chenopodium quinoa*), Amaranth (*Amaranthus* sp.) and Buckwheat (*Fagopyrum esculentum*): Influence of Cooking and Malting." *Journal of Food Composition and Analysis* 64: 181–187. https://doi.org/10.1016/J.Jfca.2017.09.003.

Motta, Carla, Isabel Castanheira, Gerard B. Gonzales, Ines Delgado, Durarte Torres, Mariana Santos, Ana S. Matos. 2019. "Impact of cooking methods and malting on amino acids content in amaranth, buckwheat and quinoa." *Journal of Food Composition and Analysis* 76: 58–65. doi:10.1016/j.jfca.2018.10.001.

Mroczek, Agnieszka. 2015. "Phytochemistry and Bioactivity of Triterpene Saponins from Amaranthaceae Family." *Phytochemistry Reviews* 14 (4): 577–605. https://doi.org/10.1007/S11101-015-9394-4.

Multari, Salvatore, Alexis Marsol-Vall, Marjo Keskitalo, Baoru Yang, and Jukka-Pekka Suomela. 2018a. "Effects of Different Drying Temperatures on the Content of Phenolic Compounds and Carotenoids in Quinoa Seeds (*Chenopodium quinoa*) from Finland." *Journal of Food Composition and Analysis* 72: 75–82. https://doi.org/10.1016/J.Jfca.2018.06.008.

Multari, Salvatore, Alexis Marsol-Vall, and Jukka-Pekka Suomela. 2018b. "Effects of Aromatic Herb Flavoring on Carotenoids and Volatile Compounds in Edible Oil from Blue Sweet Lupin (*Lupinus angustifolius*)." *European Journal of Lipid Science and Technology* 120 (10): 1800227. https://doi.org/10.1002/ejlt.201800227.

Muzquiz, Mercedes, Caralyn L. Ridout, Keith R. Price, and G. Roger Fenwick. 1993. "The Saponin Content and Composition of Sweet and Bitter Lupin Seed." *Journal of the Science of Food and Agriculture* 63 (1): 47–52. https://doi.org/10.1002/Jsfa.2740630108.

Muzquiz, Mercedes, Alejandro Varela, Carmen Burbano, Carmen Cuadrado, Eva Guillamon, and Mercedes M. Pedrosa. 2012. "Bioactive Compounds in Legumes: Pronutritive and Antinutritive Actions. Implications for Nutrition and Health." *Phytochemistry Reviews* 11 (2–3): 227–244. https://doi.org/10.1007/S11101-012-9233-9.

Nachbar, Martin S., and Joel D. Oppenheim. 1980. "Lectins in the United-States Diet—A Survey of Lectins in Commonly Consumed Foods and a Review of the Literature." *American Journal of Clinical Nutrition* 33 (11): 2338–2345. https://doi.org/10.1093/ajcn/33.11.2338.

Nickel, Julia, Luciana Pio Spanier, Fabiana Torma Botelho, Marcia Arocha Gularte, and Elizabete Helbig. 2016. "Effect of Different Types of Processing on the Total Phenolic Compound Content, Antioxidant Capacity, and Saponin Content of *Chenopodium quinoa Willd* Grains." *Food Chemistry* 209: 139–143. https://doi.org/10.1016/J.Foodchem.2016.04.031.

Oomah, B. Dave, Nathalie Tiger, Mark Olson, and Parthiba Balasubramanian. 2006. "Phenolics and Antioxidative Activities in Narrow-Leafed Lupins (*Lupinus angustifolius* L.)." *Plant Foods For Human Nutrition* 61 (2): 91–97. https://doi.org/10.1007/S11130-006-0021-9.

Otterbach, Sophie Lisa, Ting Yang, Lucilia Kato, Christian Janfelt, and Fernando Geu-Flores. 2019. "Quinolizidine Alkaloids Are Transported to Seeds of Bitter

Narrow-Leafed Lupin." *Journal of Experimental Botany* 70 (20): 5799–5808. https://doi.org/10.1093/Jxb/Erz334.

Pachari Vera, Erika, Juan Jose Alca, Giuliana Rondon Saravia, Nicolas Callejas Campioni, and Ivan Jachmanian Alpuy. 2019. "Comparison of the Lipid Profile and Tocopherol Content of Four Peruvian Quinoa (*Chenopodium quinoa* Willd.) Cultivars ('Amarilla De Maranganf', 'Blanca De Juli', Inia 415 'Roja Pasankalla', Inia 420 'Negra Collana') During Germination." *Journal of Cereal Science* 88: 132–137. https://doi.org/10.1016/J.Jcs.2019.05.015.

Pastor-Cavada, Elena, Rocio Juan, Julio E. Pastor, Manuel Alaiz, and Javier Vioque. 2010. "Antioxidant Activity in the Seeds of Four Wild Lupinus Species from Southern Spain." *Journal of Food Biochemistry* 34: 149–160. https://doi.org/10.1111/J.1745-4514.2009.00320.X.

Paucar-Menacho, Luz, Cristina Martinez-Villaluenga, Montserrat Duenas, Juana Frias, and Elena Penas. 2018. "Response Surface Optimisation of Germination Conditions to Improve the Accumulation of Bioactive Compounds and the Antioxidant Activity in Quinoa." *International Journal of Food Science and Technology* 53 (2): 516–524. https://doi.org/10.1111/Ijfs.13623.

Pellegrini, Marika, Raquel Lucas-Gonzales, Antonella Ricci, Javier Fontecha, Juana Fernandez-Lopez, Jose A. Perez-Alvarez, and Manuel Viuda-Martos. 2018. "Chemical, Fatty Acid, Polyphenolic Profile, Techno-Functional and Antioxidant Properties of Flours Obtained from Quinoa (*Chenopodium quinoa* Willd) Seeds." *Industrial Crops and Products* 111: 38–46. https://doi.org/10.1016/J.Indcrop.2017.10.006.

Pereira, Eliana, Christian Encina-Zelada, Lillian Barros, Ursula Gonzales-Barron, Vasco Cadavez, and Ferreira, Isabel C. F. R. 2019. "Chemical and Nutritional Characterization of *Chenopodium quinoa* Willd (Quinoa) Grains: A Good Alternative to Nutritious Food." *Food Chemistry* 280: 110–114. https://doi.org/10.1016/J.Foodchem.2018.12.068.

Petterson, David S., and Anne H. Fairbrother. 1996. "Lupins as a Raw Material for Human Foods and Animal Feeds." *Indonesian Food and Nutrition Progress* 3 (2): 35–41. https://doi.org/10.22146/Jifnp.52.

Pilco-Quesada, Silvia, Ye Tian, Baoru Yang, Ritva Repo-Carrasco-Valencia, and Jukka-Pekka Suomela. 2020. "Effects of Germination and Kilning on the Phenolic Compounds and Nutritional Properties of Quinoa (*Chenopodium quinoa*) and Kiwicha (*Amaranthus caudatus*)." *Journal of Cereal Science* 94: 102996. https://doi.org/10.1016/J.Jcs.2020.102996.

Prusinski, Janusz. 2017. "White Lupin (*Lupinus albus* L.)—Nutritional and Health Values in Human Nutrition—A Review." *Czech Journal of Food Sciences* 35 (2): 95–105. https://doi.org/10.17221/114/2016-Cjfs.

Ranilla, Lena Galvez, Maria Ines Genovese, and Franco Maria Lajolo. 2009. "Isoflavones and Antioxidant Capacity of Peruvian and Brazilian Lupin Cultivars." *Journal of Food Composition and Analysis* 22 (5): 397–404. https://doi.org/10.1016/J.Jfca.2008.06.011.

Repo-Carrasco Valencia, Ritva, C. Espinoza, and S. E. Jacobsen. 2003. "Nutritional Value and Use of the Andean Crops Quinoa (*Chenopodium Quinoa*) and Kañiwa (*Chenopodium pallidicaule*)." *Food Reviews International* 19 (1–2): 179–189. https://doi.org/10.1081/Fri-120018884.

Repo-Carrasco-Valencia, Ritva, Jarkko K. Hellström, Juha-Matti Pihlava, and Pirjo H. Mattila. 2010. "Flavonoids and Other Phenolic Compounds in Andean Indigenous Grains: Quinoa (*Chenopodium quinoa*), Kañiwa (*Chenopodium pallidicaule*) and Kiwicha (*Amaranthus caudatus*)." *Food Chemistry* 120 (1): 128–133. https://doi.org/10.1016/J.Foodchem.2009.09.087.

Resta, Donatella, Giovanna Boschin, Alessandra D'Agostina, and Anna Arnoldi. 2008. "Evaluation of Total Quinolizidine Alkaloids Content in Lupin Flours, Lupin-Based Ingredients, and Foods." *Molecular Nutrition and Food Research* 52 (4): 490–495. https://doi.org/10.1002/Mnfr.200700206.

Ruales, Jenny, and Baboo Nair. 1993. "Saponins, Phytic Acid, Tannins and Protease Inhibitors in Quinoa (*Chenopodium quinoa*, Willd) Seeds." *Food Chemistry* 48 (2): 137–143. https://doi.org/10.1016/0308-8146(93)90048-K.

Rumiyati, Vijay Jayasena, And Anthony P. James. 2013. "Total Phenolic and Phytosterol Compounds and the Radical Scavenging Activity of Germinated Australian Sweet Lupin Flour." *Plant Foods for Human Nutrition* 68 (4): 352–357. https://doi.org/10.1007/S11130-013-0377-6.

Ryan, E., K. Galvin, T. P. O'Connor, A. R. Maguire, and N. M. O'Brien. 2007. "Phytosterol, Squalene, Tocopherol Content and Fatty Acid Profile of Selected Seeds, Grains, and Legumes." *Plant Foods for Human Nutrition* 62 (3): 85–91. https://doi.org/10.1007/S11130-007-0046-8.

Rybinski, Wojciech, Wojciech Swiecicki, Jan Bocianowski, Andreas Boerner, Elzbieta Starzycka-Korbas, and Michal Starzycki. 2018. "Variability of Fat Content and Fatty Acids Profiles in Seeds of A Polish White Lupin (*Lupinus albus* L.) Collection." *Genetic Resources and Crop Evolution* 65 (2): 417–431. https://doi.org/10.1007/S10722-017-0542-0.

Santana, Fmc, T. Pinto, A. M. Fialho, I. Sa-Correia, and Jma Empis. 2002. "Bacterial Removal of Quinolizidine Alkaloids and Other Carbon Sources from A *Lupinus albus* Aqueous Extract." *Journal of Agricultural and Food Chemistry* 50 (8): 2318–2323. https://doi.org/10.1021/Jf011371h.

Santos, Claudia N., Ricardo B. Ferreira, and Artur R. Teixeira. 1997. "Seed Proteins of *Lupinus mutabilis*." *Journal of Agricultural and Food Chemistry* 45 (10): 3821–3825. https://doi.org/10.1021/Jf970075v.

Schrenk, Dieter, Laurent Bodin, James Kevin Chipman, Jesus Del Mazo, Bettina Grasl-Kraupp, Christer Hogstrand, Laurentius (Ron) Hoogenboom, et al. 2019. "Scientific Opinion on The Risks for Animal and Human Health Related to the Presence of Quinolizidine Alkaloids in Feed and Food, in Particular in Lupins and Lupin-Derived Products." *EFSA Journal* 17 (11): 5860. https://doi.org/10.2903/J.Efsa.2019.5860.

Siger, Aleksander, Jarosław Czubinski, Piotr Kachlicki, Krzysztof Dwiecki, Eleonora Lampart-Szczapa, and Malgorzata Nogala-Kalucka. 2012. "Antioxidant Activity and Phenolic Content in Three Lupin Species." *Journal of Food Composition and Analysis* 25 (2): 190–197. https://doi.org/10.1016/J.Jfca.2011.10.002.

Sirtori, Cesare R., Maria R. Lovati, Cristina Manzoni, Silvia Castiglioni, Marcello Duranti, Chiara Magni, Sheila Morandi, Alessandra D'Agostina, and Anna Arnoldi. 2004. "Proteins of White Lupin Seed, a Naturally Isoflavone-poor Legume, Reduce Cholesterolemia in Rats and Increase LDL Receptor Activity in HepG2 Cells." *Journal of Nutrition* 134 (1): 18–23. https://doi.org/10.1093/jn/134.1.18.

Sosulski, Frank W., and Kazimierz J. Dabrowski. 1984. "Composition of Free and Hydrolyzable Phenolic-Acids in the Flours and Hulls of 10 Legume Species." *Journal of Agricultural and Food Chemistry* 32 (1): 131–133. https://doi.org/10.1021/Jf00121a033.

Tang, Yao, Xihong Li, Bing Zhang, Peter X. Chen, Ronghua Liu, and Rong Tsao. 2015. "Characterisation of Phenolics, Betanins and Antioxidant Activities in Seeds of Three *Chenopodium quinoa* Willd. Genotypes." *Food Chemistry* 166: 380–388. https://doi.org/10.1016/J.Foodchem.2014.06.018.

Tang, Yao, Bing Zhang, Xihong Li, Peter X. Chen, Hua Zhang, Ronghua Liu, and Rong Tsao. 2016. "Bound Phenolics of Quinoa Seeds Released by Acid, Alkaline, and Enzymatic Treatments and Their Antioxidant and Alpha-Glucosidase and Pancreatic Lipase Inhibitory Effects." *Journal of Agricultural and Food Chemistry* 64 (8): 1712–1719. https://doi.org/10.1021/Acs.Jafc.5b05761.

Taraseviciene, Zivile, Akvile Virsile, Honorata Danilcenko, Pavelas Duchovskis, Aurelija Paulauskiene, and Marek Gajewski. 2019. "Effects of Germination Time on the Antioxidant Properties of Edible Seeds." *Cyta-Journal of Food* 17 (1): 447–454. https://doi.org/10.1080/19476337.2018.1553895.

Terruzzi, Ileana, Pamela Senesi, Christian Magni, Anna Montesano, Alessio Scarafoni, Livio Luzi, and Marcello Duranti. 2011. "Insulin-Mimetic Action of Conglutin-Gamma, a Lupin Seed Protein, in Mouse Myoblasts." *Nutrition Metabolism and Cardiovascular Diseases* 21 (3): 197–205. https://doi.org/10.1016/J.Numecd.2009.09.004.

Ujiroghene, Obaroakpo Joy, Lu Liu, Shuwen Zhang, Jing Lu, Cai Zhang, Jiaping Lv, Xiaoyang Pang, and Min Zhang. 2019. "Antioxidant Capacity of Germinated Quinoa-Based Yoghurt and Concomitant Effect of Sprouting on Its Functional Properties." *LWT-Food Science and Technology* 116: 108592. https://doi.org/10.1016/J.Lwt.2019.108592.

Valencia-Chamorro, Silvia A. 2003. "Quinoa." In Benjamin Caballero, Paul Finglas and Fidel Toldrá (Eds.), *Encyclopedia of Food Science and Nutrition*. Amsterdam: Academic Press, 4895–4902.

Vargas-Guerrero, Belinda, Pedro M. Garcia-Lopez, Alma L. Martinez-Ayala, Jose A. Dominguez-Rosales, and Carmen M. Gurrola-Diaz. 2014. "Administration of *Lupinus albus* Gamma Conglutin (C Gamma) to N5 Stz Rats Augmented Ins-1 Gene Expression and Pancreatic Insulin Content." *Plant Foods for Human Nutrition* 69 (3): 241–247. https://doi.org/10.1007/S11130-014-0424-Y.

Vega-Galvez, Antonio, Liliana Zura, Mariane Lutz, Rosa Jagus, M. Victoria Aguero, Alexis Pasten, Karina Di Scala, and Elsa Uribe. 2018. "Assessment of Dietary Fiber, Isoflavones and Phenolic Compounds with Antioxidant and Antimicrobial Properties of Quinoa (*Chenopodium quinoa* Willd.)." *Chilean Journal of Agricultural & Animal Sciences* 34 (1): 57–67. http://dx.doi.org/10.4067/S0719-38902018005000101.

Vincken, Jean-Paul, Lynn Heng, Aede De Groot, and Harry Gruppen. 2007. "Saponins, Classification and Occurrence in the Plant Kingdom." *Phytochemistry* 68 (3): 275–297. https://doi.org/10.1016/J.Phytochem.2006.10.008.

Woldemichael, Girma M., and Michael Wink. 2001. "Identification and Biological Activities of Triterpenoid Saponins from *Chenopodium quinoa*." *Journal of Agricultural and Food Chemistry* 49 (5): 2327–2332. https://doi.org/10.1021/Jf0013499.

Zhong, Liezhou, Gangcheng Wu, Zhongxiang Fang, Mark L. Wahlqvist, Jonathan M. Hodgson, Michael W. Clarke, Edwin Junaldi, and Stuart K. Johnson. 2019. "Characterization of Polyphenols in Australian Sweet Lupin (*Lupinus angustifolius*) Seed Coat by Hplc-Dad-Esi-Ms/Ms." *Food Research International* 116: 1153–1162. https://doi.org/10.1016/J.Foodres.2018.09.061.

Zhu, Fengmei, Bin Du, and Baojun Xu. 2018. "Anti-Inflammatory Effects of Phytochemicals from Fruits, Vegetables, and Food Legumes: A Review." *Critical Reviews in Food Science and Nutrition* 58 (8): 1260–1270. https://doi.org/10.1080/10408398.2016.1251390.

Zhu, Nanqun, Shuqun Sheng, Shengmin Sang, Jin-Whoo Jhoo, Naisheng Bai, Mukund V. Karwe, Robert T. Rosen, and Chi-Tang Ho. 2002. "Triterpene Saponins from Debittered Quinoa (*Chenopodium quinoa*) seeds." *Journal of Agricultural and Food Chemistry* 50 (4): 865–867. https://doi.org/10.1021/Jf011002l.

Chapter 2

Quinoa, Kañiwa, Amaranth, and Lupin as Ingredients in Gluten-Free Baking

Ritva Repo-Carrasco-Valencia and
Julio Mauricio Vidaurre-Ruiz

CONTENTS

2.1 INTRODUCTION

The Andean region of South America is an important center of domestication of food crops. At the time the Europeans arrived in South America, the indigenous people cultivated almost as many species of plants as the farmers of all of Asia and Europe. It has been estimated that Andean natives domesticated as many as 70 separate crop species (National Research Council 1989). From the coast and inter-Andean valleys up to 4 km high along the whole continent and in climates varying from tropical to polar, they grew roots, grains, legumes, vegetables, fruits, and nuts. During the time of the Incas, Peru was a very prosperous

DOI: 10.1201/9781003087618-2

farming country with a population of approximately 10 million people, and malnutrition was practically unknown. The cultivation of many of these plants was reduced dramatically following the arrival of the Europeans. These crops have long been neglected, and they have only received scientific and commercial interest in the last 10 years.

These native crops include some remarkably nutritious and tasty foods. The Andean grains quinoa (*Chenopodium quinoa*), kañiwa (*C. pallidicaule*), kiwicha (Andean amaranth; *Amaranthus caudatus*), and tarwi (Andean lupin; *Lupinus mutabilis*) are crops that have adapted perfectly to the harsh conditions of the mountain area of South America. Quinoa, kañiwa, and kiwicha are pseudocereals and could be used to replace common cereals in different types of preparations. Tarwi is a leguminous plant with remarkably high protein and oil contents. The proximate composition of the Andean grains is shown in Table 2.1.

All Andean grains have a higher protein content than common cereals such as corn, wheat, and rice. More important than the protein quantity is the protein quality, as defined by the amino acid composition. All Andean grains are rich in lysine, the first limiting amino acid in common cereals. They are excellent sources of dietary fiber and bioactive compounds such as flavonoids and phenolic acids. The starch of Andean grains has interesting rheological properties and could be used in different food and non-food applications. Quinoa, kañiwa, kiwicha, and tarwi are naturally gluten-free and offer an excellent alternative for introduction into the diet of persons suffering from celiac disease.

Gluten is a general term for the proteins found in wheat (durum, emmer, semolina, spelt, and einkorn), rye, barley, and triticale (a cross between wheat

TABLE 2.1 PROXIMATE COMPOSITION OF ANDEAN GRAINS (g/100g)

	Quinoa[a]	Kiwicha[a]	Kañiwa[a]	Tarwi[b,c]
Moisture	8.26–11.51	11.09–12.07	9.38–10.39	6.14–9.9
Protein	11.32–14.72	12.80–15.88	13.29–18.28	52.9–57.36
Fat	3.95–6.85	6.31–7.56	4.46–7.92	21.6–25.4
Crude fiber	1.81–4.23	2.68–7.49	5.33–14.37	8.8
Ash	2.27–3.12	2.16–2.50	3.13–3.67	2.61–2.80
Carbohydrates	59.4–70.8	55.5–63.7	52.4–60.3	12.8

Note: Data given in wet basis.
Sources: Adapted from
[a] Repo-Carrasco-Valencia et al. (2010);
[b] Vidaurre-Ruiz (2020);
[c] Rosell et al. (2009).

and rye). Gluten helps foods maintain their shape, acting as the glue that holds the food together. Gluten can be found in many foods, even those that would not be expected to contain it (Celiac Disease Foundation 2020). It is the main storage protein in wheat grains. Gluten is a complex mixture of hundreds of related but distinct proteins, mainly gliadin and glutenin. Similar storage proteins exist, such as secalin in rye, hordein in barley, and avenins in oats. These are referred to collectively as "gluten" (Biesiekierski 2017). The proteins gliadin and glutenin represent seed proteins insoluble in water but extractable in aqueous ethanol and are characterized by high levels of glutamine (38%) and proline (20%) residues (Wieser 2007).

Gluten and its properties are essential to determining the dough quality of bread and other baked products. Gliadin contains peptide sequences (known as epitopes) that are highly resistant to gastric, pancreatic, and intestinal proteolytic digestion in the gastrointestinal tract, escaping degradation in the human gut. This difficult digestion is due to gliadin's high content of the amino acids proline and glutamine, which many proteases are unable to cleave. These proline-rich residues create tight and compact structures that can mediate the adverse immune reactions in celiac disease (Hausch et al. 2002).

Celiac disease is a serious autoimmune disease that occurs in genetically predisposed people where the ingestion of gluten leads to damage in the small intestine. It is estimated to affect 1 in 100 people worldwide. It is characterized by an insufficient T-cell-mediated immune response that causes inflammatory injury to the small intestine (Tovoli 2015; Celiac Disease Foundation 2020). Currently, the only treatment for celiac disease is lifelong adherence to a strict gluten-free diet. People who follow a gluten-free diet must avoid foods with wheat, rye, and barley, such as bread, pastry, pasta, and beer.

The raw materials used for the production of gluten-free products are mainly nontoxic cereals (e.g., corn, rice, sorghum, and millet) and pseudocereals (e.g., amaranth, buckwheat, and quinoa). It has been demonstrated that persons on gluten-free diets often consume less than the recommended amounts of B vitamins, iron, calcium, and fiber (Thompson et al. 2005). The reason for this is that traditionally, mostly refined flours from rice and corn, whose germ and bran fractions have been removed, or even pure starches are used as base materials for gluten-free foods. This suggests that more emphasis should be placed on the nutritional quality of gluten-free diets to encourage patients to consume enriched and fortified products whenever possible (Thompson et al. 2005).

Pseudocereals such as amaranth, quinoa, and buckwheat have been recommended as nutritious ingredients in gluten-free formulations owing to their high protein quality and abundant quantities of fiber and minerals such as calcium and iron (Wieser et al. 2014). Incorporation of these gluten-free grains in an unrefined form into the gluten-free diet could not only add variety but also improve its nutritional value.

In recent decades, the quality and availability of gluten-free products on the market have improved continuously, and different formulations that increase the quality of gluten-free products such as biscuits, cookies, cakes, pasta, and pizza have been introduced (Alvarez-Jubete et al. 2010a). There is still a need to develop products with improved flavor, texture, and mouthfeel compared with their gluten-containing counterparts. Andean grains such as quinoa, kiwicha, kañiwa, and tarwi do not contain peptides similar to wheat gluten; therefore, they are raw materials appropriate for consumption by those with celiac disease. The inclusion of these grains in the formulations of gluten-free breads is promising.

2.2 QUINOA

Quinoa (*C. quinoa* Willd.) is a seed crop of the Amaranthaceae family. It played a very important role for the ancient Andean cultures of Peru and Bolivia. At present, quinoa is grown mainly in the Andean region from Colombia to the north of Argentina, Peru and Bolivia being the most important producers. The different varieties and landraces of quinoa have adapted to distinct environmental conditions (e.g., those of the "Altiplano" (high plateau), Andean valleys and coastal areas). The seeds of quinoa are small, about 2 mm in diameter, and can be of various colors: white, cream, purple, yellow, red, or black. Quinoa is considered one of the best vegetal protein sources, as its protein quality is similar to those found in milk and higher than those present in cereals such as wheat, rice, and maize (Gordillo-Bastidas et al. 2016). Quinoa protein is higher not only in lysine but also in another important essential amino acid, methionine, than any other cereal (Arendt and Zannini 2013). These amino acids are especially important for vegetarian or vegan diets because they are the limiting amino acids in vegetable proteins: cereals are limiting in lysine and legumes in methionine. The fat content of quinoa is between 5.7% and 7.3%, mainly located in the embryo and considerably higher than that in common cereal grains (Repo-Carrasco-Valencia et al. 2019). Quinoa is a good source of some important minerals such as calcium, magnesium, and iron (Kozioł 1992). Quinoa has been used by the National Aeronautics and Space Administration (NASA) due to its versatility in meeting the needs of humans during space missions (Vega-Gálvez et al. 2010).

2.3 KAÑIWA

Kañiwa (*C. pallidicaule* Aellen) is a close relative of quinoa; in fact, it was considered a variety of quinoa until 1929, when it was classified as a different species

(Gade 1970). Kañiwa is a resistant crop growing under very harsh environmental conditions, mainly in the Peruvian and Bolivian Altiplano. It is more tolerant to frost than quinoa. In its native region, the temperatures average less than 10 °C, and frost can occur for at least 9 months of the year. For highland farmers, kañiwa is very important because oftentimes it is the only crop that can survive during a frost. The most intensive production of kañiwa occurs in the Southern Andes of Peru and Bolivia in the surroundings of Lake Titicaca. Kañiwa is a small plant, and its seeds are smaller (approximately 1 mm) than those of quinoa. They are usually gray or brown, but there are some colored ecotypes as well.

The small grains of Kañiwa are even more nutritious than those of quinoa. Their protein content is, on average, between 14.4% and 16.9%, and the oil content is also relatively high (Repo-Carrasco-Valencia et al. 2019a). Kañiwa protein is remarkably high in lysine. Repo-Carrasco et al. (2003) analyzed the content of amino acids in quinoa and kañiwa samples and obtained lysine values between 5.6 and 6.4 g/100 g protein for quinoa and 5.7 g/100 g protein for kañiwa. Flavonoids are phenolic compounds with many health-promoting properties (e.g., antioxidant, anti-inflammatory, and anticarcinogenic effects). In kañiwa, the principal flavonoids are quercetin and isorhamnetin, the content of quercetin being exceptionally high (Repo-Carrasco-Valencia et al. 2019a). Regarding dietary fiber, kañiwa has the highest content when compared with the other Andean grains. The fiber is mainly insoluble, as is common in all grains. Pentosan is one of the important fiber components of the non-starch polysaccharides in cereal, which primarily consist of pentosan sugar of L-arabinose and D-xylose. In terms of the human nutritional aspects of dietary fiber components, pentosan not only affects food absorption but also decreases the absorption of lipids and cholesterol; therefore, pentosan is very useful in the human diet. It has a positive effect on food processing as well (e.g., on the rheological characteristics of dough and macaroni production processing). The content of pentosans in kañiwa and other Andean grains is higher than the content of these compounds in wheat, barley, and oats. The content of different tocopherols and phytosterols in Andean grains was studied by Repo-Carrasco-Valencia et al. (2019a). γ-Tocopherol was found in high amounts in kañiwa. Results from previous *in vitro* and *in vivo* studies suggest that γ- and δ-tocopherols could be more potent anticancer agents than α-tocopherol (Das Gupta and Suh 2016; Yang and Suh 2012). Kañiwa oil is rich in phytosterols, such as campesterol, stigmasterol, and β-sitosterol and could be a potential ingredient for the edible oil industry.

2.4 KIWICHA

Amaranth belongs to the Amaranthaceae family of dicotyledonous plants. The most important Andean species is *A. caudatus* Linnaeus, kiwicha. Kiwicha is cultivated in the Andes of Peru, Bolivia, Ecuador, and Argentina. *A. caudatus* originated in the same region in the Andean highlands as the common potato. The Spanish conquerors called it Inca wheat, but it was known long before the Incas. In ancient tombs, seeds have been found that are more than 4000 years old (National Research Council 1989). Kiwicha was one of the basic foods in the Incan Empire, nearly as important as maize and potatoes. The grains of amaranth are very small, but they occur in huge numbers, sometimes more than 100,000 to a plant (National Research Council 1989). Amaranth is drought-resistant and grows from sea level to 3600 m, mainly in inter-Andean valleys. Amaranthus species are grown successfully worldwide in the tropics and semi-arid regions because of their hardiness and their ability to grow in places where other more common grains cannot. Kiwicha is often called the "small giant" because of its small grains with high nutritive value. The protein content in different varieties is between 13% and 21%, and the colored varieties generally have a higher protein content than the cream varieties (Zheleznov et al. 1997; Repo-Carrasco-Valencia et al. 2019b). According to these authors, the proteins are highly digestible due to the presence of albumins and globulins. Not only the seeds but also the leaves of amaranth are sources of protein of very high quality. The lipid content of kiwicha seed is relatively high (7.20%–8.33%). The fat of amaranth is characterized by a high content of unsaturated fatty acids. Plant sterols, especially β-sitosterol, have been reported in amaranth (Marcone et al. 2003). According to Ogrodowska et al. (2014), the total phytosterol content in amaranth seeds was 177.66 ± 5.48 mg/100 g. Kiwicha oil seems to be low in α-tocopherol but, interestingly, high in γ-tocopherol. Kraujalis and Venskutonis (2013) observed that the tocopherol isomers α-, β-, γ-, and δ-tocopherol were distributed in approximate ratios of 1:27:6.5:5, respectively, in amaranth oil.

2.5 TARWI

Tarwi (*L. mutabilis* Sweet) belongs to the Lupinus family, which is very diverse, with more than 100 different species in the New World and a smaller number in the Mediterranean region. The Mediterranean species are agriculturally important and are used mainly as feed in Europe, the United States, Australia, and South Africa. These lupines (*L. angustifolius, L. luteus*, and *L. albus*) are sweet, and their content of bitter alkaloids is low compared with that of tarwi. Tarwi was domesticated by pre-Inca cultures more than 4000 years ago. Tarwi seeds have been found in the tombs of the Nazca culture (100–500 BC) on the coast of Peru (Antúnez de Mayolo 1981). Because of its excellent nutritional value, tarwi was a very important crop for pre-Incan and Incan cultures. Tarwi is also an important component in the crop rotations of the Andes because of its capacity to fix nitrogen in the soil. Tarwi is reported to be able to fix as much as 400 kg of nitrogen per hectare. On the traditional Andean terraces, tarwi was planted on the upper levels in order to allow soluble nitrogen to flow toward the lower levels. In addition to this, tarwi's strong roots loosen the soil; all of these abilities benefit the soil in which tarwi is grown (National Research Council 1989).

The nutritional composition of tarwi is very interesting. With its twofold advantage, which includes high amounts of both protein and oil, tarwi represents a valuable food crop to combat malnutrition and a promising cash crop for edible oil production. Tarwi has an exceptionally high protein content (more than 40%); it has more protein than soybeans and even more than other lupine species (Birk et al. 1990; Córdova-Ramos et al. 2020). Tarwi is an excellent source of oil (about 20%) with high nutritional quality, the main fatty acids being oleic and linoleic acids (Repo-Carrasco-Valencia 2020).

2.6 GLUTEN-FREE BAKING

Bread has been a staple food for many cultures around the world and was one of the first "processed" foods made and consumed by humanity (Arendt and Zannini 2013). It is typically made with wheat flour, which, thanks to its proteins that form the gluten network, confers on it characteristics such as good specific volume, uniform alveoli, crumb softness, a crispy crust, and a golden color (Brites et al. 2018). Making bread with gluten-free flours and ingredients is a technological challenge; this challenge has given rise to an increase in research on the development of this kind of product in the last decade, with the aim of achieving a product with characteristics similar to those of conventional bakery products.

The increase in gluten-free bakery products on the market is a global trend. Proof of this is provided by the large number of commercial brands available, as evidenced by Roman et al. (2019), who found 228 types of commercial gluten-free bread of 32 brands from 12 different countries around the world. In Latin America, the countries with the greatest advances in specific legislation for the gluten-free food and food-service industry, as well as in the development of gluten-free bakery products are Argentina and Brazil (Falcomer et al. 2020), while in Peru, the increase in the supply of gluten-free products is reflected in the increase in the specific sections of "gluten-free products" in supermarkets. However, the vast majority of gluten-free bakery products sold are imported from Italy and the United States. The increase in demand for gluten-free products, specifically gluten-free breads, can be attributed to better knowledge about the health problems that gluten intake can bring in people sensitive to this protein (Roman et al. 2019). In a recent study, it was reported for the first time that the prevalence of celiac disease in Peru was found to be similar to the world average (1%; Baldera et al. 2020). The increase in the consumption of gluten-free products could also be related to the belief that these products are healthier than their gluten-containing counterparts (Reilly 2016; Hartmann et al. 2018). However, this is not entirely true, since the vast majority of gluten-free bakery products are made from starches and rice flour, which are not nutritious (Calvo-Lerma et al. 2019; Rybicka et al. 2019).

The production of gluten-free breads differs significantly from the process of developing traditional bakery products with wheat flour. Specifically, in the preparation of gluten-free breads, the kneading stage is not developed; this is because gluten-free doughs tend to be fluid, the rheological and textural properties being of vital importance for the success of dough development during fermentation and baking (Brites et al. 2018; Vidaurre-Ruiz 2020). Another significant difference is the short duration of the mixing, fermentation, and baking processes (Arendt et al. 2008). All the parameters and ingredients involved in the process of formulating gluten-free breads have been of great interest for the development of research, which seeks to evaluate, for the most part, the influence of different types of gluten-free flours, starches, gums, hydrocolloids, enzymes, proteins, and fiber, the water content and the types of microorganisms (yeasts, lactic acid bacteria) on the physical, nutritional, and sensory quality of gluten-free breads (Ziobro et al. 2016; Sciarini et al. 2012; Horstmann et al. 2018; Carbó et al. 2020; Föste et al. 2014; Horstmann et al. 2019; Rybicka et al. 2019; Demirkesen et al. 2010). Likewise, the conditions of the process such as the mixing speed, geometry of the batter, and type of aeration (mechanical or biological) have been of interest (Elgeti et al. 2017), as have the durations of processes such as mixing and fermentation, which must be controlled to avoid weakening the bread structure (Vidaurre-Ruiz et al. 2019). Figure 2.1 shows the typical flow diagram of the process of making gluten-free bread with some considerations related to each stage of the process.

The raw material widely used in the development of gluten-free breads is starch, the three most commonly used being corn starch, potato starch, and cassava starch. The gelatinization of the starch during the cooking stage is crucial in gluten-free breads because it is essential to the structure of the bread (Brites et al. 2018). It is known that, during dough development, starch absorbs up to 45% water, based on its own weight, and is considered to act as an inert filler in the continuous dough matrix (Horstmann et al. 2017a). An important factor that can modify the properties of starch is the availability of water. When water levels are lowered, starch gelatinization is delayed; however, if water is abundant, the process will take place rapidly, affecting gelation and starch retrogradation after cooling (Brites et al. 2018). During the baking process of the bread, the starch granules gelatinize (i.e., they swell and partially solubilize but still maintain their granular identity; Horstmann et al. 2017a). Starch acts as the matrix of gluten-free bakery products, retaining carbon dioxide to allow expansion of air cells, preventing coalescence during fermentation and oven rise and stabilizing the final structure after cooling (Houben et al. 2012).

Hydrocolloids are ingredients widely used in the production of gluten-free breads; these come from a wide range of materials, such as algae (carrageenan and alginates), bacteria (xanthan gum), citrus and apples (pectin), seed extracts (guar gum, locust bean gum, tara gum), plant exudates (gum Arabic, pitch gum), and cellulose derivatives (carboxymethyl cellulose [CMC], hydroxypropyl methylcellulose [HPMC], and microcrystalline cellulose [MCC]). All hydrocolloids are of high molecular weight (generally carbohydrates) and have a high capacity to retain water (Brites et al. 2018; Vidaurre-Ruiz et al. 2019; Clapassón et al. 2020). Hydrocolloids are used to simulate the formation of the gluten network in gluten-free breads due to their ability to change the characteristics of the dough. The viscoelastic properties of the dough are enhanced by the viscoelastic properties of the polysaccharide chains in an aqueous medium. This effect depends on the structure and conformation of the hydrocolloid polysaccharide chain, which will determine the intermolecular associations (Tsatsaragkou et al. 2016; Horstmann et al. 2018). The addition of hydrocolloids to gluten-free doughs leads to an increase in viscosity due to the water-holding capacity of these molecules. It also improves the development and retention of gases during fermentation (Mir et al. 2016; Sabanis and Tzia 2011).

Currently, alterations are being made to the process of adding compounds to improve the physical, nutritional, or sensory quality of gluten-free breads. One of them is the use of sourdough; this method represents an alternative for increasing the quality of gluten-free breads. Acidification of flour by sourdough fermentation can replace gluten function to some extent and improve the swelling properties of polysaccharides, leading to better bread structure. It also improves bread volume and crumb structure, flavor, nutritional value, and shelf life (Arendt et al. 2011). Additionally, lactic acid bacteria can produce long-chain sugar polymers called exopolysaccharides, which can be produced as prebiotics and hydrocolloids to enhance the technological and nutritional properties of gluten-free breads. Lactic acid bacteria can also produce antifungal, bioactive, and aromatic compounds that have the ability to improve the overall quality of bread (Moroni et al. 2009).

The use of certain enzymes has also been investigated to improve the physical quality of gluten-free breads; for example, the use of transglutaminase has a severe effect on the water absorption of the dough, modifying its viscoelastic behavior and improving thermal stability. The use of proteases has also been shown to improve the appearance of the crumb, increase the volume, soften the texture, and slow the aging rate, which could be due to modifications in the protein-starch interactions resulting from the proteolytic activity (Hamada et al. 2013; Kawamura-Konishi et al. 2013; Azizi et al. 2020). Another enzyme that is gaining importance in the preparation of gluten-free breads is maltogenic amylase. Its greater relevance is due to the fact that it has the ability to delay the retrogradation of starch and therefore lengthen the shelf life of baked products. Recent studies on maltogenic amylase were dedicated to seeking greater thermal stability at baking temperatures, promoting its encapsulation with different matrices (Haghighat-Kharazi et al. 2020b).

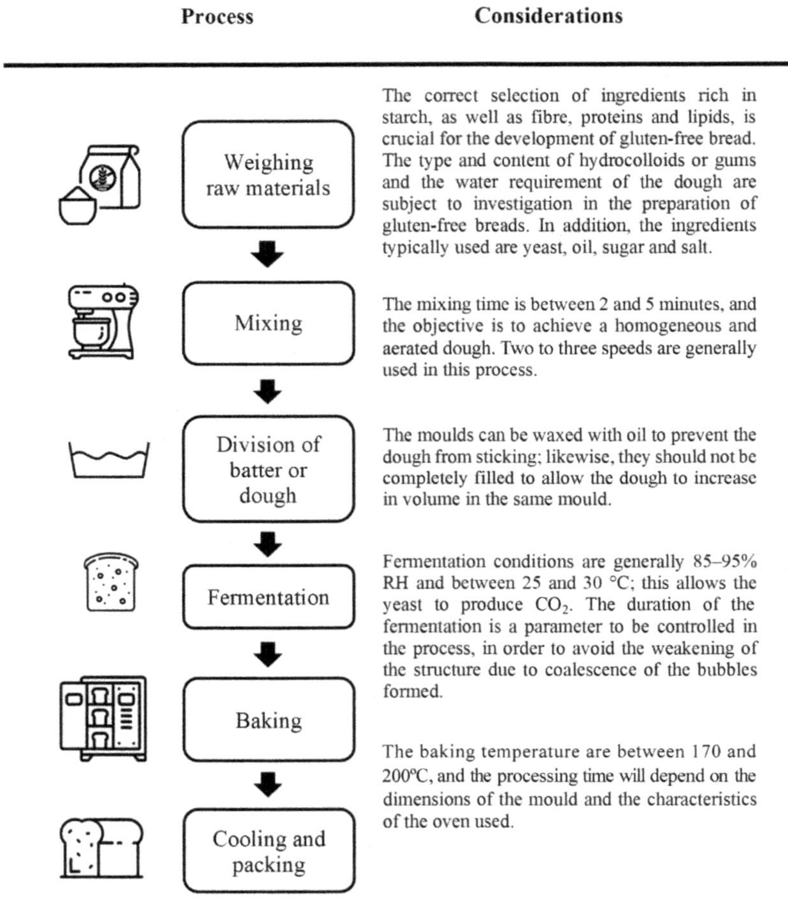

Figure 2.1 Flow diagram of the conventional production process for gluten-free bread.

Source: Adapted from Arendt et al. (2008) and Brites et al. (2018).

The problem with gluten-free bakery products is that they are of a lower sensorial quality than their gluten-containing counterparts. They often have compact and dry crumbs, a low specific volume, an unpleasant sour taste, and color and a very short shelf life. Additionally, since they are made with a high proportion of starches or white flours, such as rice flour, they present low nutritional quality, especially because they are deficient in proteins, minerals, and fiber (Rybicka et al. 2019; Benavent-Gil and Rosell 2019).

2.7 THE USE OF ANDEAN GRAINS IN GLUTEN-FREE BAKING

The use of Andean grains in the production of gluten-free breads has increased in recent years, always rescuing the nutritional and technological contribution that these flours can offer. One of the first reports on the use of kiwicha in the preparation of gluten-free breads was reported by Gambuś et al. (2002), who evaluated the viability of kiwicha flour as a gluten-free alternative to improve the nutritional quality of gluten-free breads. By substituting 10% corn starch with kiwicha flour, they found an increase in protein and fiber levels by 32% and 152%, respectively, while sensory quality was not affected. In a subsequent large study, Alvarez-Jubete et al. (2009) and Alvarez-Jubete et al. (2010a) evaluated kiwicha, quinoa, and buckwheat flours as possible healthy ingredients to improve the nutritional quality of gluten-free breads. These grains and the gluten-free breads made from them were evaluated in terms of protein, starch, total fat, dietary fiber, ash, and mineral content, as well as their fatty acid composition. These works show that gluten-free breads, formulated from these grains, have significantly higher levels of protein, fat, fiber, and minerals than control bread. The attributes of these breads conform to the nutritional recommendations of gluten-free diet experts (Thompson 2000; Thompson et al. 2005). These results suggest that kiwicha, quinoa, and buckwheat may represent a healthy alternative as ingredients in the production of gluten-free products. Since 2010, there has been a growing interest in research on Andean grains in the gluten-free product industry, highlighting the contributions made by Schoenlechner et al. (2010), Föste et al. (2014), Elgeti et al. (2014), and Mäkinen et al. (2013), among others.

The inclusion of Andean grains such as quinoa and kiwicha in the formulation of gluten-free breads has also been studied from a techno-functional point of view, since it has been shown that their components contribute to improving the water retention capacity of the dough (Encina-Zelada et al. 2018) and the texture of the crumb (Alvarez-Jubete et al. 2010b) and can contribute to increasing the specific volume of bread (Alencar et al. 2015). With regard to tarwi flour, few investigations have introduced this raw material in the production of bakery products with and without gluten (Rosell et al. 2009; Vidaurre-Ruiz 2020). However, there is evidence of the use of lupine flours and protein concentrates to improve the rheological properties of gluten-free doughs and breads (Ziobro et al. 2013, 2016; Foschia et al. 2017).

Few investigations have evaluated the independent influence of Andean grain flours in the formulation of gluten-free breads; the vast majority of Andean grains are included in the formulations between 7.5% and 50%, while the other ingredients tend to be starches, rice flour, or corn flour. Recently, it has been reported that the moderate inclusion of quinoa flour (10% and 30%), kiwicha flour (10% and 30%) and tarwi flour (10%) contribute positively to the rheological (consistency index) and textural (firmness, consistency, cohesiveness, and viscosity index) of gluten-free doughs based on potato starch with a constant water content (75%). This phenomenon is due to the lipids of Andean grains, which provide fluidity to the dough, as well as to proteins with high water-absorption capacity. The percentage of Andean grain flours can be increased in gluten-free bread formulations by optimizing the gum and water content in the formulation (Vidaurre-Ruiz et al. 2020).

The correct proportion of ingredients such flour, starch, gum, and water content is essential for the development of a dough with a structure that allows for gas retention during fermentation. One of the gums most used in the formulations of gluten-free breads from Andean grain flours is xanthan gum (XG). This gum is a high-molecular-weight extracellular heteropolysaccharide (~1000 kDa) secreted by the microorganism *Xanthomonas campestris* (Horstmann et al. 2018). It consists of repeating units of D-glucose linked to form the β-1,4-D-glucan cellulosic backbone. Recent studies in gluten-free doughs containing 20% quinoa flour show that the hydration level of the dough is closely linked to the percentage of xanthan gum used in the formulation, as well as the protein content of the quinoa flour (Encina-Zelada et al. 2018), because the 11S globulins of quinoa (*chenopodin*) could have the ability to absorb cold water, and this could contribute to the increase in the firmness of the doughs. According to Abugoch et al. (2009) and Ruiz et al. (2016), the 11S globulin from quinoa has a similar structure to that of glycinin from soybeans. Xanthan gum has been shown to trap starch granules, thus preventing complete starch swelling resulting in decreased pulp properties of systems (Horstmann et al. 2018). This effect is related to the chemical nature of xanthan gum, since this gum is an anionic hydrocolloid with two negatively charged carboxyl groups in its side chains that allows it to interact with starch and water through hydrogen bonds. The hydroxyl groups in the xanthan gum backbone are known to associate with water and starch, as well as other xanthan gum chains, forming a rigid gel at high concentrations (Crockett et al. 2011). The interaction of xanthan gum with starch and water can be seen in the lower moisture loss during baking in gluten-free breads, which maintains crumb moisture (Vidaurre-Ruiz et al. 2019).

Another gum with the potential for incorporation into gluten-free bread formulations is tara gum (TG). This gum is an indigenous hydrocolloid from Peru (Wu et al. 2015) and is used as a thickener and stabilizer in different semi-liquid products. TG is an energy reserve polysaccharide in seeds of endosperm leguminous plants (Mirhosseini and Amid 2012). It has a high molecular weight (1000 kDa; Ma et al. 2017) and consists of a linear main chain of (1–4)-β-D-mannopyranose units attached by (1–6) linkages with α-D-galactopyranose units, with a 3:1 ratio of mannose to galactose (Wu et al. 2015). It has been reported that TG, due to its neutral nature, does not trap starch, so it forms a laminar structure in the continuous phase, thus allowing the starch granules to swell freely, evidenced in the increase in the pasting properties of the gluten-free formulations (Vidaurre-Ruiz et al. 2019). There are few investigations that report the use of TG in the formulation for gluten-free breads, but it is common to find reports of other galactomannans, such as guar gum and locust bean gum. The use of guar gum has recently been reported to obtain gluten-free breads (based on rice flour 50%, corn flour 30%, and quinoa flour 20%), showing good appearance in terms of high specific volume, low crumb hardness, and high crumb elasticity in formulations with 110% water with 2.5% or 3.0% guar gum (Encina-Zelada et al. 2019).

The synergistic effect of TG with XG in gluten-free breads based on potato starch and quinoa flour (12.6%), as well as in breads based on potato starch and kiwicha flour (11.6%), has been recently reported in the research of Vidaurre-Ruiz (2020). In this investigation, it is pointed out that the mixture of both gums notably improves the textural and rheological properties of gluten-free doughs (Figure 2.2). The synergistic effect of xanthan gum with galactomannans is related to the mannose to galactose (M:G) ratio. Guar gum, which has an M:G ratio of about 2:1, is known to exhibit weak synergy with xanthan gum, while locust bean gum, with an M:G ratio of about 4:1, reacts strongly with xanthan gum (Urlacher and Noble 1997). Likewise, Vidaurre-Ruiz (2020) also finds that it is possible to achieve desirable textural properties of the doughs using a higher proportion of quinoa flours (44.7%) or kiwicha flour (46.2%). In this case, a smaller quantity of the mixture of gums (0.5%) and a moderate water content (75%–77%) were required. This result verifies that components of Andean grain flours, such as lipids (monoglycerides, free fatty acids), proteins, and the low amylose content (8%–9%) of their starch, work as natural emulsifiers in the dough, producing bread with soft crumbs (Alvarez-Jubete et al. 2010b; Steffolani et al. 2013; Vidaurre-Ruiz 2020).

The inclusion of Andean grain flours has a marked effect on the pasta properties of gluten-free doughs. It has been reported that the peak viscosity (PV) of quinoa and kiwicha flours were related to the volume of gluten-free breads, showing that at higher PV values, a greater volume was achieved in the final product (Alvarez-Jubete et al. 2010b). Likewise, Fuentes et al. (2019) point out that the low values of break viscosity (BD) of starches from Andean grains indicate that their starches have a high capacity to resist shear forces and heating to elevated temperatures. This characteristic is favorable for baked products. BD is closely related to the process of retrogradation of starches during cooling (Ragaee and Abdel-Aal 2006) and has been reported to be a parameter correlated with bread staling (Rodriguez-Sandoval et al. 2015); therefore, it can be expected that formulations with a high content of Andean grains such as quinoa or kiwicha will take a longer time to retrograde than formulations that contain a higher proportion of other starches (Vidaurre-Ruiz 2020).

The specific volume and texture properties of the crumbs of gluten-free breads with Andean grain flours in their composition is widely varied. This refers to the different proportions of gluten-free or starch-free flours that are used, the amount and type of gum, as well as the water content. Table 2.2 summarizes the quality characteristics of formulations with quinoa, kiwicha, or tarwi flour. An Andean grain still little studied in its application in the preparation of gluten-free breads is kañiwa. It has recently been reported that it is possible to obtain gluten-free breads with 100% kañiwa flour with 0.9% xanthan gum and 140% water (Repo-Carrasco-Valencia et al. 2020). This formulation produced breads with high specific volume but with exceptionally low values of crumb hardening.

The inclusion of Andean grains in gluten-free bread formulations not only improves the dough rheology or crumb texture, but it would also substantially improve the nutritional value of gluten-free breads by increasing the content of protein, fiber, and minerals (Alvarez-Jubete et al. 2010a). An alternative to

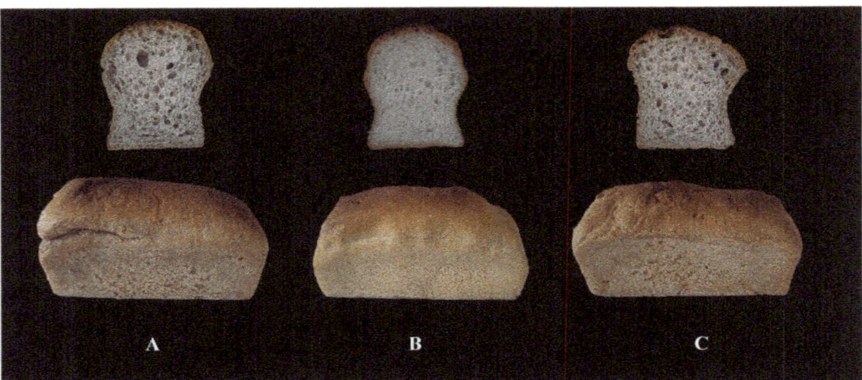

Figure 2.2 (A) Gluten-free bread with potato starch and quinoa flour. (B) Gluten-free bread with potato starch and kiwicha flour. (C) Gluten-free bread with potato starch and kañiwa flour.

Source: Photo taken by J. M. Vidaurre-Ruiz (2020).

TABLE 2.2 PHYSICAL CHARACTERISTICS OF GLUTEN-FREE BREADS WITH QUINOA, KIWICHA, KAÑIWA, OR TARWI FLOUR IN THEIR FORMULATION

Flour or starch base	Specific volume (mL/g)	Crumb firmness (N)	Reference
Rice flour (49.32%), potato starch (10.95%), cassava starch (16.43%), sour tapioca starch (3.3%), whole amaranth flour (20%)	2.30–2.65	3.87–5.36	Alencar et al. (2015)
Rice flour (49.32%), potato starch (10.95%), cassava starch (16.43%), sour tapioca starch (3.3%), whole quinoa flour (20%)	2.36–2.88	2.54–5.56	Alencar et al. (2015)
Rice flour (50%), corn flour (30%), and quinoa flour (20%)	2.41–2.92	21.28–29.30	Encina-Zelada et al. (2019)
Rice flour (10%–45%), corn flour (5%–22.5%), corn starch (25%), whole quinoa flour (7.5%–60%)	1.35–1.87	10.67–32.04	Föste et al. (2014)
Rice flour (50%), quinoa flour (50%)	1.4	30.4	Alvarez-Jubete et al. (2010b)
Rice flour (50%), kiwicha flour (50%)	1.31	17.85	Alvarez-Jubete et al. (2010b)
Rice flour (25%), potato starch (25%), quinoa flour (12.5%–50%), buckwheat flour (0–37.5%)	1.73–1.85	16.4–35	Turkut et al. (2016)
Potato starch (89.3%–52.9%), quinoa flour (10.7%–47.1%)	1.80–2.49	1.90–5.36	Vidaurre-Ruiz (2020)

(*Continued*)

TABLE 2.2 (CONTINUED)

Flour or starch base	Specific volume (mL/g)	Crumb firmness (N)	Reference
Potato starch (90%–53.8%), kiwicha flour (10%–46.2%)	2.04–2.79	3.48–4.42	Vidaurre-Ruiz (2020)
Tarwi flour (12%), quinoa flour (46%), potato starch (42%)	2.13	2.3	Repo-Carrasco-Valencia (2020)
Tarwi flour (10%), kiwicha flour (46%), potato starch (44%)	2.07	2.82	Vidaurre-Ruiz (2020)
Kañiwa flour (100%)	2.73	1.4	Repo-Carrasco-Valencia (2020)

increase the protein content in breads is the use of flours or protein isolates from legumes, since gluten-free breads usually have a lower protein content than wheat breads (Bravo-Núñez et al. 2019).

Few investigations have used lupine flour or protein concentrate in making gluten-free breads. Horstmann et al. (2017a) used 2% lupine protein concentrate (39% protein) in a formulation based on potato starch with 80% water. Ziobro et al. (2013) used 10% lupine protein concentrate (59% protein) in a formulation with 72% corn starch and 18% potato starch with 103.4% water, and in a later investigation, Ziobro et al. (2016) also reported the use of 10% lupine flour in the same formulation but with a lower water content (80.2%). Vidaurre-Ruiz (2020) reported that the inclusion of 12% tarwi flour in a formulation with 44.7% quinoa flour and 43.3% potato starch can contain 10.1% protein, 18.6% dietary fiber, and 3.3% minerals. Likewise, the formulation containing 10% tarwi flour, 46.2% kiwicha flour, and 43.8% potato starch contains 11.8% protein, 15.2% dietary fiber, and 3.2% minerals. The protein content of these breads is between two and three times higher than that of commercial gluten-free breads (Cornicelli et al. 2018).

The blend of quinoa or kiwicha flour with tarwi flour would also improve the amino acid balance in the product, since legumes are limiting in sulfur amino acids (methionine and cysteine; Villarino et al. 2016). In contrast, quinoa and kiwicha contain around 3.1 to 2.8 g/16 g of N-methionine and 5.6 to 6.0 g/16 g of N-lysine, respectively (Repo-Carrasco et al. 2003), making them good complements to the tarwi proteins.

The dietary fiber content in breads developed with Andean grain flours is also highlighted compared with the fiber in commercial breads (Vidaurre-Ruiz 2020). According to Tsatsaragkou et al. (2017), the fiber content in commercial gluten-free breads is above 3%, and gluten-free breads with up to 7.2% dietary fiber are reported in the literature. According to Wild et al. (2010), the gluten-free diet in celiac patients is deficient in fiber due to the low nutritional quality of the ingredients with which gluten-free products for daily consumption are made.

Regarding the content of minerals such as calcium and iron, bread with Andean grain flours can have three times more calcium and iron than control gluten-free bread that is made with 100% potato starch (Vidaurre-Ruiz 2020). According to Wild et al. (2010), the diet of celiac patients is deficient in calcium and iron, and whenever possible, doctors should make recommendations to increase the consumption of foods that contain these minerals. In summary, the composition of gluten-free breads with Andean grain flours presented a promising nutritional profile that can contribute to improving the diet of celiac patients and can be compared, or surpassed in some components, with that of wheat bread.

Subsequent studies should evaluate the digestibility of proteins and the bioavailability of minerals, as well as the possible functional effects that some of their bioactive compounds could have. For example, it is known that globulins are the main storage proteins in lupine grain, which are classified into four families: α-conglutin (11S globulin), β-conglutin (7S globulin), γ-conglutin (7S basic globulin), and conglutin δ (2S albumin rich in sulfur), highlighting γ-conglutin for its reported bioactivity, which can reduce blood glucose and therefore have properties beneficial to health (Villarino et al. 2016). Thus, the inclusion of tarwi flour in gluten-free bread formulations would not only improve the physical characteristics and increase the protein content in the breads but could also confer functional properties on the products. Although this aspect requires further study, the presence of γ-conglutin has already been reported in wheat breads with partial replacement of lupine flour (Villarino et al. 2015). Likewise, in the case of quinoa, the presence of flavonoids such as quercetin glycosides and kaempferol glycosides has been identified after the baking process of gluten-free breads formulated with 50% quinoa flour and 50% potato starch (Alvarez-Jubete et al. 2010c).

2.8 CONCLUSIONS

In the last decade, there has been a growing interest in the inclusion of flours of Andean grains in the formulations of gluten-free products. The inclusion of the flours of these native grains in gluten-free breads has been studied from a techno-functional point of view, and it has been shown that they can contribute to increasing the water retention capacity of the dough, improving the texture of the crumb and increasing the specific volume of the bread. The use of quinoa and kañiwa flours increases the content of good quality proteins, dietary fiber, and bioactive compounds content in the final product. The blend of quinoa or kiwicha flour with tarwi flour would optimize the amino acid balance in the bread, since legumes are limiting in sulfur amino acids. The starch of kiwicha with low amylose content improves the volume of the final product and contributes to retard the bread staling. In general, the use of Andean grain flours in gluten-free bread formulations not only improves the textural and sensorial properties, but it also substantially enhances the nutritional value of the final product.

ACKNOWLEDGMENTS

This study was supported partly by the Protein2Food project, which received funding from the European Union's Horizon 2020 Research and Innovation Framework Programme under grant agreement No 635727. We want to thank the Project CYTED 119RT0567, Spain.

REFERENCES

Abugoch, Lilian, Eduardo Castro, Cristian Tapia, María Cristina Añón, Pilar Gajardo, and Andrea Villarroel. 2009. "Stability of Quinoa Flour Proteins (*Chenopodium quinoa* Willd.) During Storage." *International Journal of Food Science and Technology* 44 (10): 2013–2020. https://doi.org/10.1111/j.1365-2621.2009.02023.x.

Alencar, Machado, Caroline Joy, Izabela Dutra, Elisa Carvalho, De Morais, Helena Maria, and Andre Bolini. 2015. "Addition of Quinoa and Amaranth Flour in Gluten-Free Breads : Temporal profile and Instrumental Analysis." *LWT—Food Science and Technology* 62: 1011–1018. https://doi.org/10.1016/j.lwt.2015.02.029.

Alvarez-Jubete, Laura, Elke K. Arendt, and Eimear Gallagher. 2009. "Nutritive Value and Chemical Composition of Pseudocereals as Gluten-Free Ingredients." *International Journal of Food Sciences and Nutrition* 60 (1): 240–257. https://doi.org/10.1080/09637480902950597.

Alvarez-Jubete, Laura, Elke K. Arendt, and Eimear Gallagher. 2010a. "Nutritive Value of Pseudocereals and Their Increasing Use as Functional Gluten-Free Ingredients." *Trends in Food Science and Technology* 21 (2). Elsevier Ltd: 106–113. https://doi.org/10.1016/j.tifs.2009.10.014.

Alvarez-Jubete, Laura, Mark Auty, Elke K. Arendt, and Eimear Gallagher. 2010b. "Baking Properties and Microstructure of Pseudocereal Flours in Gluten-Free Bread Formulations." *European Food Research and Technology* 230 (3): 437–445. https://doi.org/10.1007/s00217-009-1184-z.

Alvarez-Jubete, Laura, Hilde Wijngaard, Elke K. Arendt, and Eimear Gallagher. 2010c. "Polyphenol Composition and *in Vitro* Antioxidant Activity of Amaranth, Quinoa

Buckwheat and Wheat as Affected by Sprouting and Baking." *Food Chemistry* 119 (2): 770–778. https://doi.org/10.1016/j.foodchem.2009.07.032.

Antúnez de Mayolo, Santiago Eric. 1981. *La Nutricion En El Antiguo Perú: Banco Central de Reserva Del Perú*. Lima, Perú: Banco Central de Reserva.

Arendt, Elke K., Alice Moroni, and Emanuele Zannini. 2011. "Medical Nutrition Therapy: Use of Sourdough Lactic Acid Bacteria as a Cell Factory for Delivering Functional Biomolecules and Food Ingredients in Gluten Free Bread." *Microbial Cell Factories* 10 (Suppl 1). BioMed Central Ltd: 1–9. https://doi.org/10.1186/1475-2859-10-S1-S15.

Arendt, Elke K., Andrew Morrissey, Michelle M. Moore, and Fabio Dal Bello. 2008. "Gluten-Free Breads." In E. Arendt and F. Dal Bello (Eds.), *Gluten-Free Cereal Products and Beverages*. London: Academic Press, 289–319.

Arendt, Elke K., and Emanuele Zannini. 2013. *Cereal Grains for the Food and Beverage Industries*. Cambridge: Woodhead Publishing Limited.

Azizi, Saadat, Mohammad Hossein Azizi, Roxana Moogouei, and Peyman Rajaei. 2020. "The Effect of Quinoa Flour and Enzymes on the Quality of Gluten-free Bread." *Food Science & Nutrition* 8 (5): 2373–2382. https://doi.org/10.1002/fsn3.1527.

Baldera, Katherine, David Chaupis-Meza, César Cárcamo, King Holmes, and Patricia García. 2020. "Seroprevalencia Poblacional de La Enfermedad Celíaca En Zonas Urbanas Del Perú." *Revista Peruana de Medicina Experimental y Salud Pública* 37 (1): 63–66. https://doi.org/10.17843/rpmesp.2020.371.4507.

Benavent-Gil, Yaiza, and Cristina M. Rosell. 2019. "Technological and Nutritional Applications of Starches in Gluten-Free Products." *Starches for Food Application*: 333–358. https://doi.org/10.1016/B978-0-12-809440-2.00009-5.

Biesiekierski, Jessica R. 2017. "What Is Gluten?" *Journal of Gastroenterology and Hepatology* 32 (Suppl. 1): 78–81. https://doi.org/10.1111/jgh.13703.

Birk, Y., A. Dovrat, M. Waldmann, and C. Uzureau. 1990. *Lupin Production and Bio-Processing for Feed, Food and Other by-Products*. Brussels-Luxembourg: Proceedings of the Joint CEC-NCRD Workshop.

Bravo-Núñez, Ángela, Marta Sahagún, and Manuel Gómez. 2019. "Assessing the Importance of Protein Interactions and Hydration Level on Protein-Enriched Gluten-Free Breads: A Novel Approach." *Food and Bioprocess Technology* 12 (5): 820–828. https://doi.org/10.1007/s11947-019-02258-2.

Brites, Lara T. G. F., Marcio Schmiele, and Caroline J. Steel. 2018. "Gluten-Free Bakery and Pasta Products." *Alternative and Replacement Foods*: 385–410. https://doi.org/10.1016/B978-0-12-811446-9.00013-7.

Calvo-Lerma, Joaquim, Paula Crespo-Escobar, Sandra Martínez-Barona, Victoria Fornés-Ferrer, Ester Donat, and Carmen Ribes-Koninckx. 2019. "Differences in the Macronutrient and Dietary Fibre Profile of Gluten-Free Products as Compared to Their Gluten-Containing Counterparts." *European Journal of Clinical Nutrition* 73 (6): 930–936. https://doi.org/10.1038/s41430-018-0385-6.

Carbó, Rosa, Elena Gordún, Antía Fernández, and Marta Ginovart. 2020. "Elaboration of a Spontaneous Gluten-Free Sourdough with a Mixture of Amaranth, Buckwheat, and Quinoa Flours Analyzing Microbial Load, Acidity, and PH." *Food Science and Technology International* 26 (4): 344–352. https://doi.org/10.1177/1082013219895357.

Celiac Disease Foundation. 2020. "What Is Celiac Disease?" https://celiac.org/about-celiac-disease/what-is-celiac-disease/.

Clapassón, Priscila, Noelia B. Merino, Mercedes E. Campderrós, Maria F. Pirán Arce, and Ana N. Rinaldoni. 2020. "Assessment of Brea Gum as an Additive in the Development of a Gluten-Free Bread." *Journal of Food Measurement and Characterization* 14 (3): 1665–1670, Springer. https://doi.org/10.1007/s11694-020-00414-3.

Córdova-Ramos, Javier S., Patricia Glorio-Paulet, Felix Camarena, Andrea Brandolini, and Alyssa Hidalgo. 2020. "Andean Lupin (*Lupinus mutabilis* Sweet): Processing Effects on Chemical Composition, Heat Damage, and in Vitro Protein Digestibility." *Cereal Chemistry* 97 (4): 827–835. https://doi.org/10.1002/cche.10303.

Cornicelli, Miriam, Michela Saba, Nicoletta Machello, Marco Silano, and Susanna Neuhold. 2018. "Nutritional Composition of Gluten-Free Food versus Regular Food Sold in the Italian Market." *Digestive and Liver Disease* 50 (12). Editrice Gastroenterologica Italiana: 1305–1308. https://doi.org/10.1016/j.dld.2018.04.028.

Crockett, Rachel, Pauline Ie, and Yael Vodovotz. 2011. "How Do Xanthan and Hydroxypropyl Methylcellulose Individually Affect the Physicochemical Properties in a Model Gluten-Free Dough?" *Journal of Food Science* 76 (3): 274–282. https://doi.org/10.1111/j.1750-3841.2011.02088.x.

Das Gupta, Soumyasri, and Nanjoo Suh. 2016. "Tocopherols in Cancer: An Update." *Molecular Nutrition & Food Research* 60 (6): 1354–1363. https://doi.org/10.1002/mnfr.201500847.

Demirkesen, Ilkem, Behic Mert, Gulum Sumnu, and Serpil Sahin. 2010. "Rheological Properties of Gluten-Free Bread Formulations." *Journal of Food Engineering* 96 (2): 295–303. https://doi.org/10.1016/j.jfoodeng.2009.08.004.

Elgeti, Dana, Sebastian D. Nordlohne, Maike Föste, Marina Besl, Martin H. Linden, Volker Heinz, Mario Jekle, and Thomas Becker. 2014. "Volume and Texture Improvement of Gluten-Free Bread Using Quinoa White Flour." *Journal of Cereal Science* 59 (1): 41–47. https://doi.org/10.1016/j.jcs.2013.10.010.

Elgeti, Dana, Lu Yu, Andreas Stüttgen, Mario Jekle, and Thomas Becker. 2017. "Interrelation Between Mechanical and Biological Aeration in Starch-Based Gluten-Free Dough Systems." *Journal of Cereal Science* 76: 28–34. https://doi.org/10.1016/j.jcs.2017.05.008.

Encina-Zelada, Christian R., Vasco Cadavez, Fernando Monteiro, José A. Teixeira, and Ursula Gonzales-Barron. 2018. "Combined Effect of Xanthan Gum and Water Content on Physicochemical and Textural Properties of Gluten-Free Batter and Bread." *Food Research International* 111: 544–555. https://doi.org/10.1016/j.foodres.2018.05.070.

Encina-Zelada, Christian R., Vasco Cadavez, Fernando Monteiro, José A. Teixeira, and Ursula Gonzales-Barron. 2019. "Physicochemical and Textural Quality Attributes of Gluten-Free Bread Formulated with Guar Gum." *European Food Research and Technology* 245 (2): 443–458. https://doi.org/10.1007/s00217-018-3176-3.

Falcomer, Ana Luísa, Bruna Araújo Luchine, Hanna Ramalho Gadelha, José Roberto Szelmenczi, Eduardo Yoshio Nakano, Priscila Farage, and Renata Puppin Zandonadi. 2020. "Worldwide Public Policies for Celiac Disease: Are Patients Well Assisted?" *International Journal of Public Health* 65 (6): 937–945. https://doi.org/10.1007/s00038-020-01451-x.

Foschia, Martina, Stefan W. Horstmann, Elke K. Arendt, and Emanuele Zannini. 2017. "Legumes as Functional Ingredients in Gluten-Free Bakery and Pasta Products." *Annual Review of Food Science and Technology* 8 (1): 75–96. https://doi.org/10.1146/annurev-food-030216-030045.

Föste, Maike, Sebastian D. Nordlohne, Dana Elgeti, Martin H. Linden, Volker Heinz, Mario Jekle, and Thomas Becker. 2014. "Impact of Quinoa Bran on Gluten-Free Dough and Bread Characteristics." *European Food Research and Technology* 239 (5): 767–775. https://doi.org/10.1007/s00217-014-2269-x.

Fuentes, Catalina, Daysi Perez-Rea, Björn Bergenståhl, Sergio Carballo, Malin Sjöö, and Lars Nilsson. 2019. "Physicochemical and Structural Properties of Starch from Five Andean Crops Grown in Bolivia." *International Journal of Biological Macromolecules* 125: 829–838, March. https://doi.org/10.1016/j.ijbiomac.2018.12.120.

Gade, Daniel W. 1970. "Ethnobotany of Cañihua (*Chenopodium pallidicaule*), Rustic Seed Crop of the Altiplano." *Economic Botany* 24 (1): 55–61. https://doi.org/10.1007/BF02860637.

Gambuś, Halina, Florian Gambuś, and Renata Sabat. 2002. "The Research on Quality Improvement of Gluten-Free Bread by Amaranthus Flour Addition." *Zywnosc* 2 (31): 99–112.

Gordillo-Bastidas, E., D. A. Díaz-Rizzolo, E. Roura, T. Massanés, and R. Gomis. 2016. "Quinoa (Chenopodium Quinoa Willd), from Nutritional Value to Potential Health Benefits: An Integrative Review." *Journal of Nutrition and Food Sciences* 6 (3): 1–10. https://doi.org/10.4172/2155-9600.1000497.

Haghighat-Kharazi, Sepideh, Mohammad Reza Kasaai, Jafar Mohammadzadeh Milani, and Khosro Khajeh. 2020a. "Optimization of Encapsulation of Maltogenic Amylase into a Mixture of Maltodextrin and Beeswax and Its Application in Gluten-free Bread." *Journal of Texture Studies* 51 (4): 631–641. https://doi.org/10.1111/jtxs.12516.

Haghighat-Kharazi, Sepideh, Mohammad Reza Kasaai, Jafar Mohammadzadeh Milani, and Khosro Khajeh. 2020b. "Antistaling Properties of Encapsulated Maltogenic Amylase in Gluten-free Bread." *Food Science and Nutrition*: 1–10, September. https://doi.org/10.1002/fsn3.1865.

Hamada, Shigeki, Keitaro Suzuki, Noriaki Aoki, and Yasuhiro Suzuki. 2013. "Improvements in the Qualities of Gluten-Free Bread After Using a Protease Obtained from Aspergillus Oryzae." *Journal of Cereal Science* 57 (1): 91–97. https://doi.org/10.1016/j.jcs.2012.10.008.

Hartmann, Christina, Sophie Hieke, Camille Taper, and Michael Siegrist. 2018. "European Consumer Healthiness Evaluation of 'Free-from' Labelled Food Products." *Food Quality and Preference* 68: 377–388, September. https://doi.org/10.1016/j.foodqual.2017.12.009.

Hausch, Felix, Lu Shan, Nilda A. Santiago, Gary M. Gray, and Chaitan Khosla. 2002. "Intestinal Digestive Resistance of Immunodominant Gliadin Peptides." *American Journal of Physiology-Gastrointestinal and Liver Physiology* 283 (4): G996–G1003. https://doi.org/10.1152/ajpgi.00136.2002.

Horstmann, Stefan, Jonas Atzler, Mareile Heitmann, Emanuele Zannini, Kieran Lynch, and Elke Arendt. 2019. "A Comparative Study of Gluten-Free Sprouts in the

Gluten-Free Bread-Making Process." *European Food Research and Technology* 245 (3): 617–629. https://doi.org/10.1007/s00217-018-3185-2.

Horstmann, Stefan, Claudia Axel, and Elke Arendt. 2018. "Water Absorption as a Prediction Tool for the Application of Hydrocolloids in Potato Starch-Based Bread." *Food Hydrocolloids* 81: 129–138. https://doi.org/10.1016/j.foodhyd.2018.02.045.

Horstmann, Stefan, Martina Foschia, and Elke Arendt. 2017a. "Correlation Analysis of Protein Quality Characteristics with Gluten-Free Bread Properties." *Food and Function: Royal Society of Chemistry* 8 (7): 2465–2474. https://doi.org/10.1039/C7FO00415J.

Horstmann, Stefan, Kieran Lynch, and Elke Arendt. 2017b. "Starch Characteristics Linked to Gluten-Free Products." *Foods* 6 (4): 29. https://doi.org/10.3390/foods6040029.

Houben, Andreas, Agnes Höchstötter, and Thomas Becker. 2012. "Possibilities to Increase the Quality in Gluten-Free Bread Production: An Overview." *European Food Research and Technology* 235 (2): 195–208. https://doi.org/10.1007/s00217-012-1720-0.

Kawamura-Konishi, Yasuko, Kazuo Shoda, Hironori Koga, and Yuji Honda. 2013. "Improvement in Gluten-Free Rice Bread Quality by Protease Treatment." *Journal of Cereal Science* 58 (1): 45–50. https://doi.org/10.1016/j.jcs.2013.02.010.

Kozioł, Michael J. 1992. "Chemical Composition and Nutritional Evaluation of Quinoa (Chenopodium Quinoa Willd.)." *Journal of Food Composition and Analysis* 5 (1): 35–68. https://doi.org/10.1016/0889-1575(92)90006-6.

Kraujalis, Paulius, and Petras Rimantas Venskutonis. 2013. "Supercritical Carbon Dioxide Extraction of Squalene and Tocopherols from Amaranth and Assessment of Extracts Antioxidant Activity." *The Journal of Supercritical Fluids* 80: 78–85. https://doi.org/10.1016/j.supflu.2013.04.005.

Ma, Qianyun, Lin Du, Yang Yang, and Lijuan Wang. 2017. "Rheology of Film-Forming Solutions and Physical Properties of Tara Gum Film Reinforced with Polyvinyl Alcohol (PVA)." *Food Hydrocolloids* 63: 677–684, February. https://doi.org/10.1016/j.foodhyd.2016.10.009.

Mäkinen, Outi E., Emanuele Zannini, and Elke K. Arendt. 2013. "Germination of Oat and Quinoa and Evaluation of the Malts as Gluten Free Baking Ingredients." *Plant Foods for Human Nutrition* 68 (1): 90–95. https://doi.org/10.1007/s11130-013-0335-3.

Marcone, Massimo F., Yukio Kakuda, and Rickey Y. Yada. 2003. "Amaranth as a Rich Dietary Source of β-Sitosterol and Other Phytosterols." *Plant Foods for Human Nutrition* 58 (3): 207–211. https://doi.org/10.1023/B:QUAL.0000040334.99070.3e.

Mir, Shabir Ahmad, Manzoor Ahmad Shah, Haroon Rashid Naik, and Imtiyaz Ahmad Zargar. 2016. "Influence of Hydrocolloids on Dough Handling and Technological Properties of Gluten-Free Breads." *Trends in Food Science and Technology* 51: 49–57. https://doi.org/10.1016/j.tifs.2016.03.005.

Mirhosseini, Hamed, and Bahareh Tabatabaee Amid. 2012. "A Review Study on Chemical Composition and Molecular Structure of Newly Plant Gum Exudates and Seed Gums." *Food Research International* 46 (1): 387–398. https://doi.org/10.1016/j.foodres.2011.11.017.

Moroni, Alice V., Fabio Dal Bello, and Elke K. Arendt. 2009. "Sourdough in Gluten-Free Bread-Making: An Ancient Technology to Solve a Novel Issue?" *Food Microbiology* 26 (7): 676–684. https://doi.org/10.1016/j.fm.2009.07.001.

National Research Council. 1989. *Lost Crops of the Incas: Little-Known Plants of the Andes with Promise for Worldwide Cultivation.* Washington, DC: National Academy Press. https://doi.org/10.2307/3673751.

Ogrodowska, Dorota, Ryszard Zadernowski, Sylwester Czaplicki, Dorota Derewiaka, and Beata Wronowska. 2014. "Amaranth Seeds and Products—The Source of Bioactive Compounds." *Polish Journal of Food and Nutrition Sciences* 64 (3): 165–170. https://doi.org/10.2478/v10222-012-0095-z.

Ragaee, Sanaa, and El-Sayed M. Abdel-Aal. 2006. "Pasting Properties of Starch and Protein in Selected Cereals and Quality of Their Food Products." *Food Chemistry* 95 (1): 9–18. https://doi.org/10.1016/j.foodchem.2004.12.012.

Reilly, Norelle R. 2016. "The Gluten-Free Diet: Recognizing Fact, Fiction, and Fad." *The Journal of Pediatrics* 175: 206–210. https://doi.org/10.1016/j.jpeds.2016.04.014.

Repo-Carrasco-Valencia, Ritva. 2020. "Nutritional Value and Bioactive Compounds in Andean Ancient Grains." *Proceedings* 53 (1): 1–5. https://doi.org/10.3390/proceedings2020053001.

Repo-Carrasco, Ritva, Clara Espinoza, and Sven-Erik Jacobsen. 2003. "Nutritional Value and Use of the Andean Crops Quinoa (*Chenopodium quinoa*) and Kañiwa (*Chenopodium pallidicaule*)." *Food Reviews International* 19 (1–2): 179–189. https://doi.org/10.1081/fri-120018884.

Repo-Carrasco-Valencia, Ritva, Jarkko K. Hellström, Juha-Matti Pihlava, and Pirjo H. Mattila. 2010. "Flavonoids and Other Phenolic Compounds in Andean Indigenous Grains: Quinoa (Chenopodium Quinoa), Kañiwa (Chenopodium Pallidicaule) and Kiwicha (*Amaranthus caudatus*)." *Food Chemistry* 120: 128–133. https://doi.org/10.1016/j.foodchem.2009.09.087.

Repo-Carrasco-Valencia, R., Silvia Melgarejo-Cabello, and Juha Matti Pihlava. 2019a. "Nutritional Value and Bioactive Compounds in Quinoa (*Chenopodium quinoa* Willd.), Kañiwa (*Chenopodium pallidicaule* Aellen) and Kiwicha (*Amaranthus caudatus* L.)." In P. G. Peiretti and F. Gai (Eds.), *Quinoa: Cultivation, Nutritional Properties and Effects on Health.* New York: Nova Science Publishers, Inc., 83–113.

Repo-Carrasco-Valencia, Ritva, and Julio Mauricio Vidaurre-Ruiz. 2019b. "Quinoa and Other Andean Ancient Grains: Super Grains for the Future." *Cereal Foods World* 64 (5): 1–10. https://doi.org/10.1094/CFW-64-5-0053.

Repo-Carrasco-Valencia, Ritva, Julio Vidaurre-Ruiz, and Genny Isabel Luna-Mercado. 2020. "Development of Gluten-Free Breads Using Andean Native Grains Quinoa, Kañiwa, Kiwicha and Tarwi." *Proceedings* 53 (1): 15. https://doi.org/10.3390/proceedings2020053015.

Rodriguez-Sandoval, Eduardo, Misael Cortes-Rodriguez, and Katherine Manjarres-Pinzon. 2015. "Effect of Hydrocolloids on the Pasting Profiles of Tapioca Starch Mixtures and the Baking Properties of Gluten-Free Cheese Bread." *Journal of Food Processing and Preservation* 39 (6): 1672–1681. https://doi.org/10.1111/jfpp.12398.

Roman, Laura, Mayara Belorio, and Manuel Gomez. 2019. "Gluten-Free Breads: The Gap Between Research and Commercial Reality." *Comprehensive Reviews in Food Science and Food Safety* 18 (3): 690–702. https://doi.org/10.1111/1541-4337.12437.

Rosell, Cristina M., Gladys Cortez, and Ritva Repo-Carrasco. 2009. "Breadmaking Use of Andean Crops Quinoa, Kañiwa, Kiwicha, and Tarwi." *Cereal Chemistry* 86 (4): 386–392. https://doi.org/10.1094/CCHEM-86-4-0386.

Ruiz, Geraldine Avila, Wukai Xiao, Martinus Van Boekel, Marcel Minor, and Markus Stieger. 2016. "Effect of Extraction PH on Heat-Induced Aggregation, Gelation and Microstructure of Protein Isolate from Quinoa (Chenopodium Quinoa Willd)." *Food Chemistry* 209: 203–210. https://doi.org/10.1016/j.foodchem.2016.04.052.

Rybicka, Iga, Karolina Doba, and Olga Bińczak. 2019. "Improving the Sensory and Nutritional Value of Gluten-free Bread." *International Journal of Food Science & Technology* 54 (9): 2661–2667. https://doi.org/10.1111/ijfs.14190.

Sabanis, Dimitrios, and Constantina Tzia. 2011. "Effect of Hydrocolloids on Selected Properties of Gluten-Free Dough and Bread." *Food Science and Technology International* 17 (4): 279–291. https://doi.org/10.1177/1082013210382350.

Schoenlechner, Regine, Ioanna Mandala, Alexandra Kiskini, Athanasios Kostaropoulos, and Emmerich Berghofer. 2010. "Effect of Water, Albumen and Fat on the Quality of Gluten-Free Bread Containing Amaranth." *International Journal of Food Science and Technology* 45 (4): 661–669. https://doi.org/10.1111/j.1365-2621.2009.02154.x.

Sciarini, Lorena, Pablo Ribotta, Alberto Edel León, and Gabriela Teresa Pérez. 2012. "Incorporation of Several Additives into Gluten Free Breads: Effect on Dough Properties and Bread Quality." *Journal of Food Engineering* 111 (4): 590–597. https://doi.org/10.1016/j.jfoodeng.2012.03.011.

Steffolani, María Eugenia, Alberto Edel León, and Gabriela Teresa Pérez. 2013. "Study of the Physicochemical and Functional Characterization of Quinoa and Kañiwa Starches." *Starch/Staerke* 65 (11–12): 976–983. https://doi.org/10.1002/star.201200286.

Thompson, Tricia. 2000. "Folate, Iron, and Dietary Fibre Contents of the Gluten-Free Diet." *Journal of the American Dietetic Association* 100 (11): 1389–1396. https://doi.org/10.1016/S0002-8223(00)00386-2.

Thompson, Tricia, Melinda Dennis, Laurie Higgins, Anne R. Lee, and Mary K. Sharrett. 2005. "Gluten-Free Diet Survey: Are Americans with Coeliac Disease Consuming Recommended Amounts of Fibre, Iron, Calcium and Grain Foods?" *Journal of Human Nutrition and Dietetics* 18 (3): 163–169. https://doi.org/10.1111/j.1365-277X.2005.00607.x.

Tovoli, Francesco. 2015. "Clinical and Diagnostic Aspects of Gluten Related Disorders." *World Journal of Clinical Cases* 3 (3): 275. https://doi.org/10.12998/wjcc.v3.i3.275.

Tsatsaragkou, Kleopatra, Theodora Kara, Christos Ritzoulis, Ioanna Mandala, and Cristina M Rosell. 2017. "Improving Carob Flour Performance for Making Gluten-Free Breads by Particle Size Fractionation and Jet Milling." *Food and Bioprocess Technology*: 831–841. https://doi.org/10.1007/s11947-017-1863-x.

Tsatsaragkou, Kleopatra, Styliani Protonotariou, and Ioanna Mandala. 2016. "Structural Role of Fibre Addition to Increase Knowledge of Non-Gluten Bread." *Journal of Cereal Science* 67: 58–67. https://doi.org/10.1016/j.jcs.2015.10.003.

Turkut, Gulsum M., Hulya Cakmak, Seher Kumcuoglu, and Sebnem Tavman. 2016. "Effect of Quinoa Flour on Gluten-Free Bread Batter Rheology and Bread Quality." *Journal of Cereal Science* 69: 174–181. https://doi.org/10.1016/j.jcs.2016.03.005.

Urlacher, B., and O. Noble. 1997. "Xanthan Gum." In *Thickening and Gelling Agents for Food*. Boston, MA: Springer, 284–311. https://doi.org/10.1007/978-1-4615-2197-6_13.

Vega-Gálvez, Antonio, Margarita Miranda, Judith Vergara, Elsa Uribe, Luis Puente, and Enrique A. Martínez. 2010. "Nutrition Facts and Functional Potential of Quinoa

(Chenopodium Quinoa Willd.), an Ancient Andean Grain: A Review." *Journal of the Science of Food and Agriculture* 90 (15): 2541–2547. https://doi.org/10.1002/jsfa.4158.

Vidaurre-Ruiz, Julio. 2020. "Desarrollo de Panes Libres de Gluten con Harinas de Granos Andinos." Ph.D. diss., Universidad Nacional Agraria La Molina, Lima, Peru.

Vidaurre-Ruiz, Julio, Shessira Matheus-Diaz, Francisco Salas-Valerio, Gabriela Barraza-Jauregui, Regine Schoenlechner, and Ritva Repo-Carrasco-Valencia. 2019. "Influence of Tara Gum and Xanthan Gum on Rheological and Textural Properties of Starch-Based Gluten-Free Dough and Bread." *European Food Research and Technology* 245 (7): 1347–1355. https://doi.org/10.1007/s00217-019-03253-9.

Villarino, Casiana Blanca Jucar, Vijay Jayasena, Ranil Coorey, Sumana Chakrabarti-Bell, Rhonda C. Foley, Kent Fanning, and Stuart Johnson. 2015. "The Effects of Lupin (*Lupinus angustifolius*) Addition to Wheat Bread on Its Nutritional, Phytochemical and Bioactive Composition and Protein Quality." *Food Research International* 76 (P1): 58–65. https://doi.org/10.1016/j.foodres.2014.11.046.

Villarino, Casiana Blanca Jucar, Vijay Jayasena, Ranil Coorey, Sumana Chakrabarti-Bell, and Stuart Johnson. 2016. "Nutritional, Health, and Technological Functionality of Lupin Flour Addition to Bread and Other Baked Products: Benefits and Challenges." *Critical Reviews in Food Science and Nutrition* 56 (5): 835–857. https://doi.org/10.1080/10408398.2013.814044.

Wieser, Herbert. 2007. "Chemistry of Gluten Proteins." *Food Microbiology* 24 (2): 115–119. https://doi.org/10.1016/j.fm.2006.07.004.

Wieser, Herbert, Peter Koehler, and Katharina Konitzer. 2014. *Celiac Disease and Gluten: Multidisciplinary Challenges and Opportunities*. London: Elsevier. https://doi.org/10.1016/C2013-0-13353-3.

Wild, D., G. G. Robins, V. J. Burley, and P. D. Howdle. 2010. "Evidence of High Sugar Intake, and Low Fibre and Mineral Intake, in the Gluten-Free Diet." *Alimentary Pharmacology and Therapeutics* 32 (4): 573–581. https://doi.org/10.1111/j.1365-2036.2010.04386.x.

Wu, Yanbei, Wei Ding, Lirong Jia, and Qiang He. 2015. "The Rheological Properties of Tara Gum (*Caesalpinia spinosa*)." *Food Chemistry* 168: 366–371. https://doi.org/10.1016/j.foodchem.2014.07.083.

Yang, Chung S., and Nanjoo Suh. 2012. "Cancer Prevention by Different Forms of Tocopherols." *Peptide-Based Materials* 310: 21–33. https://doi.org/10.1007/128_2012_345.

Zheleznov, A. V., L. P. Solonenko, and N. B. Zheleznova. 1997. "Seed Proteins of the Wild and the Cultivated Amaranthus Species." *Euphytica* 97 (2): 177–182. https://doi.org/10.1023/A:1003073804203.

Ziobro, Rafał, Lesław Juszczak, Mariusz Witczak, and Jarosław Korus. 2016. "Non-Gluten Proteins as Structure Forming Agents in Gluten Free Bread." *Journal of Food Science and Technology* 53 (1): 571–580. https://doi.org/10.1007/s13197-015-2043-5.

Ziobro, Rafał, Teresa Witczak, Lesław Juszczak, and Jarosław Korus. 2013. "Supplementation of Gluten-Free Bread with Non-Gluten Proteins: Effect on Dough Rheological Properties and Bread Characteristic." *Food Hydrocolloids* 32 (2): 213–220. https://doi.org/10.1016/j.foodhyd.2013.01.006.

Chapter 3

Biodiversity of Andean Maize (*Zea mayz*). Nutritional, Functional, and Technological Properties

Norma Sammán, Alejandra Gimenez,
Cristina Segundo, and Manuel Lobo

CONTENTS

3.1 ORIGINS AND BIODIVERSITY OF MAIZE

Mexico is the region of origin, domestication, and diversification of maize (*Zea mays* L.) from its wild progenitor, teosinte (*Zea mays* ssp. *parviglumis*). Development bases of great Ancient civilizations—the Olmecs, Mayans,

DOI: 10.1201/9781003087618-3

Teotihuacanes, Toltecas, and Mixtecas, among others—were seated on maize cultivations, where teosinte was domesticated to improve the morphological and genetic diversity (Serratos-Hernández 2012; Matsuoka et al. 2002).

The word *corn* has been used over time for different meanings; however, after the discovery of America, Europeans called it *Indian corn*; it was then named *mahis* in the Native American language, which was later called *maize* (Salvador-Reyes and Pedrosa Silva Clerici 2020).

From the beginning of its domestication, maize was taken by indigenous communities while moving through different regions of the American continent. One of the transfer trails headed toward the lowlands of southern Mexico and reached the central Andes (Matsuoka et al. 2002; Vigouroux et al. 2008). In the Andean region, where the Incas had achieve advanced agricultural techniques (sowing in terraces and furrows, irrigation, and fertilization), maize found suitable conditions to settle down and diversify in races throughout Ecuador, Peru, Bolivia, Chile, and Argentina, taking advantage of the different ecosystems (Galinat 1992; Salvador Reyes and Pedrosa Silva Clerici 2020). The magnitude of development achieved in Peru turned it into the second center of domestication and diversification of maize (Grobman et al. 1961).

Maize diversity is characterized by numerous varieties, which acquire different morphological and genetic characteristics when adapting to diverse ecosystems, crossing and human selection resulting in a large number of subvarieties or races (Vigouroux et al. 2008; McClintock et al. 1981).

The ancient civilizations of Mesoamerica and South America reached a high domain of agriculture techniques, such as seed adaptation to different environments, sowing, fertilizing, and harvesting. Furthermore, ancestral knowledge regarding nutritional properties of the varieties used in food and beverage preparation was passed over generations until the arrival of the Spanish conquerors, who mitigated or eliminated these traditions with the loss of essential information of maize biodiversity and the conservation (León-Portilla 2002; Pope et al. 2001).

During the late 19th and mid-20th centuries, the first characterizations of maize biodiversity were carried out. Initially, the plant morphological characteristics were used, especially the shape, size, and color of the cob. Then, the sowing place began to be taken into consideration, due to the phenotypic differences observed in the same breeds grown in different places. Later assays used physiological, agronomic, and genetic features. Currently, the large volume of information is organized through multivariate statistics into dendrograms to define differences and affinities between races (Serratos et al. 1987; Carvalho et al. 2004; Labate et al. 2003).

In Mexico, most of the researchers agree with a multicentric theory of maize origin, identifying five domestication centers. This situation strongly determines the biodiversity characterization in the largest center of origin, domestication and diversification of maize. Therefore, the maize diversity in Mexico is described by racial complexes, characterized by the racial richness of territory and genetic identification with chromosomal nodes. In this way, the racial complexes Zapalote, Pepitilla, Mesa Central, Altos de Guatemala, and Tuxpeño were defined with the identification of around 60 races (Kato et al. 2009).

In Peru, according to the morphological and cytogenetic description, 55 races of maize were identified, which can be divided into five groups: primitive (early plants with small cobs and grains); first-step derivation (plants between the primitive and immediate derivatives); second-step derivation (plants with greater vegetative development and better grain yield); introduced races (foreign races crossed with Peruvian natives); and incipient races (in development in small plots); (Salvador-Reyes and Pedrosa Silva Clerici 2020).

3.2 STRUCTURE OF THE MAIZE GRAIN

The chemical characteristics and size of the main structural components (endosperm, germ, and pericarp) of maize grain are affected by several features such as the genotype, type of endosperm, environment, and agronomic conditions (Vázquez-Carrillo et al. 2017). The main parts of the maize kernel are shown in Figure 3.1.

Pericarp, endosperm, germ, and pedicel are the main anatomical parts of a mature maize kernel. The pericarp is the outer covering; aleurone is a set of protein granules in the external part of the endosperm; the endosperm is the site where the seed reserves are located; starch granules and proteins are the main components. The germ or embryo contains fatty matter and especially the biological machinery for germination; the pedicel is the remaining tissue

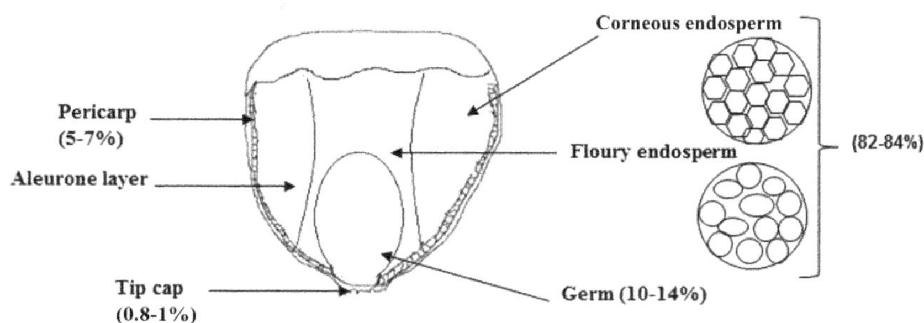

Figure 3.1 A cross section of the maize grain and percentage in weight of its parts.

TABLE 3.1 AVERAGE CHEMICAL COMPOSITION (g/100g D.B.) OF THE MAIN MAIZE STRUCTURES

Component	Endosperm	Germ	Pericarp	Pedicel
Starch	87	8.3	7.3	5.3
Proteins	8.0	18.4	3.7	9.1
Lipids	0.8	33.2	1.0	3.8
Ash	0.3	10.5	0.8	1.6
Others*	3.9	29.6	87.2	80.2

Sources: Pérez de la Cerda et al. (2007), Shukla and Cheryan (2001).
* By difference, it includes fiber and soluble sugars.

where the grain connects to the cob. The average chemical composition of the main parts of maize grains is shown in Table 3.1.

3.3 MORPHOLOGICAL CHARACTERISTICS. VARIABILITY BETWEEN RACES

The vast colors and size variability among some of the morphological characteristics of the native maize races in Mexico and throughout Latin America are mainly due to the diversity of climatic and geographical conditions to which this crop is able to adapt. Also, the reproductive system that depends on cross-pollination and seed exchange between producers results in the characteristics of different Andean maize races (Pardey-Rodríguez et al. 2016). Figure 3.2 shows variability in grain and cob size and color in Andean maize from the north of Argentina, planted in the same ecosystem.

The morphological characterization takes into consideration both qualitative descriptors (e.g., plant color, grain color, and texture) and quantitative descriptors slightly influenced by the environment (e.g., the height of the plant, number of leaves per plant, and the number of spike ramifications) (Coral-Valenzuela and Andrade- Bolaños 2019).

3.3.1 Color Variability of Maize Grains

The color spectrum of maize grains varies from white through yellow, red, brown, purple, and blue to black. Pigments can be accumulated in the aleurone, pericarp, and/or endosperm. The endosperm can be light yellow or white and can be seen when the aleurone layer is colorless (Jompuk et al. 2020).

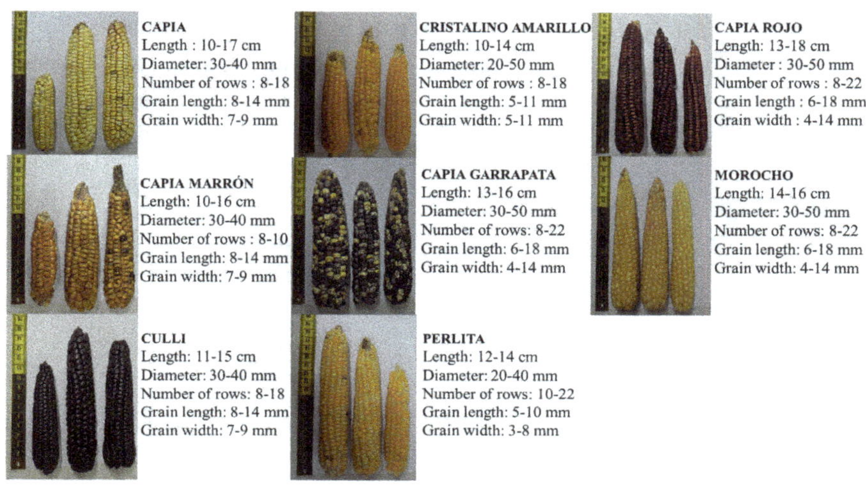

Figure 3.2 Variability of native maize from northwest Argentina.

Flavonoids, mainly anthocyanin and specifically flobaphen, represent the principal pigments in grains. Blue, purple, and red tones come from anthocyanin compounds, while flobaphenes primarily contribute to red and purple tones (Hu et al. 2016). Magenta-tone grains concentrate pigments in the pericarp and the aleurone layer, and blue/purple grains concentrate pigments only in the aleurone layer (Salinas-Moreno et al. 2013). The endosperm tones come from two classes of pigments: carotenes and xanthophylls. Generally, the main carotenoids are α and β carotenes, while β cryptoxanthin, lutein, and zeaxanthin constitute the majority of xanthophylls (Žilić et al. 2012).

In some races, populations with up to three colored grains were found (Salinas Moreno et al. 2012). Pigments can be located in different grain structures determining the possible use of different maize varieties. For example, if the pigment is concentrated in aleurone, the grain can be used for nixtamalization to produce blue-toned foods, but if pigments are accumulated in the pericarp in sufficient quantity, the grain could be used for pigment extraction (Bello-Pérez et al. 2016). Table 3.2 shows the color variability of some maize grains and the location of the pigment in the grain structure.

Food use of the different races is specifically paired with the physicochemical properties of the grain. In this aspect, hardness, size, and water absorption capacity are the most important characteristics (Fernández Suárez et al. 2013). These features are influenced by the composition and structure of the grain.

TABLE 3.2 VARIABILITY OF MAIZE COLORS AND PIGMENT LOCATION

Races	Color	Origin	Pigment location	Reference
Chullpi	Beige	Peru	Endosperm (yellow)	Salvador-Reyes and Silva Cleresi (2019)
Piscorunto	White and black		Pericarp	
Sacsa	White and red		Pericarp	
Morado	Purple		Pericarp (purple)	
Olotillo	Blue Red	Mexico	Aleuronic layer Pericarp	Salinas Moreno et al. (2012)
Tepecintle	Blue		Aleuronic layer	
Tuxpeño	Blue Red Magenta		Aleuronic layer Pericarp	
Pepitilla	Blue black	Mexico	Aleuronic layer	Broa Rojas et al. (2019)
Elotes Occidentales	Red		Pericarp y aleuronic layer	

3.4 NUTRITIONAL COMPOSITION

The nutritional composition of maize grains shows high variability due to the great genetic diversity of the species as well as the influence of the environment during cultivation. This variability is related with the different proportion of components distributed among maize types. Table 3.3 shows the nutritional composition of different native races in Latin America.

Starch represents 63%–73% of the grain weight. It is located mainly in the endosperm and acts as an energy reserve for the embryo. Starch is made up of two macromolecules: amylose and amylopectin. Maize starch digestion decreases while increasing amylose content and the length of the branched-chain of amylopectin (Rincon-Londoño et al. 2016). The amylose content of non-pigmented maize races is high, while in pigmented races the amylose starch content was found to be between 13% and 28% (De la Rosa-Millan et al. 2010). Utrilla-Coello et al. (2009) observed that the native starch granules of the blue maize race are more likely to present damage during the purification process compared to the non-pigmented races, which is possibly due to the differences in the enzymes that participate in the starch synthesis. Proteins are the second most abundant component in the maize grain. The protein content in Latin American maize races ranges from 5.7% to 12.5% of their dry weight (Table 3.4). The protein content in soft endosperm races is lower than in hard endosperm races due to the lower presence of protein bodies (prolamins) surrounding the

TABLE 3.3 PROXIMATE COMPOSITION OF MAIZE RACES FROM LATIN AMERICA (g/100g D.B.)

Maize Origin	Protein	Lipids	Ash	Carbohydrates	Starch	References
Mexico	9.6–11.6	4.2–5.6	–	–	–	Salinas Moreno et al. (2013)
Mexico	6.6–9.4	4.5–5.9	1.4–1.7	–	78.5–89.0	Agama Acevedo et al. (2011)
Mexico	9.4–12.5	3.4–4.5	1.1–1.8	–	57.9–70.2	Cazares Sánchez et al. (2015)
Mexico	7.6–11.5	4.5–5.9	–	–	–	Broa Rojas et al. (2019)
Mexico	9.3–11.3	4.1–5.3	1.2–1.4	81.0–85.2	71.9–77.7	Vazquez Carrillo et al. (2017)
Peru	5.7–7.8	2.7–4.8	1.3–1.5	–	69.5–74.8	Salvador-Reyes and Silva Clerici (2020)
Bolivia	9.0–10.7	4.6–5.7	1.3–1.5	74.0–74.5	–	Narváez González et al. (2006)
Ecuador	8.2–11.0	5.1–5.8	1.2–1.4	73.5–75.4	–	Narváez González et al. (2006)
Argentina	7.0–12.0	3.5–6.1	1.2–1.7	81.8–87.5	–	Gimenez et al. (2019)

TABLE 3.4 CONTENT OF TOTAL PHENOLIC COMPOUNDS AND BETA CAROTENES IN MAIZE OF DIFFERENT COLORS

Maize	Total phenolic (mg/100 g d.b.)	Anthocyanin (mg/100 g d.b.)	β Carotene (μg/100 g d.b.)	Xanthophylls* (μg/100 g d.b.)
White	170.0–260.7	1.33–1.33	4.92 ± 018	13.0 ± 0.2
Yellow	551.0–285.8	0.57–70.2	33.6 ± 1.2	778 ± 23.1
Red	243.8–465.0	9.75–85.2	20.2 ± 1.9	246.7 ± 12.1
Blue	266.2–343.0	85.2–99.5	23.1 ± 2.1	22.9 ± 1.0
Purple	465.0 ± 9.8	93.2 ± 1.1		
Black	454.0 ± 7.4	76.2 ± 2.2		

* Include cryptoxanthin and lycopene.
Source: De la Parra et al. (2007).

starch granules (Bello-Pérez et al. 2016). Differences in the protein content of some native breeds may be related with the difference in crop management, the use of fertilizers, and soil nitrogen content (Agama-Acevedo et al. 2011). Cázares-Sánchez et al. (2015) and Broa-Rojas et al. (2019) found that some maize races were characterized by presenting higher content of proteins, lysine, and tryptophan among the local maize from Yucatán and Morelos (Mexico).

Proteins are distributed mainly in the germ and endosperm, and their characteristics differ significantly. According to their solubility, four classes of proteins can differentiate: albumins, globulins, glutelins, and zeins. Prolamins or zeins are the most abundant in the grain, they are located in the endosperm and can be fractionated into α-, β-, γ-, and δ-zeins polypeptides. These zeins have different characteristics depending on their solubility, molecular weight, and structure (Ortiz-Martinez et al. 2017). The most abundant amino acids in zeins are glutamine (21%–26%), leucine (20%), and proline (10%). Their lysine and tryptophan content is relatively low, which is why zeins are considered of low nutritional quality (Tapia-Hernández et al. 2019). Excess leucine antagonizes the use of isoleucine and, therefore, can produce its deficiency (Ortiz-Martinez et al. 2017). On the other hand, the essential amino acid pattern of the germ proteins is better balanced.

The lipid content of native maize varies between 2.70% and 6.14%, according to Navarro-Cortez et al. (2016). The lipid content in maize depends on the variety and races as well as climatic factors, such as drought. Lipids are found mainly in the germ (76%–83%), then in the aleuronic layer (13%–15%), and finally the pericarp and endosperm (1%–2%). Maize oil contains approximately 79% triglycerides, 9% polar lipids (phospholipids and glycolipids), 5%

sterols, 4% mono- and diglycerides, 3% hydrocarbon esters and sterols, 1% free fatty acids, and small amounts of waxes, tocopherols, and carotenoids (Ai and Jane 2016).

Maize oil is rich in tocopherols, mainly in the gamma form. Due to the high levels of natural antioxidants and a low percentage of linolenic acid (C18:3 <1%), maize oil is stable. In native maize oils, palmitic (11.4%–15.0%), stearic (2.2%–3.5%), oleic (31.4%–46.6%), and linoleic (39.0%–53.0%) acid contents were found (Guzman-Maldonado et al. 2015). The total content of unsaturated fatty acids is greater than 80%, which may provide positive health effects (Serna-Saldivar 2017).

In maize, complex carbohydrates such as fiber are found mainly in the pericarp and pedicel but also in the cell walls of the endosperm and to a lesser extent in the germ (Nuss and Tanumihardjo 2010). The dietary fiber content in native maize races varies between 8.47 and 19.5%. Differences were found in the mean content of dietary fiber between pigmented races (13.95%) and white races (18.37%; Ortiz 2006). The importance of dietary fiber in health is widely known, and it is supported by numerous studies (Wang et al. 2015). The reduction in nutrient absorption by fibers may be due to the thickening of the intestinal lumen contents, so the migration of nutrients to the intestine slows down. Although data from literature confirm that the consumption of insoluble dietary fiber slightly alters mineral absorption, indigestible oligosaccharides probably stimulate intestinal microflora to produce vitamins and short-chain fatty acids, which can promote mineral absorption (Kiewlicz and Rybicka 2020). Maize bran is widely used in products such as breakfast cereals to increase dietary fiber content. The seed coat, a by-product of wet milling of maize, has recently been used to produce functional soluble dietary fiber. The product, known as maize fiber gum, is composed mainly of arabinoxylan and has a strong emulsifying property, which makes it a good substitute for Arabic gum in beverages and confectionery products (Ai and Jane 2016).

Maize grains are rich in vitamins and minerals. Yellow grains contain high amounts of carotenoids with provitamin A activity. A higher content of carotenoids was found in maize with hard or vitreous endosperm. Vitamin E plays an important role as an antioxidant, improving the immune response and preventing the oxidation of unsaturated fatty acids (Shah et al. 2016). In addition, maize contains niacin (B_3), biotin (B_7), thiamine (B_3), and pyridoxine (B_6). The presence of vitamins A, C, and K, together with beta-carotene and selenium, helps to improve the functioning of the thyroid gland and immune system (Siyuan et al. 2018).

The ash content in different races of native maize fluctuates between 1.13 and 1.84%. Minerals found in the highest proportion are phosphorus, potassium, magnesium, sodium, calcium, and to a lesser extent iron, copper, manganese, and zinc, which are mainly found in the germ (Robles-Ramírez et al. 2012). The content of some minerals is variable depending on the type of maize, for instance.

Iron content can vary from 11.28 to 83.35 mg/kg (Kumar et al. 2019). The presence of phytates in whole maize, specifically in the germ, interferes with the bioavailability of certain minerals, especially calcium, magnesium, zinc, and iron (Loy and Lundy 2019).

3.5 BIOACTIVE COMPOUNDS. HEALTHY EFFECTS

Besides nutrients, maize also contains other compounds with health-promoting properties in general, or in specific tissues or cells due to their ability to modulate metabolic processes, which are called bioactive compounds. For instance, the presence of antioxidant compounds—such as phenolic acids, vitamin C, and carotenoids—is associated with the colors of the Andean and Mesoamerican maize races. Table 3.2 shows the distribution of the colored compounds within the grain structure. Some of them played an important role in the diet and traditional medicine of ancient cultures (Adom and Liu 2002). Currently, research focused on the use of pigmented maize with antioxidant and bioactive properties has increased around the world due to the health benefits of maize. Urias-Lugo et al. (2015) reported the antiproliferative effect of anthocyanin and phenolic acids of blue maize on breast, liver, colon, and prostate cancer cells. Phenolic compounds are colored substances that exist in the plant kingdom and present a wide spectrum of colors in nature. Several health-promoting and pharmacological properties to phenolic compounds are used with therapeutic purposes in some cases (Cuevas-Montilla et al. 2008). Among all the cultivated maize races, investigations regarding these compounds' content and functions are mainly oriented to purple maize, and to a lesser degree to the red and black. The total content of phenolic compounds, anthocyanins, and carotenoids of various types of maize is shown in Table 3.4.

The β-carotenes and cryptoxanthin have provitamin A activity and, together with lycopene, are strong scavengers of free radicals. The wide range of antioxidants of native colored maize and their activity levels indicate the importance of promotion the development of varieties or races with high amounts of health-benefit compounds (Harakotr et al. 2015). Table 3.5 shows the different phenolic compounds contained in purple maize.

Anthocyanin are water-soluble and thermosensitive polyphenols that differ from others because of the sugar content within their functional groups, which in most cases implies several OH– groups. There is a great variability in the anthocyanin profile in colored maize; the main ones were identified as cyanidin-3-glucoside, pelargonidin-3-glucoside, peonidin-3-glucoside, and

TABLE 3.5 PHENOLIC COMPOUNDS CONTENT IN PURPLE MAIZE

Phenolic compounds	Compounds	Content
anthocyanin (mg/g water soluble fraction)	Cyanidin-3-glucoside	15.43 ± 1.59
	Pelargonidin 3 glucoside	2.33 ± 0.85
	Peonidin 3 glucoside	4.44 ± 1.09
	Acylated cyanidin 3 glucoside	10.37 ± 3.83
	Acylated pelargonidin 3 glucoside	2.83 ± 0.84
	Acylated peonidin 3 glucoside	4.85 ± 0.71
nonanthocyanin (mg/g ethyl acetate soluble fraction)	Protocatechuic acid	14.61 ± 0.08
	Vanillic acid	8.46 ± 0.09
	Unknown	23.10 ± 0.65
	p-coumaric acid Traces	4.00 ± 2.31
	Quercetin derivative	45.38 ± 1.44
	Hydroxycinnamic acid derivative	22.74 ± 0.37

Source: Pedreschi and Cisneros-Zevallos (2007).

their respective acylated counterparts. Other phenolic compounds present in purple maize and (to a lesser degree) in other, less colored ones were p-coumaric, vanillic, protocatechuic acids, derivatives of hesperidin and quercetin, and bound hydroxycinnamic acid forms composed of p-coumaric and ferulic acid. Quercetin derivatives seem to be the major no-anthocyanin phenolic compounds present in Andean purple maize, followed by ferulic and p-coumaric acid derivatives. There are some phenolic compounds that still remain unidentified. The presence of significant amounts diverse non-anthocyanin phenolic compounds should be taken into consideration when studying purple maize bioactive properties.

Anthocyanins are pigments with high biological capacity, which present their maximum color expression at acidic pH but are colorless at neutral and alkaline pH. At acid pH, anthocyanins, particularly the majority (cyanidin-3-glucoside), are red and therefore used in the manufacture of slightly acidic foods. The color stability depends on total anthocyanin concentration rather than the qualitative pigment composition and is related with other factors, including pigment structure temperature, light intensity, presence of co-pigments, metallic ions, enzymes, oxygen, ascorbic acid, sugars, and their degradation products and sulfur dioxide, among others. Currently, maize pigments are demanded by the food industry for replacing synthetic colorants (Cevallos-Casals and Cisneros-Zevallos 2004).

In Peru, purple maize is responsible for a third of maize exports; this growth brought with it an increase in its production and added value, as well as in the commercialization of concentrates and formulated products, opening up new research perspectives for the food sector (Cristianini and Guillén-Sánchez 2020). Purple maize could be a good source for the production of natural colorants in the market food (Cevallos-Casals and Cisneros-Zevallos 2004). This maize is used as a colorant for typical drinks and desserts, such as *chicha morada* and *mazamorra morada* (Saldaña et al. 2018).

In this way, anthocyanins stand out not only as food colorants but also as health food ingredients. Polyphenols show numerous beneficial health effects—anti-inflammatory, anti-allergic, antitumor, antimicrobial, vasorelaxant, and antioxidant properties, among others. Several reports highlight their protective role against cancer cell proliferation and heart disease, and also the prevention of lipid damage in food (Li et al. 2012; Ramos-Escudero et al. 2012; Colín-Chavez 2020). Galvez-Ranilla et al. (2017) reported the potential effects of free and bound phenolic fractions of purple maize on the gut microbiome. Guzman-Geronimo et al. (2017) concluded that blue maize extracts enhanced high-density lipoprotein cholesterol and decreased systolic blood pressure, serum triglycerides, total cholesterol, and epididymal adipose tissue weight in laboratory rats. Chen et al. (2017) reported that the colored maize pericarp derived from the dry milling process contains anthocyanins with inhibitory effects against pro-inflammatory enzymes, and anti-inflammatory and antidiabetic potentials in induced insulin-resistant 3T3-L1 adipocytes by TNF-α rich. Luna-Vital (2020) found evidence that anthocyanins from colored maize could attenuate the inflammation associated with obesity in the context of interactions between adipocytes and macrophages in adipose tissue. Treatment with purple maize decreased abdominal adiposity, improved glucose metabolism, and lowered blood pressure in healthy people.

Other authors investigated the effect of different types of pigmented maize on the *in vitro* digestibility of cooked or gelatinized starch (Rocchetti et al. 2018) and compared it with that of commercial yellow maize. Raw maize flours were characterized by differences in chemical composition and antioxidant activity. *In vitro* starch digestion data after cooking revealed that pigmented maize varieties generally had higher slowly digestible starch and resistant starch contents and lower rapidly digestible starch and hydrolysis index starch values compared with yellow maize. Significant changes in phenolic profiles were also observed during *in vitro* digestion, suggesting the participation of anthocyanin in modulating the enzymatic activity and starch digestibility (Camelo-Mendez 2018). Therefore, the use of pigmented maize flours could be an effective strategy to create healthier gluten-free foods since they are characterized by having higher antioxidant properties and higher levels of slowly digestible starch and resistant starch than commercial yellow maize flour.

3.6 ENDOSPERM HARDNESS

The hardness of the endosperm contributes the mechanical resistance of maize grain, which is an important parameter in any industrial process since it determines the ability of the grain to absorb and retain water during cooking (Salinas-Moreno et al. 2010). Two kinds of endosperm are recognized in the same maize grain, as shown in Figure 3.1: the corneous (hard and translucent) and the floury (soft and opaque).

The physicochemical basis of endosperm hardness can be attributed to complex interactions between zeins and starch. The strong zein-starch interaction provides a greater degree of compaction in the corneal endosperm (Narváez-González et al. 2006). Both endosperm types vary in protein concentration as well as in the shape and properties of starch granules. The α-zeins (Zeins 1, Z1) are concentrated in the nucleus, while the β- and γ-zeins (Zeins 2, Z2) lay on the surface. The hardness of the grain has been particularly correlated with the Z2 zeins (Gerde et al. 2017). Starch granules are polygonal in shape and rarely contain air space in the vitreous endosperm. In the floury endosperm, the starch granules are spherical in shape with a large amount of air space (Yu et al. 2015). Therefore, the vitreous endosperm has a hard and compact structure, while the floury has a soft and loose structure (Navarro-Cortez et al. 2016).

The terms *hard* and *soft* are used to designate the relationship between the floury or soft and the crystalline, vitreous, or corneal areas. Hard maize grains are those with more than 50% of flint endosperm, while soft grains have a predominant portion of floury endosperm (Zhang and Xu 2019). The proportions of each endosperm vary with the type and races of maize. Based on the grain characteristics, there are five general classes of maize: dent, flint, sweet, popcorn, and floury. In floury, the soft endosperm predominates. Many native races of Latin America are characterized as floury maize (García-Lara et al. 2019).

The molecular structure of the starch is another factor that contributes to the hardness of the endosperm. Increased accumulation of the starch synthase I (GBSSI) enzyme was found in vitreous endosperm starch granules, therefore a higher percentage of amylose is associated with greater hardness in grains (Agama-Acevedo et al. 2013). However, contradictory results were found since a greater tendency to increase the endosperm hardness with amylose content was observed only in floury and semi-hard grains, in hard endosperm maize races, the relation with amylose content is no verified (Robutti et al. 2000). Hard maize starch shows higher gelatinization temperatures and lower enthalpy due to its highly compacted endosperm (Santiago-Ramos et al. 2017).

Generally, maize with hard endosperm is used for the production of starch and animal feed (Salinas-Moreno et al. 2013), and maize with soft endosperm is used for human food in dishes and drinks since starch easily hydrates (Salinas-Moreno et al. 2010).

TABLE 3.6 ENDOSPERM HARDNESS INDICATORS

Hardness	FI	Hectolitric weight	RC/F	NIR
Very hard	0–12	–	–	–
Hard	13–37	78	4.7–6.2	595–691
Intermediate	38–62	74–75	3.0–5.6	406–568
Soft	63–87	73	1.8–4.1	310–421
Very soft	88–100	–	–	–

Abbreviations: FI, flotation index; RC/F, coarse/fine ratio; NIR, near-infrared reflectance.

The common parameters used to estimate endospermic hardness are hectolitric weight, flotation index (FI), near-infrared reflectance (NIR) hardness, and coarse/fine ratio (RC/F; Arriaga-Pérez et al. 2019). Table 3.6 shows the scales used for the mentioned parameters.

3.7 TECHNO-FUNCTIONAL PROPERTIES OF MAIZE STARCH

Maize starch is a valuable ingredient used to thicken, gelatinize, retain moisture, and improve the texture of numerous foodstuffs. The physicochemical and functional properties of starch depend on the structure and morphology of the granules. Several factors alter starch functional behavior, such as small changes in the amylose/amylopectin ratio, as well as the structure of the amylopectin molecules (chain length and branching degree), the lipid content, and the granules architecture (size, amorphous/crystalline structure ratio; BeMiller 2019). Starch physicochemical properties of both endosperm types are different (Narváez-González et al. 2006). Starch granules rich in amylopectin have a more regular shape, and those rich in amylose show a smoother surface (Rincon-Londoño et al. 2016). High amylopectin starches increase viscosity therefore are commonly used as texturizing and thickening agents in foods such as soups, sauces, and yogurts. Starches with amylose contents of 50% and 85% have strong gelling and film-forming properties and are widely used for products such as doughs and noodles (Ai and Jane 2016). Starch gelatinization is the loss of crystalline structure, as crystallinity is mainly a property of amylopectin molecules; a starch with high amylopectin content would be expected to have a high gelatinization enthalpy. The increase in the gelatinization enthalpy is also associated with the variation in the distribution of the amylopectin chain length since a higher amount of energy is needed to dissociate longer chains (Tazrart et al. 2019).

Peak temperatures and gelatinization enthalpies of 67–75.8 °C and 3.5–13.4 J/g were found in starches from different Latin American maize races (Narváez-González et al. 2006; Gimenez et al. 2020). Furthermore, lipids play an important role in gelatinization and retrogradation properties since they form complexes in associations with amylose chains (Santiago-Ramos et al. 2017). These interactions significantly influence starch properties, reduce starch swelling and solubility, delay starch gelatinization and retrogradation, affect gel formation and texture, slow starch digestion, and enhance palatability (Niu et al. 2019).

On the other hand, maize can be considered one of the safest grains for people with celiac disease, since is a gluten-free crop. Nevertheless, the use of maize for pasta or baked products production represents a challenge due to the absence of prolamins that form the viscoelastic gluten network. Zeins do not form a viscoelastic mass at room temperature. However, they provide a cohesive and extensible mass above their glass transition temperature—approximately 35 °C—with moisture content above 20% (Ozturk and Mert 2018). However, in gluten-free bread and pasta, it was found that pure starch and zein mixture conferred an undesirable texture. On the contrary, when zein was used in combination with hydrocolloids, it served as a structural crumb enhancer (Sosa et al. 2019).

3.7.1 Nixtamalization

Nixtamalization is a maize alkaline-cooking process that has been practiced in Mexico and some Central American countries since pre-Columbian times. The cooked maize (nixtamal) obtained is left overnight, then it was washed and ground to obtain a dough, which can be used to produce *tortillas*. Or it can be dried and ground to make products such as those typical of Mexico and Central America (tortillas, tamales, pozole, atoles, tortilla chips, sandwiches, tacos, tostadas, and nachos). Today many of these products are sold all over the world. Recently, food industries that produce nixtamalized flours have been installed in Russia, Korea, China, Australia, and several European countries (Serna-Saldivar and Chuck-Hernandez 2019).

The classic nixtamalization process (CNP) used by the Mayans consisted in cooking maize with wood ash. Later, the Aztecs substituted ash with lime, creating the traditional nixtamalization process (TNP); (Santiago-Ramos et al. 2017; Mariscal-Moreno et al. 2015). A third process, the ecological nixtamalization process (ENP), was patented by Figueroa-Cárdenas et al. (2011), but faced some challenges regarding environmental and nutritional feature. The tortillas made with ENP have a lower glycemic index than tortillas made with TNP. However, ENP demands large water volumes with a high level of soluble solids

(3%–15%), which are discarded during cooking and washing, producing high levels of contamination (Mariscal-Moreno et al. 2015). Alternative processes to TNP such as extrusion, high-pressure cooking, and processes based on electrothermal technologies such as ohmic heating, moderate electric field, and pulsed electric field, among others, have been studied (Ramírez-Jiménez et al. 2019).

Nixtamalization produces the hydrolysis and solubilization of the pericarp components. The released arabinoxylans act as hydrocolloids imparting desirable textural properties to the dough. The ease of pericarp removing is closely related to the maize genotype, type, and strength of alkali, cooking time, and temperature (Serna-Saldivar and Chuck Hernandez 2019; Bressani 1990). The thermo-alkaline process produces simultaneously starch partial gelatinization and annealing; some lipids are saponified and proteins are solubilized and polymerized (Santiago-Ramos et al. 2017).

Calcium is absorbed due to fatty acids saponification, the neutralization of the acid groups of hemicelluloses (mainly uronic acids) and the interaction between calcium ions and amylose or amylopectin chains through Ca^{2+} bridges formation (Santiago-Ramos et al. 2018). Another important change is the amylose-lipid complexes establishment, which affects the functional properties of nixtamalized flours (Santiago-Ramos et al. 2017).

After nixtamalization, a part of the starch retains its characteristic structure with a decrease in crystallinity, indicating the existence of resistant starch (Villada et al. 2017). Starch spherulite formation has also been demonstrated; as they have different optical and structural properties from those of native starch, these characteristics are used to identify nixtamalization in archaeological records (Johnson and Marston 2020). Nixtamalized grains also shows higher calcium content, higher niacin and iron bioavailability, and reduced content of mycotoxins and phytic acid. However, the process negatively affects the bio-functional value, since cooking increases the leaching of phenolic compounds and other important antioxidants (Palacios-Rojas et al. 2020; Bressani et al. 2004).

The quality of the maize grain for alkaline processing is determined by its physical characteristics and chemical composition. This quality is important for grain processors at the industrial level, although not for artisanal elaboration in rural areas, where maize type and processing conditions are selected according to consumer particular preferences. Among the main maize characteristics for nixtamalization, the grain hardness stands out. The Mexican standards for maize destined for the nixtamalization process establish a minimum hectolitric weight of 74 kg/L, a value corresponding to intermediate hardness maize endosperm (Serna-Saldivar and Chuck-Hernandez 2019; Bressani 1990). Osorio-Díaz et al. (2011) reported that tortillas made with hard maize (vitreous endosperm) had a stiffer texture and lower available starch content than tortillas made with soft maize (floury endosperm). Authors attributed these results to a higher rate of retrogradation. In the same way, Cruz-Vazquez et al. (2019) found that hard maize grains produce softer and more adhesive Mexican tamales.

The type of calcium-providing compound, used as an alternative of calcium hydroxide during cooking, has also an important effect on the viscoelastic behavior of the nixtamalized dough, since it not only influences the gelatinization process but also pericarp hydrolysis, calcium, and water absorption, lipid saponification, protein solubilization, and the amylose-lipid complex formation affecting the strength and elasticity of the dough (Santiago-Ramos et al. 2018). The use of other calcium sources that generates less alkaline solution is important to minimize the damage of anthocyanins in the nixtamalization of pigmented maize (Sánchez-Madrigal et al. 2015).

3.7.2 Toasted

Among the different ways to consume maize toasting is a rapid processing method that implies dry heat for short periods (Mrad et al. 2014). Toasted grains exhibit improved digestibility, color, shelf life, flavor and texture, greater volume and crispiness, and reduced antinutrient factors (Sandhu et al. 2017). A typical ancient Aztec food is *pinole*, made with toasted maize flour, sugar and cinnamon. In rural populations, *pinole* is still used as a travel food, usually in powder form, that can be used to prepare a drink and a baked product. Characteristics such as a fast and totally natural energy booster are the reasons why many athletes choose it. As urban migration in Mexico grows, fewer households prepare or eat traditional foods, such as *pinole*. Therefore, the Puebla region promotes its local variety made with native blue maize (Littaye 2016).

The toasting process is carried out at high temperatures (250–270 °C) in the absence of humidity, causing important physicochemical transformations in the macronutrients. The grain structure protects the starch granule during heat treatment, preventing thermal fractionation and a large degree of melting. Therefore, it may be considered as an alternative heat treatment to obtain starch with some desirable properties, such as greater crystallinity (Carrera et al. 2015). According to Mendez-Albores et al. (2004), maize toasting causes the inactivation of mycotoxins, and furthermore, it is more effective than boiling in reducing aflatoxins. In toasted white and yellow maize varieties, an increase in phenolic compounds and antioxidant capacity has been reported after roasting (Oboh et al. 2010). Sandhu et al. (2017) suggested that toasting could release more phenolic acids from the decomposition of cellular components, and the higher antioxidant properties of toasted maize could be due to the formation of Maillard products such as HMF (5-hydroxymethyl-2-furaldeh yde). Toasting also increases the available and fast-digesting starch fraction as well as the glycemic index. These improvements are more noticeable in the yellow maize race than in the purple one (Carrera et al. 2015).

3.7.3 Fermentation

The preservation of food through fermentation is a millenary practice that has allowed extending the shelf life of products, due to the presence of metabolites such as alcohol and lactic and acetic acids, which appear in the fermented product. In addition, fermentation generates an important nutrient and health-promoting compound, destroys undesirable compounds, inhibits food spoilage and pathogenic microorganisms, and improves food digestibility (Pérez-Armendáriz and Cardoso-Ugarte 2020).

Traditional fermented maize-based products are an essential element of the cultural and culinary heritage of many Latin American countries, such as *chicha, champus, tesguino*, sour *atole*, sourdough, and *pozole* (Soro-Yao et al. 2014). Some traditional techniques allow the use of a pretreatment such as grain nixtamalization or germination. These processes cause physical and chemical changes in the grains, thus acting as selective agents of the microbiota that guides the fermentation processing the desired way. The fermentation processes modify the grains through several steps, which involve endogenous enzymes (amylases, proteases, phytases) and microbial enzymes (generally from lactic acid bacteria and yeasts). Proper selection of microorganism leads to different product quality and highly variable sensory characteristics (Chaves-López et al. 2020).

In various fermented maize products, the coexistence of lactic acid bacteria (LAB), yeast, fungi, acetic acid bacteria (AAB), and *Bacillus* species is common. LAB also increase the content of free amino acids and B vitamins, improving the bioavailability of iron, zinc, and calcium by breaking down phytic acid. In addition, LAB produce volatile compounds that may contribute to the sensory properties of the products (Mosso et al. 2019).

Yeasts, as well as LAB, provide growth factors such as vitamins and soluble nitrogen in addition to extracellular enzymes (lipases, esterases, amylases, and phytases), some of which participate in flavor and aroma production. On the other hand, spore-forming aerobic bacteria (*Bacillus* spp.) secrete a wide range of enzymes that can also produce antimicrobial compounds, such as bacillysin, capable of inhibiting molds and bacteria, and iturine and chloromethane, which inhibit bacteria (Chaves-López et al. 2020).

From a technological point of view, LAB play an important role in the production of baked products. These bacteria have great potential to produce exopolysaccharides with properties similar to those observed in commercial hydrocolloids (Seitter et al. 2020). In this regard, Falade et al. (2014) found better elastic and viscous properties in maize sourdough than in native maize dough.

Among the traditional fermented products, *pozole* is a fermented drink, originally from the Mayan culture. Nixtamalized maize is used for *pozole* elaboration, so the microorganisms present are completely different from other maize grains used for dough preparation. It is generally prepared with Mexican maize varieties such as Cacahuacintle, Tabloncillo, Western Elotes, and Maíz Ancho. These races have large cobs with broad grains and a higher proportion of soft endosperm (Vázquez-Carrillo et al. 2017). The main microorganisms identified included yeasts, molds and bacteria—in particular *Lactobacillus* such as *L. casei, L. fermentum, L. plantarum*, and *L. paracasei*. A bacterial isolate extracted from pozole identified as a member of the *Bacillus* genus showed an antimicrobial compound production that inhibits a wide spectrum of microorganisms, including bacteria, molds, and yeasts (Pérez-Armendáriz and Cardoso-Ugarte 2020).

In *chicha*, another traditional fermented beer-like drink, strains of *Leuconostoc* and *L. plantarum* were identified and characterized as species with metabolic pathways for riboflavin and folate biosynthesis. Among the isolated microorganisms, an aminoglycoside resistance gene linked to streptomycin and spectinomycin resistance was also identified; also ampicillin resistance was observed *in vitro* but no related genes were found (Rodrigo-Torres et al. 2019). Piló-Barbosa et al. (2018) showed that *chicha* is fermented by a high number of different strains of *Saccharomyces cerevisiae*, while other species provided a minor contribution to the fermentation process.

3.8 CONCLUSIONS

Native maize originated in Mexico and distributed throughout the Americas by different migratory currents. The varieties, by adaptation in different ecosystems, acquired diverse morphological and genetic characteristics, resulting in the use of the term *races* to include the cultivation site. Several ancient civilizations from Mesoamerica and South America acquired advanced knowledge on maize adaptation, planting, and harvesting as well as on its nutritional importance and technologies for food and beverages production.

Maize contains numerous bioactive compounds, associated with the grain color, capable of modulating metabolic processes in cells and tissues. The purple maize is the most studied race, mainly by its anthocyanin content associated with anticancer, anti-inflammatory, anti-allergic, antimicrobial and anticholesterolemic properties. Besides, purple maize could be a good source of natural colorants for the food industry.

The endosperm characteristics (corneus and floury) and the starch (amylose/amylopectin ratio) are fundamental to determine the maize industrial application. Nixtamalization, toasting, and fermentation are processes that have been practiced in Latin America since pre-Columbian times. Currently, some of those processes were revalorized to pretreat maize and develop the suitable technological properties for novel gluten-free foods and beverage formulation.

ACKNOWLEDGMENTS

This work was supported by grants laValSe-Food-CYTED (Ref. 119RT0567); Consejo Nacional de Investigaciones Científicas y Técnicas (CONICET) and Universidad Nacional de Jujuy, Argentina.

REFERENCES

Adom, Kafui Kwami, and Ha Rui Liu. 2002. "Antioxidant Activity of Grains." *Journal of Agricultural and Food Chemistry* 50 (21): 6182–6187. https://doi.org/10.1021/jf0205099.

Agama-Acevedo, Edith, Erika Juárez-García, Silvia Evangelista-Lozano, Olga L. Rosales-Reynoso, and Luis A. Bello-Pérez. 2013. "Características del almidón de maíz y relación con las enzimas de su biosíntesis." *Agrociencia* 47 (1): 1–12. www.scielo.org.mx/scielo.php?script=sci_arttext&pid=S1405.

Agama-Acevedo, Edith, Yolanda Salinas-Moreno, Glenda Pacheco-Vargas, and Luis Arturo Bello-Pérez. 2011. "Características físicas y químicas de dos razas de maíz azul: Morfología del almidón." *Revista Mexicana de Ciencias Agrícolas* 2 (3): 317–329. www.scielo.org.mx/scielo.php?script=sci_arttext&pid=S2007.

Ai, Yongfeng, and Jane Jaylin. 2016. "Macronutrients in Corn and Human Nutrition." *Comprehensive Reviews in Food Science and Food Safety* 15 (3): 581–598. https://doi.org/10.1111/1541-4337.12192.

Arriaga-Pérez, Wendy, Marcela Gaytán-Martínez, and María L. Reyes-Vega. 2019. "Métodos para medir la dureza del grano de maíz: Revisión." *Digital Ciencia@ Uaqro* 12 (2): 67–78. http://ciencia.uaq.mx/index.php/ojs/article/view/45. ISSN: 2395–8847.

Bello-Pérez, Arturo L., Gustavo A. Camelo-Mendez, Edith Agama-Acevedo, and Rubí G. Utrilla-Coello. 2016. "Aspectos nutracéuticos de los maíces pigmentados: Digestibilidad de los carbohidratos y antocianinas." *Agrociencia* 50 (8): 1041–1063. www.scielo.org.mx/scielo.php?script=sci_arttext&pid=S1405.

BeMiller, James N. 2019. "Corn Starch Modification in *Corn* (Third Edition)." *Chemical and Technology* Chapter 19: 537–549. https://doi.org/10.1016/B978-0-12-811971-6.00019-X.

Bressani, Ricardo. 1990. "Chemistry, Technology, and Nutritive Value of Maize Tortillas." *Food Reviews International* 6 (2): 225–264. https://doi.org/10.1080/87559129009540868.

Bressani, Ricardo, Juan Carlos Turcios, Ana Silvia Colmenares de Ruiz, and Patricia Palacios de Palomo. 2004. "Effect of Processing Conditions on Phytic Acid, Calcium, Iron, and Zinc Contents of Lime-Cooked Maize." *Journal of Agricultural and Food Chemistry* 52 (5): 1157–1162. https://doi.org/10.1021/jf030636k.

Broa-Rojas, Elizabeth, María Gricelda Vázquez-Carrillo, Néstor Gabriel Estrella-Chulím, José Hilario Hernández-Salgado, Benito Ramírez-Valverde, and Gregorio Bahena-Delgado. 2019. "Características fisicoquímicas y calidad de la proteína de maíces nativos pigmentados de Morelos en dos años de cultivo." *Revista Mexicana de Ciencias Agrícolas* 10 (3): 683–697. https://doi.org/10.29312/remexca.v10i3.481.

Camelo-Mendez, Gustavo A., Pamela C. Flores-Silva, Edith Agama-Acevedo, Juscelino Tovar, and Luis A. Bello-Pérez. 2018. "Incorporation of Whole Blue Maize Flour Increases Antioxidant Capacity and Reduces in Vitro Starch Digestibility of Gluten-Free Pasta." *Starch/Stärke* 70: 1700126. https://doi.org/10.1002/star.201700126.

Carrera, Y., Rubi Utrilla-Coello, Angel Bello-Pérez, José, Alvarez-Ramirez, E. Jaime, and Vernon-Carter. 2015. "In Vitro Digestibility, Crystallinity, Rheological, Thermal, Particle Size and Morphological Characteristics of Pinole, a Traditional Energy Food Obtained from Toasted Ground Maize." *Carbohydrate Polymers* 123: 246–255. https://doi.org/10.1016/j.carbpol.2015.01.044.

Carvalho Valdemar, P., Claudete F. Ruas, Josué M. Ferreira, Rosângela M. P. Moreira, and Paulo M. Rúas. 2004. "Genetic Diversity Among Maize (*Zea mays L.*) Landrace Assessed by RAPD Markers." *Genetics and Molecular Biology* 27 (2): 228–236. http://doi.org/10.1590/S1415-47572004000200017.

Cázares-Sánchez, Esmeralda, José L. Chávez-Servia, Yolanda Salinas-Moreno, Fernando Castillo-González, and Porfirio Ramírez-Vallejo. 2015. "Variación en la composición del grano entre poblaciones de maíz (*Zea mays* L.) nativas de Yucatán, México." *Agrociencia* 49 (1): 15–30. www.redalyc.org/articulo.oa?id=30236850002.

Cevallos-Casals, Bolivar A., and Luis Cisneros-Zevallos. 2004. "Stability of Anthocyanin-Based Aqueous Extracts of Andean Purple Corn and Red-Fleshed Sweet Potato Compared to Synthetic and Natural Colorants." *Food Chemistry* 86: 69–77. https://doi.org/10.1016/j.foodchem.2003.08.011.

Chaves-López Clemencia, Chiara Rossi, Francesca Maggio, Antonello Paparella, and Annalisa Serio. 2020. "Changes Occurring in Spontaneous Maize Fermentation: An Overview." *Fermentation* 6: 36. https://doi.org/10.3390/fermentation6010036.

Chen, Cheng, Somava Pavelt Yija Singh, and Elvira Gonzalez de Mejia. 2017. "Chemical Characterization of Proanthocyanidins in Purple, Blue, and Red Maize Coproducts from Different Milling Processes and Their Antiinflammatory Properties." *Industrial Crops and Products*: 464–475. https://doi.org/10.1016/j.indcrop.2017.08.046.

Colín-Chavez Citlali, José J. Virgen-Ortiza, Luis E. Serrano-Rubio, Miguel A. Martínez-Téllez, and Marta Astier. 2020. "Comparison of Nutritional Properties and Bioactive Compounds Between Industrial and Artisan Fresh Tortillas from Maize Landraces." *Current Research in Food Science* 3: 189–194. https://doi.org/10.1016/j.crfs.2020.05.004.

Coral-Valenzuela, Jenny V., and Héctor J. Andrade-Bolaños. 2019. "Caracterización morfológica y agronómica de dos genotipos de maíz (Zea mays L.) en la zona media de la Parroquia Malchinguí." *Avances en Ciencias e Ingeniería*. 11 (17): 40–49. http://dx.doi.org/10.18272/aci.v11i1.1091.

Cristianini, Marcelo, and Jhoseline S. Guillén-Sánchez. 2020. "Extraction of Bioactive Compounds from Purple Corn Using Emerging Technologies: A Review." *Journal of Food Science* 85 (4): 862–869. https://doi.org/10.1111/1750-3841.15074.

Cruz-Vazquez, Celia, Adriana Villanueva-Carvajal, Gaspar Estrada-Campuzano, and Aurelio Dominguez-Lopez. 2019. Tamales texture properties as a function of corn endosperm type. *International Journal of Gastronomy and Food Science* 16(2019): 100153. https://doi.org/10.1016/j.ijgfs.2019.100153.

Cuevas-Montilla, Elyana, Alejandro Antezana, and Peter Winterhalter. 2008. "Análisis y caracterización de antocianinas en diferentes variedades de maíz (*Zea mays*) boliviano." Memorias del Encuentro Final Alfa Lagrotech, Cartagena, Colombia, 21–26.

De la Rosa-Millan Julián, Edith Agama-Acevedo, Antonio R. Jiménez-Aparicio, and Luis A. Bello-Pérez. 2010. "Starch Characterization of Different Blue Maize Varieties." *Starch/Stärke* 62: 549–557. https://doi.org/10.1002/star.201000023.

Falade, Adediwura T., Naushad Emmambux, Elna M. Buys, and John R. Taylor. 2014. "Improvement of Maize Bread Quality Through Modification of Dough Rheological Properties by Lactic Acid Bacteria Fermentation." *Journal of Cereal Science* 60: 471–476. https://doi.org/10.1016/j.jcs.2014.08.010.

Fernández Suárez, Rocío, Luis A. Morales Chávez, and Mariscal Amanda Gálvez. 2013. "Importancia de los maíces nativos de México en la dieta nacional: Una revisión indispensable." *Revista fitotecnia Mexicana* 36: 275–283. www.scielo.org.mx/scielo.php?script=sci_arttext&pid=S0187-73802013000500004&lng=es&tlng=es.

Figueroa-Cárdenas, Juan de Dios, Antonio Rodríguez-Chong, and José Juan Veles-Medina. 2011. "Proceso ecológico de nixtamalización para la producción de harinas, masa y tortillas integrales." *Patente Mexicana Número*: 289339.

Galinat, Walton C. 1992. "Evolution of Corn." *Advances in Agronomy* 47: 203–231. https://doi.org/10.1016/S0065-2113(08)60491-5.

Galvez-Ranilla, Lena, Ashish Christopher, Dipayan Sarkar, Kalidas Shetty, Rosana Chirinos, and David Campos. 2017. "Phenolic Composition and Evaluation of the Antimicrobial Activity of Free and Bound Phenolic Fractions from a Peruvian Purple Corn (*Zea mays* L.)." *Accession Journal of Food Science* 82 (12): 2968–2976. https://doi.org/10.1111/1750-3841.13973.

García-Lara, Silverio, Cristina Chuck-Hernandez, and Sergio O. Serna-Saldivar. 2019. "Development and Structure of the Corn Kernel." *Corn* 6: 147–163. AACC International Press. https://doi.org/10.1016/B978-0-12-811971-6.00006-1.

Gerde, José A., Joel Spinozzi, and Lucas Borrás. 2017. "Maize Kernel Hardness, Endosperm Zein Profiles, and Ethanol Production." *Bioenergy Research* 10: 760–771. https://doi.org/10.1007/s12155-017-9837-4.

Gimenez, Maria A., Cristina N. Segundo, Manuel O. Lobo, and Norma C. Samman. 2020. "Physicochemical and Techno-Functional Characterization of Native Corn Reintroduced in the Andean Zone of Jujuy, Argentina." *Proceedings, II Congreso Internacional de Ia ValSe-Food Network* 53 (1): 7. https://doi.org/10.3390/proceedings2020053007.

Grobman, Alexander, Wilfredo Salhuana, Ricardo Sevilla, and Paul C. Mangelsdorf. 1961. *Races of Maize in Peru*. Washington, DC: National Academy of Sciences and National Research Council, Pub, 915.

Guzman, Geronimo, Rosa Isela, Tania Margarita Alarcón-Zavaleta, Rosa María Oliart-Ros, José Enrique Meza-Alvarado, Socorro Herrera-Meza, and José Luis Chavez-Servia. 2017. "Blue Maize Extract Improves Blood Pressure, Lipid Profiles, and Adipose Tissue in High-Sucrose Diet-Induced Metabolic Syndrome in Rats." *Journal of Medicinal Food* 20 (2): 110–115. https://doi.org/10.1089/jmf.2016.0087.

Guzmán-Maldonado, Salvador, Gricelda Vázquez-Carrillo, Alfonso Aguirre-Gómez and Isela Serrano-Fujarte. 2015. "Contenido de ácidos grasos, compuestos fenólicos y calidad industrial de maíces nativos de Guanajuato". *Revista Fitotecnia Mexicana* 38(2): 213–222.

Harakotr, Bhornchai, Bhalang Suriharn, Paul Scott Marvin, and Kamol Lertrat. 2015. "Genotypic Variability in Anthocyanins, Total Phenolics, and Antioxidant Activity Among Diverse Waxy Corn Germplasm." *Euphytica* 203: 237–248. https://doi.org/10.1007/s10681-014-1240-z.

Hu, Chaoyang, Quanlin Li, Xuefang Shen, Sheng Quan, Hong Lin, Lei Duan, Yifa Wang, Qian Luo, Guorun Qu, Qing Han, Yuan Lu, Dabing Zhang, Zheng Yuan, and Jianxin Shi. 2016. "Characterization of Factors Underlying the Metabolic Shifts in Developing Kernels of Colored Maize." *Scientific Reports* 6: 35479. https://doi.org/10.1038/srep35479.

Johnson, Emily S., and John M. Marston. 2020. "The Experimental Identification of Nixtamalized Maize Through Starch Spherulites." *Journal of Archaeological Science* 113: 105056. https://doi.org/10.1016/j.jas.2019.105056.

Jompuk, Choosak, Chadamas Jitlaka, Peeranuch Jompuk, and Peter Stamp. 2020. "Combining Three Grain Mutants for Improved-Quality Sweet Corn." *Agricultural & Environmental Letters* 5: 20010. https://doi.org/10.1002/ael2.20010.

Kato, Takeo Angel, Cristina Mapes, Luz María Mera-Ovando, José Antonio Serratos-Hernández, and Robert A. Bye Boettler. 2009. *Origen y diversificación del maíz: una revisión analítica*. México: Universidad Nacional Autónoma de México, Comisión Nacional para el Conocimiento y Uso de la Biodiversidad, 116.

Kiewlicz. Justyna, and Iga Rybicka. 2020. "Minerals and Their Bioavailability in Relation to Dietary Fiber, Phytates and Tannins from Gluten and Gluten-Free Flakes." *Food Chemistry* 305: 125452. https://doi.org/10.1016/j.foodchem.2019.125452.

Kumar, Pardeep, Mukesh Choudhary, Firoz Hossain, N. K. Singh, Poonam Choudhary, Mamta Gupta, Vishal Singh, G. K. Chikappa, Ramesh Kumar, Bhupender Kumar, S. L. Jat, and Sujay Rakshit. 2019. "Nutritional Quality Improvement in Maize (*Zea mays*): Progress and Challenges." *Indian Journal of Agricultural Sciences* 89 (6): 895–911. http://krishi.icar.gov.in/jspui/handle/123456789/21430.

Labate, Joanne A., Kendall R. Lamkey, Sharon E. Mitchell, Stephen Kresovich, Hillary Sullivan, and John S. C. Smith. 2003. "Molecular and Historical Aspects of Corn Belt Dent Diversity." *Crop Science* 43: 80–91. https://doi.org/10.2135/cropsci2003.8000.

León-Portilla, Miguel. 2002. "Mitos de los orígenes en Mesoamérica." *Arqueología Mexicana* X: 56: 20–29.

Li Jing, Soon Sung Lim, Jae-Yong Lee, Jin-Kyu Kim, Sang-Wook Kanga, and Jung-Lye Young-Hee Kang. 2012. "Purple corn anthocyanins dampened high-glucose-induced mesangial fibrosis and inflammation: possible renoprotective role in diabetic nephropathy". *Journal of Nutritional Biochemistry* 23: 320–331. https://doi.org/10.1016/j.jnutbio.2010.12.008.

Littaye, Alexandra Zelda. 2016. "The Multifunctionality of Heritage Food: The Example of Pinole, a Mexican Sweet." *Geoforum* 76: 11–19. http://doi.org/10.1016/j.geoforum.2016.08.008.

Loy, Daniel, and Erika Lundy. 2019. "Nutritional Properties and Feeding Value of Corn and Its Coproducts in Maize Chemical and technology." *Corn*, 3rd edition, Chapter 23: 633–659. https://doi.org/10.1016/B978-0-12-811971-6.00023-1.

Luna-Vital, Diego, Iván Luzardo-Ocampo, Liceth Cuellar-Nuñez, Guadalupe Loarca-Piña, and Elvira González de Mejía. 2020. "Maize Extract Rich in Ferulic Acid and Anthocyanins Prevent High-Fat-Induced Obesity in Mice by Modulating SIRT1, AMPK and IL-6 Associated Metabolic and Inflammatory Pathways." *Journal of Nutritional Biochemistry* 79: 108343. https://doi.org/10.1016/j.jnutbio.2020.108343.

Mariscal-Moreno, Rosa María, Juan de Dios Figueroa, David Santiago-Ramos, Gerónimo Arambula-Villa, Sergio Jimenez-Sandoval, Patricia Rayas-Duarte, José Juan Veles-Medina, Hector Martín, and Eduardo Flores. 2015. "The Effect of Different Nixtamalisation Processes on Some Physicochemical Properties, Nutritional Composition and Glycemic Index." *Journal of Cereal Science* 65: 140–146. http://doi.org/10.1016/j.jcs.2015.06.016.

Matsuoka, Yoshihiro, Yves Vigouroux, Major M. Goodman, Jesus Sanchez, Edward Buckler, and John Doebley. 2002. "A Single Domestication for Maize Shown by Multilocus Microsatellite Genotyping." *Proceedings of the National Academy of Science* 99 (9): 6080–6084. https://doi.org/10.1073/pnas.052125199.

McClintock, Barbara, Takeo Angel Kato, and Almiro Blumenschein. 1981. "Chromosome ConstitutionofRacesofMaize:It'sSignificanceintheInterpretationofRelationships Between Races and Varieties in the Americas." Colegio de Postgraduados, Chapingo, México, CIMMYT, Programa de Recursos Naturales." www.worldcat.org/title/chromosome-constitution-of-races-of-maize-its-significance-in-the-interpretation-of-relationships-between-races-and-varieties-in-the-americas/oclc/9181898.

Mendez-Albores, Abraham, F. De Jesús-Flores, Elsa Castañeda-Roldan, Gerónimo Arambula-Villa, and Ernesto Moreno-Martínez E. 2004. "The Effect of Toasting and Boiling on the Fate of B-Aflatoxins During Pinole Preparation." *Journal of Food Engineering* 65: 585–589. http://doi.org/10.1016/j.jfoodeng.2004.02.024.

Mosso, Ana, María E. Jiménez, Graciela Vignolo, Jean Guy Leblanc, and Norma Samman. 2018. "Increasing the Folate Content of Tuber Based Foods Using Potentially Probiotic Lactic Acid Bacteria." *Food Research International* 109: 168–174. https://doi.org/10.1016/j.foodres.2018.03.073.

Mrad, Rachelle, Espérance Debs, Rachad Saliba, Richard G. Maroun, and Nicolass Louka. 2014. "Multiple Optimization of Chemical and Textural Properties of Roasted Expanded Purple Maize Using Response Surface Methodology." *Journal of Cereal Science* 60: 397–405. http://doi.org/10.1016/j.jcs.2014.05.005.

Narváez-González, Ernesto David, Juan de Dios Figueroa-Cárdenas, Suketoshi Taba, Eduardo Castaño-Tostado, Ramón Alvar Martínez-Peniche, and Froylán Rincón-Sánchez. 2006. "Relationships Between the Microstructure, Physical Features, and Chemical Composition of Different Maize Accessions from Latin America." *Cereal Chemistry* 83: 595–604. https://doi.org/10.1094/CC-83-0595.

Navarro-Cortez, Ricardo Omar, Alberto Gómez-Aldapa, Ernesto Aguilar-Palazuelos, Efren Delgado-Licon, Juan Castro Rosas, Juan Hernández-Ávila, Aquiles Solís-Soto, Luz Araceli Ochoa-Martínez, and Hiram Medrano-Roldán. 2016. "Blue Corn (*Zea mays* L.) with Added Orange (*Citrus sinensis*) Fruit Bagasse: Novel Ingredients for Extruded Snacks." CyTA—*Journal of Food* 14 (2): 349–358. https://doi.org/10.1080/19476337.2015.1114026.

Niu, Bin, Chen Chao, Jing Cai, Yizhe Yan, Les Copeland, Shuo Wang, and Shujun Wang. 2019. "The Effect of NaCl on the Formation of Starch-Lipid Complexes." *Food Chemistry* 299: 125–133. https://doi.org/10.1016/j.foodchem.2019.125133.

Nuss, Emily T., and Sherry A. Tanumihardjo. 2010. "Maize: A Paramount Staple Crop in the Context of Global Nutrition." *Comprehensive Reviews in Food Science and Food Safety* 9: 417–436. https://doi.org/10.1111/j.1541-4337.2010.00117.x.

Oboh, Ganiyu, Adedayo O. Ademiluyi, and Afolabi A. Akindahunsi. 2010. "The Effect of Roasting on the Nutritional and Antioxidant Properties of Yellow and White Maize Varieties." *International Journal of Food Science and Technology* 45: 1236–1242. https://doi.org/10.1111/j.1365-2621.2010.02263.x.

Ortiz, Prudencio S. A. 2006. "Determinación de la composición química proximal y fibra dietaria de 43 variedades criollas de maíz de 7 municipios del Sureste del Estado de Hidalgo." Universidad Autónoma, Pachuca de Soto, México, Tesis de Licenciado en Nutrición, México, 79. http://repository.uaeh.edu.mx.

Ortiz-Martinez, Margarita, Elvira Gonzalez de Mejia, Silverio García-Lara, Oscar Aguilar, Margarita Lopez-Castillo, and José T. Otero-Pappatheodorou. 2017. "Efecto antiproliferativo de fracciones peptídicas aisladas de una proteína de calidad de maíz, un maíz híbrido blanco, y sus péptidos derivados en células HepG2 humanas de hepatocarcinoma." *Journal of Functional Foods* 34: 36–48. https://doi.org/10.1016/j.jff.2017.04.015.

Osorio-Díaz, Perla, Edith Agama-Acevedo, Luis A. Bello-Pérez, José Juan Islas-Hernández, Noel. O Gomez-Montiel, and Octavio Paredes-López. 2011. "Effect of Endosperm Type on Texture and in Vitro Starch Digestibility of Maize Tortillas." *LWT—Food Science and Technology* 44: 611–615. https://doi.org/10.1016/j.lwt.2010.09.011.

Ozturk, Oguz Kaan, and Behic Mert. 2018. "The Use of Microfluidization for the Production of Xanthan and Citrus Fiber-Based Gluten-Free Corn Breads." *LWT—Food Science and Technology*. https://doi.org/10.1016/j.lwt.2018.05.025.

Palacios-Rojas, Natalia, Laura McCulley, Mikayla Kaeppler, Tyler J. Titcomb, Nilupa S. Gunaratna, Santiago Lopez-Ridaura, and Sherry A. Tanumihardjo. 2020. "Mining Maize Diversity and Improving Its Nutritional Aspects Within Agro-Food Systems." *Comprehensive Reviews in Food Science and Food Safety*: 1–26. https://doi.org/10.1111/1541-4337.12552.

Pardey-Rodríguez, Catherine, Mario Augusto García-Dávila, and Nataly Moreno-Cortés. 2016. "Caracterización de maíz procedente del departamento del Magdalena-Colombia." *Corpoica Ciencia y Tecnología Agropecuaria* 17 (2): 167–190. www.scielo.org.co/scielo.php?pid=S0122-87062016000200003&script=sci_abstract&tlng=en.

Pedreschi, Romina, and Luis Cisneros-Zevallos. 2007. "Phenolic Profiles of Andean Purple Corn (*Zea mays* L.)." *Food Chemistry* 100: 956–963. https://doi.org/10.1016/j.foodchem.2005.11.004.

Pérez-Armendáriz, Beatriz, and Gabriel Abraham Cardoso-Ugarte. 2020. "Traditional Fermented Beverages in Mexico: Biotechnological, Nutritional, and Functional Approaches." *Food Research International*: 109307. https://doi.org/10.1016/j.foodres.2020.109307.

Pérez de la Cerda, Felipe de Jesús, Aquiles Carballo Carballo, Amalio Santacruz Varela, Adrián Hernández Livera and Juan Celestino Molina Moreno. 2007. "Calidad fisiológica en semillas de maíz con diferencias estructurales". *Agricultura Técnica en México* 33(1): 53–61.

Piló-Barbosa, Fernanda, Enrique Javier Carvajal-Barriga, María Cristina Guamán-Burneo, Patricia Portero-Barahona, Arthur Dias, Freitas Matoso, Larissa Morato, Daher de Gomes Falabella, Fátima de Cássia Oliveira, and Carlos Augusto Rosa. 2018. "Poblaciones de Saccharomyces cerevisiae y otras levaduras asociadas con cervezas indígenas (chicha) del Ecuador." *Revista Brasileña de Microbiología* 49 (4): 808–815. https://doi.org/10.1016/j.bjm.2018.01.002.

Pope, Kevin O., Mary E. D. Pohl, John G. Jones, David L. Lentz, Christopher von Nagy, Francisco J. Vega, and Irvy R. Quitmyer. 2001. "Origin and Environmental Setting of Ancient Agriculture in the Lowlands of Mesoamerica." *Science* 292: 1370–1373. https//doi.org.10.1126 / science.292.5520.1370.

Ramírez-Jiménez, Aurea K., Jorge Rangel-Hernández, Eduardo Morales-Sánchez, Guadalupe Loarca-Piña, and Marcela Gaytán-Martínez. 2019. "Changes on the Phytochemicals Profile of Instant Corn Flours Obtained by Traditional Nixtamalization and Ohmic Heating Process." *Food Chemistry* 276: 57–62. https://doi.org/10.1016/j.foodchem.2018.09.166.

Ramos-Escudero Fernando, Ana María Muñoz, Carlos Alvarado-Ortíz, Angel Alvarado, and Jaime A. Yañez. 2012. "Purple Corn (*Zea mays* L.) "Phenolic Compounds Profile and Its Assessment as an Agent Against Oxidative Stress in Isolated Mouse Organs." *Journal of Medicinal Food* 15 (2): 206–215. https://doi.org/10.1089/jmf.2010.0342.

Rincon-Londoño, Natalia, Lineth J. Vega-Roja, Margarita Contreras-Padilla, Andrés Antonio Acosta-Osorio, and Mario E. Rodríguez-García. 2016. "Analysis of the Pasting Profile in Corn Starch: Structural, Morphological, and Thermal Transformations, Part I." *International Journal of Biological Macromolecules* 91: 106–114. http://doi.org/10.1016/j.ijbiomac.2016.05.070.

Robles-Ramírez, María del Carmen, Areli Flores-Morales, and Rosalva Mora-Escobedo. 2012. "Corn Tortillas: Physicochemical, Structural and Functional Changes." In *Maize: Cultivation, Uses and Health Benefits*. New York: Nova Science Publishers, Inc., Chapter 7, 89–112. 978-1-62081-518-2 (eBook).

Robutti, José, Francisco Borras, Marcelo Ferrer, Mabel Percibaldi, and Clarence A. Knutson. 2000. "Evaluation of Quality Factors in Argentine Maize Races." *Cereal Chemistry* 77 (1): 24–26. https://doi.org/10.1094/CCHEM.2000.77.1.24.

Rocchetti, Gabriele, Gianluca Giuberti, Antonio Galloa, Jamila Bernardic, Adriano Marocco, and Luigi Lucini. 2018. "Effect of Dietary Polyphenols on the in Vitro Starch Digestibility of Pigmented Maize Varieties Under Cooking Conditions." *Food Research International* 108: 183–191. https://doi.org/10.1016/j.foodres.2018.03.049.

Rodrigo-Torres, Lidia, Alba Yepez, Rosa Aznar, and David Arhal. 2019. "Genomic Insights into Five Strains of Lactobacillus Plantarum with Biotechnological Potential Isolated from Chicha, a Traditional Maize-Based Fermented Beverage from Northwestern Argentina." *Frontiers in Microbiology* 10: 1–16. https://doi.org/10.3389/fmicb.2019.02232.

Saldaña, Eric, Juan Rios-Mera, Hubert Arteaga, Jhordin Saldaña, Cathia Samán, Miriam Mabel Selani, and Nilda Doris Montes Villanueva. 2018. "How Does Starch Affect the Sensory Characteristics of Mazamorra Morada? A Study with a Dessert Widely Consumed by Peruvians." *International Journal of Gastronomy and Food Science* 12: 22–30.

Salinas-Moreno, Yolanda, Flavio Aragón-Cuevas, Carmen Moncada-Ybarra, Jessica Aguilar-Villarreal, Bernabé Altunar-López, and Eliseo Sosa-Montes. 2013. "Caracterización física y composición química de razas de maíz de grano azul/morado de las regiones tropicales y subtropicales de Oaxaca." *Revista Fitotecnia Mexicana* 36 (1): 23–31. www.scielo.org.mx/scielo.php?script=sci_arttext&pid=S0187-73802013000100003&lng=es&tlng=es.

Salinas-Moreno, Yolanda, Francisco J. Cruz-Chávez, Silvia A. Díaz-Ortiz, and Fernando Castillo-González. 2012. "Granos de maíces pigmentados de Chiapas, características físicas, contenido de antocianinas y valor nutracéutico." *Revista Fitotecnia Mexicana* 35 (1): 33–41. www.scielo.org.mx/scielo.php?script=sci_arttext&pid=S0187-73802012000100006&lng=es&tlng=es.

Salinas-Moreno, Yolanda, Noel Orlando Gómez-Montiel, José Ernesto Cervantes-Martínez, Mauro Sierra-Macías, Artemio Palafox-Caballero, Esteban Betanzos-Mendoza, and Bulmaro Coutiño-Estrada. 2010. "Calidad nixtamalera y tortillera en maíces del trópico húmedo y sub-húmedo de México." *Revista Mexicana de Ciencias Agrícolas* 1 (4): 509–523. www.redalyc.org/articulo.oa?id=263120639005.

Salvador-Reyes, Rebeca, and María Teresa Pedrosa Silva Clerici. 2020."Peruvian Andean Maize: General Characteristics, Nutritional Properties, Bioactive Compounds, and Culinary Uses." *Food Research International* 130: 108934. https://doi.org/10.1016/j.foodres.2019.108934.

Sánchez-Madrigal, Miguel Ángel, Armando Quintero-Ramos, Fernando Martínez-Bustos, Carmen O. Meléndez-Pizarro, Martha G. Ruiz-Gutiérrez, Alejandro Camacho-Dávila, Patricia Isabel Torres-Chávez, and Benjamín Ramírez-Wong. 2015. "Effect of Different Calcium Sources on the Bioactive Compounds Stability of Extruded and Nixtamalized Blue Maize Flours." *Journal of Food Science and Technology* 52 (5): 2701–2710. https://doi.org/10.1007/s13197-014-1307-9.

Sandhu, Kawaljit Singh, Poonam Godara, Maninder Kaur, and Sneh Punia. 2017. "Effect of Toasting on Physical, Functional and Antioxidant Properties of Flour from Oat (*Avena sativa* L.) Cultivars." *Journal of the Saudi Society of Agricultural Sciences* 16: 197–203. http://doi.org/10.1016/j.jssas.2015.06.004.

Santiago-Ramos, David, Juan de Dios Figueroa-Cardenas, José Juan Veles-Medina, Rosa María Mariscal-Moreno. 2017. "Changes in the Thermal and Structural Properties of Maize Starch During Nixtamalization and Tortilla-Making Processes as Affected by Grain Hardness." *Journal of Cereal Science* 74: 72–78. http://doi.org/10.1016/j.jcs.2017.01.018.

Santiago-Ramos, David, Juan de Dios Figueroa-Cárdenas, and José Juan Véles-Medina. 2018. "Viscoelastic Behaviour of Masa from Corn Flours Obtained by

Nixtamalization with Different Calcium Sources." *Food Chemistry* 248: 21–28. https://doi.org/10.1016/j.foodchem.2017.12.041.

Seitter, Michael, Fleig, Markus, Herbert Schmidt, and Christian Hertel. 2020."Effect of Exopolysaccharides Produced by *Lactobacillus sanfranciscensis* on the Processing Properties of Wheat Doughs." *European Food Research and Technology* 246: 461–469. https://doi.org/10.1007/s00217-019-03413-x.

Serna-Saldivar, Sergio O. 2017. "History of Corn and Tortilla in Tortillas." *Wheat Flour and Corn Products*, Chapter 1: 1–28. https://doi.org/10.1016/B978-1-891127-88-5.50001-3.

Serna-Saldivar, Sergio O., and Cristina Chuck-Hernandez. 2019. "Food Uses of Lime-Cooked Corn with emphasis in Tortillas and Snacks." In *Corn (Third Edition): Chemical and Technology*, Chapter 17: 469–500. https://doi.org/10.1016/B978-0-12-811971-6.00017.

Serratos-Hernández, José Antonio. 2012. "El origen y la diversidad del maíz en el continente americano." www.researchgate.net/publication/304498935_El_origen_y_la_diversidad_del_maiz_en_el_continente_americano_2a_edicion.

Serratos-Hernández, José Antonio, John T. Arnason, Constanza Nozzolillo, John D. H. Lambert, Bernard J. R. Philogene, Gary Fulcher, Kathryn Davidson, L Peacock, Jefrey Atkinson, and Peter Morand. 1987. "Factors Contributing to Resistance of Exotic Maize Populations to Maize Weevil, Sitophilus Zeamais." *Journal of Chemical Ecology* 13: 751–762. https://doi.org/10.1007/BF01020157.

Shah, Tajamul Rouf, Kamlesh Prasad, and Pradyuman Kumar. 2016. "Maize-A Potential Source of Human Nutrition and Health: A Review." *Cogent Food and Agriculture* 2 (1): 1166995. https://doi.org/10.1080/23311932.2016.1166995.

Shukla Rishi, and Munir Cheryan. 2001. "Zein: the industrial protein from corn". *Industrial Crops and Products* 13(3): 171–192. https://doi.org/10.1016/S0926-6690(00)00064-9.

Siyuan, Sheng, Li Tong, and Liu Rui Hai. 2018. "Corn Phytochemicals and Their Health Benefits." *Food Science and Human Wellness* 7: 185–195. https://doi.org/10.1016/j.fshw.2018.09.003.

Soro-Yao, Amenan Anastasie, Kouakou Brou, Georges Amani, Philippe Thonart, and Koffi Marcelin Djè. 2014. "The Use of Lactic Acid Bacteria Starter Cultures During the Processing of Fermented Cereal-Based Foods in West Africa: A Review." *Tropical Life Sciences Research* 25: 81–100. www.ncbi.nlm.nih.gov/pmc/articles/PMC4814148/.

Sosa, Meli, Alicia Califano, and Gabriel Lorenzo. 2019. "Influence of Quinoa and Zein Content on the Structural, Rheological, and Textural Properties of Gluten-Free Pasta." *European Food Research Technology* 245: 343–353. https://doi.org/10.1007/s00217-018-3166-5.

Tapia-Hernández, José Agustín, Carmen Lizett Del Toro-Sánchez, Francisco Javier Cinco-Moroyoquia, Elias Juárez-Onofreb, Saúl Ruiz-Cruzc, Elizabeth Carvajal-Milland, Amanda Guadalupe López-Ahumada, Daniela Denisse Castro-Enriqueza, Carlos Gregorio Barreras-Urbina, and Francisco Rodríguez-Felixa. 2019. "Prolamins from Cereal by-Products: Classification, Extraction, Characterization and Its Applications in Micro- and Nanofabrication." *Trends in Food Science and Technology* 90: 111–132. https://doi.org/10.1016/j.tifs.2019.06.005.

Tazrart, Karima, Farid Zaisi, Ana Salvador, and Claudia Monika Haros. 2019. "Effect of Broad Bean (*Vicia faba*) Addition on Starch Properties and Texture of Dry and Fresh Pasta." *Food Chemistry* 278 (25): 476–481. https://doi.org/10.1016/j.foodchem.2018.11.036.

Urias-Lugo, Diana Angelina, Jose Basilius Heredia, María Dolores Muy-Rangel, José Beningno Valdez-Torres, Sergio O. Serna-Saldıvar, and Janet Alejandra Gutierrez-Uribe. 2015. "Anthocyanins and Phenolic Acids of Hybrid and Native Blue Maize (Zea mays L.) Extracts and Their Antiproliferative Activity in Mammary (MCF7), Liver (HepG2), Colon (Caco2 and HT29) and Prostate (PC3) Cancer Cells." *Plant Foods for Human Nutrition* 70: 193–199. https://doi.org/10.1007/s11130-015-0479-4.

Utrilla-Coello, Rubi G., Edith Agama-Acevedo, Ana Paulina Barba de la Rosa, José L. Martinez-Salgado, Sandra L. Rodriguez-Ambriz, and Luis A. Bello-Perez. 2009. "Blue Maize: Morphology and Starch Synthase Characterization of Starch Granule." *Plant Foods for Human Nutrition* 64: 18–24. https://doi.org/10.1007/s11130-008-0106-8.

Vázquez-Carrillo, Maria Gricelda, David Santiago-Ramos, Edith Domínguez-Rendón, and Marcos Antonio Audelo-Benites. 2017. "Effects of Two Different Pozole Preparation Processes on Quality Variables and Pasting Properties of Processed Maize Grain." *Journal of Food Quality*: 1–15. https://doi.org/10.1155/2017/8627363.

Vigouroux, Yves, Jeffrey C. Glaubitz, Yoshihiro Matsuoka, Major Matsuoka Goodman, G. Jesus Sanchez, and John Doebley. 2008. "Population Structure and Genetic Diversity of New World Maize Races Assessed by DNA Microsatellites." *American Journal of Botanic* 95 (10): 1240–1253. https://doi.org/10.3732/ajb.0800097.

Villada, Jhon Alexander, Feliciano Sanchez-Sinencio, Orlando Zelaya-Angel, Elsa Gutierrez-Cortez, and Mario Enrique Rodríguez-García. 2017. "Study of the Morphological, Structural, Thermal, and Pasting Corn Transformation During the Traditional Nixtamalization Process: From Corn to Tortilla." *Journal of Food Engineering* 212: 242–251. https://doi.org/10.1016/j.jfoodeng.2017.05.034.

Wang Yan, Yu-Li Zhou, Ying-un Cheng, Zhen-Yan Jiang, Ye Jin, Han-Si Zhan, Da Liu, Li-Rong Teng, and Gui-Rong Zhang. 2015. "Enzymo-Chemical Preparation, Physico-Chemical Characterization and Hypolipidemic Activity of Granular Corn Bran Dietary Fibre." *Journal of Food Science and Technol*ogy 52 (3): 1718–1723. https://doi.org/10.1007/s13197-013-1140-6.

Yu, Xurun, Heng Yu, Jing Zhang, Shanshan Shao, Fei Xiong, and Zhong Wang. 2015. "Endosperm Structure and Physicochemical Properties of Starches from Normal, Waxy, and Super-Sweet Maize." *International Journal of Food Properties* 18 (12): 2825–2839. https://doi.org/10.1080/10942912.2015.1015732.

Zhang, Haiyan, and Guanghai Xu. 2019. "Physicochemical Properties of Vitreous and Floury Endosperm Flours in Maize." *Food Science and Nutrition* 7 (8): 2605–2612. https://doi.org/10.1002/fsn3.1114.

Žilić, Slađana, Arda Serpen, Gül Akıllıoğlu‖, Vural Gökmen, and Jelena Vančetović. 2012. "Phenolic Compounds, Carotenoids, Anthocyanins, and Antioxidant Capacity of Colored Maize (*Zea mays* L.) Kernels." *Journal of Agricultural and Food Chemistry* 60: 1224–1231. https://doi.org/10.1021/jf204367z.

Chapter 4

Bioactive Compounds in Native Cocoa (*Theobroma cacao* L.)

Rosario Rojas, Billy Cabanillas, Rosario Portales, and Candy Ruiz

CONTENTS

DOI: 10.1201/9781003087618-4

4.1 INTRODUCTION

Cocoa (*Theobroma cacao* L.) is a superfood and the main ingredient of chocolate. Recent archaeological evidence indicates that *T. cacao* was domesticated in South America at least 1500 years before it was moved into Central America and Mesoamerica. This reveals the upper Amazon region as the oldest center of cacao domestication yet identified. Ongoing research is dedicated to explain cacao's domestication history and the mechanisms of spread between the upper Amazon region, the Pacific coast, Central America, and Mesoamerica (Zarrillo et al. 2018). The upper Amazon region is also considered the center of the genetic diversity of *T. cacao* (Motamayor et al. 2008).

Cocoa beans are rich in carbohydrates (30%–40%), proteins (10%–15%), fat (46%–54%), fiber (3%), and ash (5.6%; Oracz et al. 2019). Additionally, cocoa beans are a source of secondary metabolites like alkaloids, amines, and phenolic compounds that have been shown to promote health and assist in the prevention of certain diseases. The content of these bioactive compounds in cocoa and cocoa by-products, and how they are affected by genetic factors and post-harvest procedures, will be described in this chapter.

4.2 COMPOUNDS FROM COCOA

4.2.1 Alkaloids and Phenylamines

4.2.1.1 Methylxanthines

The methylxanthines theobromine, caffeine, and theophylline represent more than 99% of the total alkaloids detected in cocoa beans (Loureiro et al. 2017). The theobromine content reported in *T. cacao* beans can vary between 2% and 3%, while the caffeine (<1%) and theophylline (traces) concentrations are much lower (Bertazzo et al. 2013; Franco et al. 2013; Vásquez-Ovando et al. 2016). It is known that theobromine, caffeine, and phenolic compounds are the main compounds responsible for the bitter taste of cocoa (Brunetto et al. 2007; Peláez et al. 2016). The chemical composition of cocoa beans is complex and varies depending on the bean type, geographical area, and bean maturity, as well as the fermentation and drying methods (Loureiro et al. 2017; Peláez et al. 2016).

T. cacao has a wide genetic diversity; however, cocoa beans are usually classified into three morphogenetic groups or complexes: Forastero (bulk cocoa), Criollo (sweet cocoa), and Trinitario. Criollo beans are characterized by intense fruity/nutty flavor and slight bitterness. In turn, Forastero beans have a relatively bitter and acidic taste. The Trinitario group, a hybrid of the Criollo and Forastero varieties, possesses intermediate features. Because of their attributes, Criollo and Trinitario varieties are preferred by the fine-flavored chocolate industry (Gopaulchan et al. 2019).

Davrieux et al. (2005) proposed a relationship between the cocoa genotype and the theobromine (Teob)/caffeine (Caf) ratio. Samples belonging to the Criollo variety exhibit the lowest Teob/Caf values (1–2), while the highest Teob/Caf values (>9) correspond to samples of the Forastero variety. Trinitario beans show intermediate Teob/Caf values (Brunetto et al. 2007; Carrillo et al. 2014; Loureiro et al. 2017; Peláez et al. 2016).

The Teob/Caf ratio may be an indication of the "fineness" of cocoa (Vásquez-Ovando et al. 2016). Peláez et al. (2016) obtained Teob/Caf ratios between 3.1 and 6.3 for 123 native cocoa samples from Huánuco (the central region of Perú), which suggests that cacao samples from this region have organoleptic qualities similar to those of the Trinitario group. On the other hand, native cocoas of the Chuncho variety from Cusco (Southern part of Perú) have Teob/Caf ratios between 0.9 and 1.3, similar to those of the Criollo group (Rojas et al. 2017).

Alkaloids do not undergo chemical transformations during fermentation, instead, about 20%–30% of them are eliminated by diffusion to the outside of the cocoa bean. Thus, fermentation decreases the bitter taste of cacao beans, mainly by lowering Teob and Caf concentrations (Brunetto et al. 2007; Loureiro et al. 2017; Peláez et al. 2016; Trognitz et al. 2013).

Caffeine produces a stimulating effect on the central nervous system by increasing attention and the sense of alarm. Other effects include vasodilatation and diuresis (Bertazzo et al. 2013; Dang and Nguyen 2019). The main action mechanism of caffeine and theobromine consists of blocking adenosine receptors and inhibiting phosphodiesterases (Martínez-Pinilla et al. 2015). Compared to caffeine, theobromine has a weaker effect on the nervous system but is considered safer (Dang and Nguyen 2019). Other activities reported for theobromine include antitumoral, vasodilation, muscle relaxant, blood pressure reduction, antitussive, diuretic, anti-inflammatory, and cardiovascular protection, without the undesirable side effects described for caffeine (Martínez-Pinilla et al. 2015).

4.2.1.2 Tetrahydro-β-Carbolines and Tetrahydroisoquinolines

Tetrahydro-β-carbolines (THβCs) are indole alkaloids that occur in various foods like cheese, yogurt, bread, wine, and chocolate. These compounds possess various biological activities including antimalarial, antileishmanial, and antiviral. Moreover, THβCs are considered neuromodulators, since they have been demonstrated to inhibit monoamine oxidase (MAO), to alter the reuptake of biogenic amines, and to interfere with benzodiazepine receptors (Maity et al. 2019; Gutsche and Herderich 1997).

In cocoa, the following THβCs were detected at low concentrations (<10 μg/g): 6-hydroxy-1-methyl-1,2,3,4-THβC; 1,2,3,4-THβC-3-carboxylic acid; 1-methyl-1,2,3,4-THβC-3-carboxylic acid; and 1-methyl-1,2,3,4-THβC. A 30 g chocolate bar can contain up to 0.21 mg of total THβCs; this amount could play a role in craving. Since THβCs inhibit MAO, they may enhance the activities of amines occurring in chocolate (phenylethylamine, tryptamine, serotonin; Herraiz 2000).

Salsolinol is a tetrahydroisoquinoline that can be found in cocoa and chocolate up to a concentration of 25 μg/g. This alkaloid binds to the dopamine D3-receptor, which plays a role in the reward system. The consumption of 100 g of dark chocolate would be enough to reach a pharmacological effect (Tuenter et al. 2018).

4.2.1.3 Phenylamines

Tryptamine, tyramine, and 2-phenylethylamine are the most common amines in cocoa. Their concentration in cocoa beans is highly variable (0–22.0 mg/kg; Bertazzo et al. 2013). Differences in bioactive amine profiles in fresh cocoa beans may be attributed to factors such as types of cultivars, growing region, growing conditions, degree of ripeness, post-harvest procedures, and storage conditions (do Carmo Brito et al. 2017).

The 2-phenylethylamine has also been found in chocolate. This amine can activate brain receptors of dopamine and noradrenalin and is considered to be responsible for chocolate craving and mood-lifting. However, other foods like cheese and sausage have higher concentrations of 2-phenylethylamine, yet they are not craved as chocolate (Bruinsma and Taren 1999; Bertazzo et al. 2013; do Carmo Brito et al. 2017).

In rodents, 2-phenylethylamine causes neurochemical and behavioral alterations due to the production of hydroxyl radicals and oxidative stress. It may be suggested that the overconsumption of chocolate might be toxic to dopaminergic neurons; however, this toxic effect may be attenuated by chocolate's polyphenols (Borah et al. 2013).

Tyramine and tryptamine at low concentrations modulate vaso- and neuro-activities; however, at high concentrations, they may induce headache and hypertension (do Carmo Brito et al. 2017).

4.2.2 Phenolic Compounds

Cocoa (*T. cacao* L.) is one of the richest dietary sources of phenolic compounds (Pérez-Jiménez et al. 2010). In cocoa beans, polyphenols contribute to the bitter and astringent taste, and their content varies depending on the provenience and post-harvest treatment, being the fermentation process that produces the greatest loss of phenolics (Bordiga et al. 2015; Oracz et al. 2015). Published data suggest that polyphenols may represent between 3% and 18% (dry basis) of the total mass of beans (Andújar et al. 2012; Loureiro et al. 2017). The major class of phenolic compounds in beans is flavonoids (about 98%) followed by a lower number of phenolic acid derivatives. Flavonoids are polyphenolic compounds consisting of two aromatics rings (rings A and B) connected by a 3-carbon bridge that often forms a third ring (ring C). Based on the degree of oxidation and saturation and its connectivity through the carbon bridge, flavonoids are divided into subclasses (Marais et al. 2006). Five subclasses of flavonoids have been reported in cocoa: flavanols, anthocyanins, flavonols, flavones, and flavanones.

Flavanols are the most abundant subclass found in cocoa beans and foods containing cocoa. They are present in the form of monomers and oligomers. Monomers (flavan-3-ols) are less abundant compared to oligomers (procyanidins); the principal monomer is (–)-epicatechin followed by (+)-catechin (Adamson et al. 1999; Gu et al. 2006). Their diastereoisomers (+)-epicatechin and (–)-catechin can be found in cocoa products resulting from epimerization during processing (Payne et al. 2010). Furthermore, other monomers have been reported in traces (Table 4.1). Oligomers (or polymers) are phytochemicals formed by the condensation of monomeric flavanols and, depending on their degree of polymerization (DP), they are grouped as dimers, trimers, tetramers, etc. Procyanidins with DP ranging from 2 to 10 are common in cocoa, although polymers with DP >10 are also reported (Hammerstone et al. 1999; Gu et al. 2006). Dimers, trimers, and tetramers are the most abundant oligomers with DP ≤ 10, and among them, dimers represent the largest group of procyanidins in different cocoa products (Ortega et al. 2008). Many dimers have been isolated and identified in cocoa, but the most abundant are type B procyanidins (Table 4.1), in particular procyanidin B2. Likewise, literature reports procyanidin C1 as the most abundant of six trimers found in cocoa beans and liquor (Hatano et al. 2002; Porter et al. 1991). Regarding higher polymers, only a tetramer and a pentamer (cinnamtannin A2 and cinnamtannin A3, respectively) have been identified (Esatbeyoglu et al. 2015). Anthocyanins are a class of compounds only reported in fresh and fermented-like cocoa beans. These pigments give a purple coloring to the fresh beans and disappear during fermentation by the action of the glycosidases (Niemenak et al. 2006). Mayor anthocyanins found are cyanidin 3-*O*-galactoside and cyanidin 3-*O*-arabinoside. Flavanols reported in cocoa beans and cocoa products consist of quercetin and kaempferol derivatives. Other flavonoids include apigenin and luteolin derivatives (flavones) and naringenin derivatives (flavanones; Andres-Lacueva et al. 2008; Pereira-Caro et al. 2013; Sanbongi et al. 1998; Sánchez-Rabaneda et al. 2003; Stark et al. 2005).

N-phenylpropenoyl-L-amino acids (NPAs) are polyphenol/amino acid conjugates that contribute, as well as catechins and procyanidins, to the astringent taste of the cocoa beans. (Stark et al. 2005; Stark and Hofmann 2005). Some NAPs have shown diverse pharmacological activities, including strong

TABLE 4.1 COMPOUNDS PRESENT IN COCOA BEANS AND COCOA-DERIVED PRODUCTS

Compound	Occurrence	References
Flavan-3-ols		
Monomers		
(+)-catechin	CB, CL, CH	Sanbongi et al. 1998; Hatano et al. (2002); Ioannone et al. (2015); Tokusoglu and Ünal (2002)
(–)-epicatechin	CB, CL, CH	
(–)-epicatechin 8-*C*-galactoside	CL	
(–)-epigallocatechin	CB, CH	
(–)-epigallocatechin gallate	CH	
Dimers		
bis-8,8'-catechinylmethane	CL	Hatano et al. (2002); Porter et al. (1991); Pedan et al. (2016)
epicatechin-(2β→5,4β→6)-*ent*-epicatechin	CB	
3T-*O*-β-D-galactopyranosyl-*ent*-epicatechin-(2α→7, 4α→8)-epicatechin	CB, CL	
3T-*O*-α-L-arabinopyranosyl-*ent*-epicatechin-(2α→7, 4α→8)-epicatechin	CB, CL	
3T-*O*-α-L-arabinopyranosyl-*ent*-epicatechin-(2α→7, 4α→8)-catechin	CL	
Procyanidin A1	CL	
Procyanidin A2	CL	
Procyanidin B1	CB	
Procyanidin B2	CB, CL	
Procyanidin B3	CB	
Procyanidin B4	CB	
Procyanidin B5	CB, CL	
Trimers		
Procyanidin C1	CB, CL	Hatano et al. (2002); Porter et al. (1991); Żyżelewicz et al. (2016)
3T-*O*-α-L-arabinopyranosyl-cinnamtannin B1	CL	
3T-*O*-β-L-galactopyranosyl-cinnamtannin B1	CL	

Compound	Occurrence	References
Tetramers		
cinnamtannin A2	CB, CL	Esatbeyoglu et al. (2015); Hatano et al. (2002)
Pentamers		
cinnamtannin A3	CB	Esatbeyoglu et al. (2015)
Anthocyanins		
cyanidin 3-*O*-galactoside	CB	Niemenak et al. (2006)
cyanidin 3-*O*-arabinoside	CB	
Flavonols		
Quercetin	CL	Sanbongi et al. (1998); Stark et al. (2005); Andres-Lacueva et al. (2008); Rodríguez-Carrasco et al. (2018); Pereira-Caro et al. (2013)
Quercetin 3-*O*-glucoside	CB, CL	
Quercetin 3-*O*-galactoside	CB	
Quercetin 3-*O*-arabinoside	CL	
Quercetin 3-*O*-glucuronide	CP	
Quercetin 3-*O*-rhamnoside	CH	
Quercetin 3-*O*-rutinoside	CH	
Isorhamnetin	CH	
Kaempferol	CH	
Kaempferol 3-*O*-glucoside	CH	
Kaempferol 3-*O*-rutinoside	CH	
Kaempferol 7-O-neohesperidoside	CH	
Flavones		
Apigenin	CP	Rodríguez-Carrasco et al. (2018); Sánchez-Rabaneda et al. (2003); Stark et al. (2005)
Apigenin 6-*C*-glucoside (isovitexin)	CB	
Apigenin 8-*C*-glucoside (vitexin)	CB CP	
Apigenin 7-*O*-Glucoside	CH	
Apigenin 7-*O*-rutinoside	CH	
Amentoflavone	CH	
Biapigenin	CH	
Luteolin	CP	
Luteolin 8-C-glucoside (orientin)	CP	
Luteolin 6-C-glucoside (isoorientin)	CP	
Luteolin 7-*O*-glucoside	CP, CB	

(*Continued*)

TABLE 4.1 (CONTINUED)

Compound	Occurrence	References
Flavanones		
Naringenin	CP	Rodríguez-Carrasco et al. (2018); Sánchez-Rabaneda et al. (2003); Stark et al. (2005)
Naringenin-7-*O*-glucoside	CB, CP	
Naringenin 7-O-neohesperidoside	CH	
Other phenolics		
N-phenylpropenoyl-L-amino acids		
(+)-N-caffeoyl-L-aspartic acid	CB	Sanbongi et al. (1998); Stark and Hofmann (2005)
(−)-N-caffeoyl-L-glutamic acid	CB	
(−)-N-caffeoyl-3-*O*-hydroxy-L-tyrosine	CB, CL	
(−)-N-caffeoyl-L-tyrosine	CB	
(+)-N-*p*-coumaroyl-L-aspartic acid	CB	
(+)-N-*p*-coumaroyl-L-glutamic acid	CB	
(−)-N-*p*-coumaroyl-L-tyrosine	CB, CL	
(−)-N-*p*-coumaroyl-3-*O*-hydroxy-L-tyrosine	CB	
(+)-N-feruloyl-L-aspartic acid	CB	
(+)-N-cinnamoyl-L-aspartic acid	CB	
Benzoic acids		
Protocatechuic acid	CB, CL, CP	Ali et al. (2015); Barnaba et al. (2017); Hatano et al. (2002)
Gallic acid	CB, CP, CH	
Syringic acid	CB	
Cinnamic acids		
Chlorogenic acid	CP	Ali et al. (2015); Barnaba et al. (2017); Hatano et al. (2002)
Caffeic acid	CB, CH	
Coumaric acid	CB, CH	
Ferulic acid	CB, CH	
Stilbenes		
***trans*-resveratrol**	CL, CH	Counet et al. (2006); Jerkovic et al. (2010)
***trans*-piceid**	CL, CH	

Abbreviations: CB, cocoa beans; CL, cocoa liquor; CP, cocoa powder; CH, chocolate.

anti-adhesive properties against the adhesion of *Helicobacter pylori* to human stomach tissue. Reports describe ten NAPs found in cocoa beans, among them (+)-N-caffeoyl-L-aspartic acid is the major metabolite (Lechtenberg et al. 2012). Phenolic acid derivatives are usually reported in cocoa powder or chocolate, their presence would result from the decomposition of NAPs and catechins during processing (Elwers et al. 2009). Two stilbenes, *trans*-resveratrol and its glycoside *trans*-piceid, have been detected in cocoa liquor and chocolate (Counet et al. 2006). Resveratrol has shown a broad spectrum of pharmacological activities that include antioxidant, anti-inflammatory, anticancer, cardioprotective, anti-aging, and antidiabetic properties (Chan et al. 2019).

The concentration of phenolic compounds in cocoa beans is also influenced by factors ranging from the place of harvest and the state of maturity of the fruit to the post-harvest treatment (Oracz et al. 2015). A study carried out with samples of Criollo, Upper Amazon Forastero, Lower Amazon Forastero, Nacional, and Trinitario varieties from different origins did not show significant differences either in the total polyphenol content or (–)-epicatechin content. On the other hand, Criollo beans showed few or no anthocyanins but a high content of caffeic acid aspartate compared to the other varieties. The decrease in catechins found during fermentation and drying is stronger in the Criollo seed samples than that described for other genotype groups. This might be the reason for the mild flavor of Criollo chocolates (Elwers et al. 2009).

4.2.3 Lipids

Cocoa beans contain about 46%–54% fat on dry weight basis. Similar to other vegetable fats and oils, cocoa fat is mainly composed of triglycerides together with small amounts of phospholipids. Triglycerides are compounds consisting of a glycerol moiety connected to fatty acid chains through an ester bond. The major triglycerides in cocoa beans are POP (2-oleoyl-1,3-dipalmitoylgly cerol), POS (2-oleoyl-1-palmitoyl-3-stearoylglycerol), and SOS (2-oleoyl-1,3-distearoylglycerol); their proportion may vary depending on the origin and processing (Servent et al. 2018; Sirbu et al. 2018). Fatty acids in cocoa beans are mainly saturated (60%–66%), followed by monounsaturated (32%–36%) and a lower content of polyunsaturated (2%–4%). Stearic acid and palmitic acid represent over 95% of total saturated fatty acids (SFA) and nearly 60% of total fatty acids. Oleic acid is the main unsaturated fatty acid and represents about 34% of total fatty acids (Grassia et al. 2019; Torres-Moreno et al. 2015). Although the consumption of foods with saturated fats is associated with increased plasma cholesterol and risk of thrombosis, this is not the case with cocoa butter. Intake of foods rich in stearic acid does not affect blood cholesterol levels as it does with other saturated fatty acids (Steinberg et al. 2003). Phospholipids have a similar structure to triglycerides, except that one of the fat chains has been replaced by a phosphate group modified with another organic molecule. The content of phospholipids in cocoa butter is less than 1%. Phosphatidylcholine, phosphatidylinositol, and phosphatidyl-ethanolamine are the main compounds found in cocoa butter; together they accumulate around 80% of total phospholipids. The concentration of phospholipids in cocoa butter would affect its viscosity, which would be reduced as the phospholipid content increases (Parsons and Keeney 1969; Parsons et al. 1969). The unsaponifiable fraction of cocoa butter contains tocopherols (100–300 mg/kg) and tocotrienols (2–5 mg/kg). The most abundant tocopherol isomer is γ-tocopherol (Lipp et al. 2001; Żyżelewicz et al. 2014). This compound has shown health-promoting properties including antioxidant, anti-inflammatory, and anticancer (Zheng et al. 2020). Several sterols have been identified in cocoa butter; however, the two major are β-sitosterol and stigmasterol (Lipp and Anklam 1998).

4.3 BIOACTIVE COMPOUNDS IN COCOA PRODUCTS

The concentration of bioactive compounds in raw cocoa beans is different from that found in cocoa products. The different processes applied in the industry alter the chemical composition of the raw material used. Table 4.2 shows the caffeine and theobromine content of various cocoa products. Cocoa powder

TABLE 4.2 PURINE ALKALOIDS CONTENT (mg/g FRESH WEIGHT) IN COCOA BEANS AND PRODUCTS

Source	Theobromine	Caffeine	Reference
Cocoa liquor	8.2–17.3	0.6–4.2	Zoumas et al. (1980)
Cocoa powder	14.6–26.6	0.8–3.5	Zoumas et al. (1980)
	17.5–23.5	1.2–2.4	Rýdlová et al. (2020)
Milk chocolate	1.35–1.86	0.05–0.54	Zoumas et al. (1980)
	0.5–4.5	0.5–1.7	Alañón et al. (2016)
Dark chocolate	2.2–13.7	1.0–3.1	Alañón et al. (2016)
	5.3–16.4	0.3–2.4	Langer et al. (2011)
Instant cocoa beverages	2.1–6.6	0.2–0.9	Rýdlová et al. (2020)
	2.6	0.21	Craig and Nguyen (1984)
Chocolate milk	0.23	0.01	Craig and Nguyen (1984)

TABLE 4.3 PHENOLIC COMPOUNDS CONTENT (mg/g FRESH WEIGHT) IN COCOA PRODUCTS

Source	Epicatechin	Catechin	Total procyanidins	Reference
Cocoa liquor	1.76–2.01	0.52–1.17	18.76–25.20	Gu et al. (2006)
Cocoa powder	1.47–2.83 1.58–2.58	0.35–0.90 0.61–0.90	20.94–24.38 32.19–48.70	Miller et al. (2009) Gu et al. (2006)
Baking chocolate	1.00–1.22	0.26–0.73	12.57–16.32	Miller et al. (2009)
Baking chips	0.41–0.58 0.66–1.07	0.11–0.24 0.26–0.50	3.70–6.29 8.71–15.57	Miller et al. (2009) Gu et al. (2006)
Dark chocolate	0.31–0.37 0.52–1.25	0.11–0.23 0.11–0.40	2.79–4.10 8.52–19.85	Miller et al. (2009) Gu et al. (2006)
Milk chocolate	0.02–0.15 0.18–0.24	0.01–0.08 0.05–0.12	0.43–0.90 2.16–3.14	Miller et al. (2009) Gu et al. (2006)
Chocolate syrup	0.03–0.12	0.03–0.06	0.37–0.85	Miller et al. (2009)

provides the best content of theobromine and caffeine. Other products such as chocolates and beverages report a lower content, which is related to the processing conditions and the percentage of cocoa used in the commercial product.

Phenolic compounds are also affected during industrial processing. The highest concentration of polyphenols is found in cocoa powder (Table 4.3). Cocoa products show lower levels of polyphenols as a result of the processes carried out on the cocoa beans before and after their industrial transformation (Urbańska et al. 2019). A study reveals that major phenolic epicatechin is affected in a similar proportion to other polyphenols during processing (Miller et al. 2009). In recent years, the studies showing the potential of cocoa polyphenols in the prevention of certain diseases has generated interest in developing procedures focused on reducing the degradation of polyphenols during fermentation and thus obtaining enriched raw material for the preparation of cocoa products (Cienfuegos-Jovellanos et al. 2009; Schinella et al. 2010; Tomas-Barberan et al. 2007).

4.4 COCOA AND HEALTH

4.4.1 Bioavailability of Cocoa Compounds

Health benefits attributed to compounds depend on the amount consumed and their bioavailability. The bioavailability of a drug compound is usually a fraction of an administered drug that reaches the systemic circulation in its unaltered form (not metabolite). Bioavailability is usually determined by measuring its concentration in plasma or urine after the ingestion of a given amount of a foodstuff or pure compound (Jalil and Ismail 2008). The structural characteristics and chemical composition of a compound influence its bioavailability. The diversity of compounds reported in cocoa has contributed in part to the differences observed in absorption and bioavailability. Matrix components (fat, carbohydrates, protein) would also affect the absorption of bioactive compounds (Arranz et al. 2013; Oracz et al. 2020). Methylxanthines are quickly absorbed in the gastrointestinal tract and metabolized mainly in the liver to subsequently eliminate the metabolites in urine. Published results indicate that caffeine is absorbed faster in tissues than theobromine, reaching its maximum plasma concentration between 30 and 70 minutes after consumption, while theobromine reaches its maximum concentration in approximately 2 hours (Oracz et al. 2020). Phenolics in cocoa include anthocyanins, flavan-3-ols, flavonols, and phenolic acids. Monomeric flavan-3-ols report better plasma bioavailability than oligomers. They are rapidly absorbed in the small intestine and reach their maximum concentration 2–3 hours after consumption. Epicatechin is much better absorbed than catechin probably due to stereochemical factors (Rusconi and Conti 2010). Procyanidins are less absorbed (under 0.5%) by the small intestine and reach the colon, where they are degraded by microbiota. Similarly, anthocyanins are poorly absorbed and are degraded in the colon. Glycosylated flavonols have exhibited better bioavailability compared to their aglycones, probably due to their better water solubility. The main flavonols in cocoa beans and cocoa products are quercetin glycosides. They have not been found in plasma; after entering the gut they are quickly absorbed and then metabolized in the small intestine, kidneys, liver, and colon. Phenolic acids are also rapidly absorbed in the small intestine; in particular, gallic acid is extremely well absorbed and rapidly metabolized in comparison to other polyphenols (Oracz et al. 2020). Bioavailability studies of stilbenes in plasma have shown variable results. These compounds would be absorbed by the small intestine and quickly metabolized, which is the reason why they have only been detected in traces in plasma (Oracz et al. 2020).

4.4.2 Antioxidant Effect

Cocoa has a high antioxidant activity that is superior to other foods and drinks such as red wine, green tea, black tea, and apples. This property is attributed to its content of methylxanthines and flavonoids, particularly catechins and procyanidins. In cocoa products, procyanidins contribute more to the antioxidant activity. Flavonoids act by neutralizing free radicals, chelating metals (Fe^{2+} and Cu^{+}) that enhance reactive oxygen species, inhibiting enzymes, and by upregulating antioxidant defenses. Epicatechin and catechin are very effective in neutralizing various types of radicals, while quercetin is good at chelating metal ions (Katz et al. 2011; Rusconi and Conti 2010). The antioxidant capacity of cocoa flavonoids has been proven by *in vitro* tests. They decrease oxidant-induced erythrocyte hemolysis, protect intestinal Caco-2 cell monolayers from lipophilic oxidants, inhibits superoxide anion formation and xanthine oxidase activity in leukemia HL-60 cells, inhibits lipid oxidation phosphatidylcholine liposomes, and inhibits ultraviolet-induced DNA oxidation (Andújar et al. 2012; Kim et al. 2014; Ramiro-Puig and Castell 2009). *In vivo* studies using polyphenol-rich cocoa extracts exhibited a reduction of lipid peroxide in rats fed diets deficient in vitamin E. Also, consumption of a 10% cocoa diet resulted in increased antioxidant activity in the thymus, spleen, and liver. Human studies have shown that consumption of cocoa powder increases the antioxidant capacity of plasma after 1–2 hours and decreases 6 hours after ingestion, probably due to the rapid metabolization of phenolics in the intestine. Furthermore, it has been found that cocoa consumption reduces the concentration of low-density lipoprotein (LDL) cholesterol in human plasma. Cocoa phenols act as a potent inhibitor of LDL oxidation (Kim et al. 2014; Ramiro-Puig and Castell 2009).

4.4.3 Cardioprotective Effects

Several studies have shown that cocoa has a positive impact on cardiovascular health. Cocoa compounds prevent the development of atherosclerosis by decreasing platelet adhesion and activation in blood vessel walls and by reducing LDL concentration in plasma (Franco et al. 2013; Zięba et al. 2019). Cocoa flavonols induce vasodilation and consequent reduction of blood pressure. Studies conducted on healthy people, patients with cardiovascular risk factors, and patients with arterial hypertension have shown variable results, some of them contradictory. A recent review analyzed the action of cocoa intake in 35 trials through a meta-analysis. The results indicate only a slight decrease in blood pressure (systolic and diastolic). Patients with high initial blood pressure and with a longer period of chocolate consumption showed the best blood pressure reduction (Garcia et al. 2018; Zięba et al. 2019). Atherosclerosis involves an endothelial dysfunction produced by inflammation. Cocoa polyphenols have been shown to reduce inflammation in human organisms through various mechanisms (Arranz et al. 2013). Other properties associated with moderate chocolate consumption lead to a low risk of coronary heart disease, heart failure, cerebrovascular accidents, and peripheral vascular disease. High consumption of chocolate increases the risk of heart failure (Garcia et al. 2018), as is usually expected with the consumption in excess of several food items.

4.4.4 Anticancer Effect

Cancer is a disease characterized by the growth and proliferation of abnormal cells. Its presence in humans is related to internal factors such as oxidative stress, hypoxia, and genetic mutations, as well as external factors such as stress, pollution, smoking, and radiation. Several studies have reported the capacity of food flavonoids to suppress cancer cell proliferation and prevent tumor development (Kopustinskiene et al. 2020). *In vitro* tests have shown that cocoa procyanidins can inhibit the growth of human colonic and prostate cancer cells. *In vivo* studies in rats indicate that cocoa procyanidins inhibit the growth of lung carcinomas and thyroid adenomas and also block breast and pancreatic tumorigenesis. Theobromine and caffeine from cocoa would act as inhibitors of the angiogenesis by inhibiting the expression of vascular endothelial growth factor (VEGF); cocoa polyphenols would also contribute to this activity (Kim et al. 2014; Montagna et al. 2019). Currently, there are no clinical studies that indicate an association between cocoa consumption and cancer prevention or treatment. A recent systematic review of the scientific literature indicates that there is no risk of colorectal cancer from cocoa consumption (Morze et al. 2020).

4.4.5 Effect on Obesity

Cocoa oligomeric procyanidins have shown a hypercholesterolemic effect in rats. According to the results, the procyanidins would act by inhibiting the intestinal absorption of cholesterol and bile acids through the decrease of the solubility of micellar cholesterol (Kim et al. 2014). In human beings, it has been found that the aroma of dark chocolate could suppress the appetite. Likewise, the regular consumption of cocoa powder or dark chocolate could contribute to decreasing LDL cholesterol levels and to the reduction of abdominal circumference in obese patients (Montagna et al. 2019; Kim et al. 2014).

4.4.6 Effects on Mood

Chocolate craving is a phenomenon still not completely understood. Although it seems that the sensory properties of chocolate (creamy texture, sweetness, and aroma) might be the main factor for craving and mood enhancement, certain pharmacologically active compounds found in chocolate would also be responsible. Among these compounds are 2-phenylethylamine (known as the "love drug") and the endocannabinoid anandamide; however, both of them are present in chocolate in very low concentrations (Herraiz 2000; Di Tomaso et al. 1996).

On the other hand, the methylxanthines theobromine and caffeine (the main alkaloids in chocolate), possess psychostimulant effects that reinforce chocolate consumption (Smit et al. 2004). Another interesting compound is the tetrahydroisoquinoline salsolinol, an alkaloid with an important dopaminergic activity (Tuenter et al. 2018).

To explain the relationship between cocoa or chocolate and mood, Tuenter et al. (2018) proposed the "mood pyramid" model. At the base of the pyramid are the flavonoids (cocoa flavanols) responsible for the enhancement of cognition. At the second level are theobromine and caffeine, with their effects on cognition and alertness. At the third level are the tetrahydroisoquinoline alkaloids, in particular salsolinol with its dopaminergic activities. Last, at the fourth and most important level are the orosensory properties of chocolate, which make it one of the most palatable foods (Tuenter et al. 2018; Smit et al. 2004).

4.5 CONCLUSIONS

Cocoa (*T. cacao* L.) is considered a superfood. The chemical composition of its beans is very complex and depends on the genotype, geographical origin, maturity, and post-harvesting procedures (fermentation and drying).

The main bioactive compounds in cocoa are alkaloids and phenolics. The methylxanthines theobromine, caffeine, and theophylline are the most abundant alkaloids in cocoa beans. Caffeine has a stimulating effect on the central nervous system by increasing attention and the sense of alarm. Caffeine and theobromine act by blocking adenosine receptors and by inhibiting phosphodiesterases.

Cocoa is one of the richest dietary sources of phenolic compounds; they represent 3%–18% of the total weight. Cocoa has a high antioxidant activity that is superior to other foods and drinks such as red wine, green tea, black tea, and apples. This property is attributed to its high content of flavonoids, particularly catechins and procyanidins. These flavanols are the most abundant subclass of flavonoids found in cocoa beans and chocolate. They are appreciated for their potential in the prevention of oxidative stress and inflammatory processes.

Several investigations have shown that cocoa and chocolate have a positive impact on cardiovascular health; however, some studies are contradictory. Moderate chocolate consumption may lead to a low risk of coronary heart disease, heart failure, cerebrovascular accidents, and peripheral vascular disease. However, high consumption of chocolate increases the risk of heart failure. The regular consumption of cocoa powder or dark chocolate may decrease LDL cholesterol levels and reduce abdominal circumference in obese patients.

Chocolate craving and its positive influence on mood are probably explained by the effects of cocoa flavanols, theobromine, and caffeine on cognition and alertness, as well as by the dopaminergic activities of tetrahydroisoquinoline alkaloids (salsolinol)—but more importantly because of its palatable properties (creamy texture, sweetness, and aroma).

REFERENCES

Adamson, Gary E., Sheryl A. Lazarus, Alyson E. Mitchell, Ronald L. Prior, Guohua Cao, Pieter H. Jacobs, Bart G. Kremers, John F. Hammerstone, Robert B. Rucker, Karla A. Ritter, and Harold H. Schmitz. 1999. "HPLC Method for the Quantification of Procyanidins in Cocoa and Chocolate Samples and Correlation to Total Antioxidant Capacity." *Journal of Agricultural and Food Chemistry* 47 (10): 4184–4188. https://doi.org/10.1021/jf990317m.

Alañón, María Elena, Sophie M. Castle, P. J. Siswanto, Tania Cifuentes-Gómez, and Jeremy P. Spencer. 2016. "Assessment of Flavanol Stereoisomers and Caffeine and Theobromine Content in Commercial Chocolates." *Food Chem*istry 208: 177–184. https://doi.org/10.1016/j.foodchem.2016.03.116.

Ali, Faisal, Yazan Ranneh, Amin Ismail, and Norhaizan Mohd Esa. 2015. "Identification of Phenolic Compounds in Polyphenols-Rich Extract of Malaysian Cocoa Powder Using the HPLC-UV-ESI—MS/MS and Probing Their Antioxidant Properties." *Journal of Food Science and Technology* 52 (4): 2103–2111. https://doi.org/10.1007/s13197-013-1187-4.

Andres-Lacueva, Cristina, María Monagas, Nasiruddin Khan, María Izquierdo-Pulido, Mireia Urpi-Sarda, Joan Permanyer, and Rosa M. Lamuela-Raventós. 2008. "Flavanol and Flavonol Contents of Cocoa Powder Products: Influence of the Manufacturing Process." *Journal of Agricultural and Food Chemistry* 56 (9): 3111–3117. https://doi.org/10.1021/jf0728754.

Andújar, Isabel, María Carmen Recio, Rosa M. Giner, and José L. Ríos. 2012. "Cocoa Polyphenols and Their Potential Benefits for Human Health." *Oxidative Medicine and Cellular Longevity* 9: 906252. https://doi.org/10.1155/2012/906252.

Arranz, Sara, Palmira Valderas-Martinez, Gemma Chiva-Blanch, Rosa Casas, Mireia Urpi-Sarda, Rosa M. Lamuela-Raventós, and Ramon Estruch. 2013. "Cardioprotective Effects of Cocoa: Clinical Evidence from Randomized Clinical Intervention Trials in Humans." *Molecular Nutrition and Food Research* 57 (6): 936–947. https://doi.org/10.1002/mnfr.201200595.

Barnaba, Chiara, Tiziana Nardin, A. Pierotti, Mario Malacarne, and Roberto Larcher. 2017. "Targeted and Untargeted Characterisation of Free and Glycosylated Simple Phenols in Cocoa Beans Using High Resolution-Tandem Mass Spectrometry (Q-Orbitrap)." *Journal of Chromatography A* 1480: 41–49. https://doi.org/10.1016/j.chroma.2016.12.022.

Bertazzo, Antonella, Stefano Comai, Francesca Mangiarini, and Su Chen. 2013. "Composition of Cacao Beans." In Ronald Ross Watson, Victor R. Preedy, and Sherma Zibadi (Eds.), *Chocolate in Health and Nutrition, Nutrition and Health.* Wien: Springer. https://doi.org/10.1007/978-1-61779-803-0.

Borah, Anupom, Rajib Paul, Muhammed K. Mazumder, and Nivedita Bhattacharjee. 2013. "Contribution of β-phenethylamine, a Component of Chocolate and Wine, to Dopaminergic Neurodegeneration: Implications for the Pathogenesis of Parkinson's Disease." *Neuroscience Bulletin* 29 (5): 655–660. https://doi.org/10.1007/s12264-013-1330-2.

Bordiga, Matteo, Monica Locatelli, Fabiano Travaglia, Jean Daniel Coïsson, Giuseppe Mazza, and Marco Arlorio. 2015. "Evaluation of the Effect of Processing on Cocoa Polyphenols: Antiradical Activity, Anthocyanins and Procyanidins Profiling from Raw Beans to Chocolate." *International Journal of Food Science & Technology* 50 (3): 840–848. https://doi.org/10.1111/ijfs.12760.

Brunetto, María del Rosario, Lubin Gutiérrez, Yelitza Delgado, Máximo Gallignani, Alexis Zambrano, Álvaro Gómez, Gladys Ramos, and Carlos Romero. 2007. "Determination of Theobromine, Theophylline and Caffeine in Cocoa Samples by a High-Performance Liquid Chromatographic Method with On-Line Sample Cleanup in a Switching-Column System." *Food Chemistry* 100 (2): 459–467. https://doi.org/10.1016/j.foodchem.2005.10.007.

Bruinsma, Kristen, and Douglas L. Taren. 1999. "Chocolate: Food or Drug?" *Journal of the American Dietetic Association* 99 (10): 1249–1256. https://doi.org/10.1016/S0002-8223(99)00307-7.

Carrillo, Luis C., Julián Londoño-Londoño, and Andrés Gil. 2014. "Comparison of Polyphenol, Methylxanthines and Antioxidant Activity in *Theobroma cacao* Beans from Different Cocoa-Growing Areas in Colombia." *Food Research International* 60: 273–280. https://doi.org/10.1016/j.foodres.2013.06.019.

Chan, Eric Wei Chiang, Chen Wai Wong, Yong Hui Alvin Tan, Jenny Pei Yan Foo, Siu Kuin Wong, and Hung Tuck Chan. 2019. "Resveratrol and Pterostilbene: A Comparative Overview of Their Chemistry, Biosynthesis, Plant Sources and Pharmacological Properties." *Journal of Applied Pharmaceutical Science* 9 (7): 124–129. https://doi.org/10.7324/JAPS.2019.90717.

Cienfuegos-Jovellanos, Elena, María del Mar Quiñones, Begoña Muguerza, Leila Moulay, Marta Miguel, and Amaya Aleixandre. 2009. "Antihypertensive Effect of a Polyphenol-Rich Cocoa Powder Industrially Processed to Preserve the Original Flavonoids of the Cocoa Beans." *Molecular Nutrition and Food Research* 57 (14): 6156–6162. https://doi.org/10.1021/jf804045b.

Counet, Christine, Delphine Callemien, and Sonia Collin. 2006. "Chocolate and Cocoa: New Sources of Trans-Resveratrol and Trans-Piceid." *Food Chemistry* 98 (4): 649–657. https://doi.org/10.1016/j.foodchem.2005.06.030.

Craig, Winston J., and Thuy T. Nguyen. 1984. "Caffeine and Theobromine Levels in Cocoa and Carob Products." *Journal of Food Science* 49 (1): 302–303. https://doi.org/10.1111/j.1365-621.1984.tb13737.x.

Dang, Yen K. T., and Hà V. H. Nguyen. 2019. "Effects of Maturity at Harvest and Fermentation Conditions on Bioactive Compounds of Cocoa Beans." *Plant Foods for Human Nutrition* 74 (1): 54–60. https://doi.org/10.1007/s11130-018-0700-3.

Davrieux, Fabrice, Sophie Assemat, Darin Sukha, Denis Bastianelli, Boulanger Renaud, and Emile Cros. 2005. "Genotype Characterization of Cocoa into Genetic Groups Through Caffeine and Theobromine Content Predicted by NIRS." Paper presented at the 12th International Conference for Near Infrared Spectroscopy. https://agritrop.cirad.fr/530989/. Accessed November 5, 2020.

di Tomaso, Emmanuelle, Massimiliano Beltramo, and Daniele Piomelli. 1996. "Brain Cannabinoids in Chocolate." *Nature* 382 (6593): 677–678. https://doi.org/10.1038/382677a0.

do Carmo Brito, Brenda de Nazaré, Renan Campos Chisté, Rosinelson da Silva Pena, Maria Beatriz Abreu Gloria, and Alessandra Santos Lopes. 2017. "Bioactive Amines and Phenolic Compounds in Cocoa Beans Are Affected by Fermentation." *Food Chemistry* 228: 484–490. https://doi.org/10.1016/j.foodchem.2017.02.004.

Elwers, Silke, Alexis Zambrano, Christina Rohsius, and Reinhard Lieberei. 2009. "Differences Between the Content of Phenolic Compounds in Criollo, Forastero and Trinitario Cocoa Seed (*Theobroma cacao* L.)." *European Food Research and Technology* 229 (6): 937–948. https://doi.org/10.1007/s00217-009-1132-y.

Esatbeyoglu, Tuba, Victor Wray, and Peter Winterhalter. 2015. "Isolation of Dimeric, Trimeric, Tetrameric and Pentameric Procyanidins from Unroasted Cocoa beans (*Theobroma cacao* L.) Using Countercurrent Chromatography." *Food Chemistry* 179: 278–289. https://doi.org/10.1016/j.foodchem.2015.01.130.

Franco, Rafael, Ainhoa Oñatibia-Astibia, and Eva Martínez-Pinilla. 2013. "Health Benefits of Methylxanthines in Cacao and Chocolate." *Nutrients* 5 (10): 4159–4173. https://doi.org/10.3390/nu5104159.

Garcia, José P., Adrian Santana, Diego Lugo Baruqui, and Nicholas Suraci. 2018. "The Cardiovascular Effects of Chocolate." *Reviews in Cardiovascular Medicine* 19 (4): 123–127. https://doi.org/10.31083/j.rcm.2018.04.3187.

Gopaulchan, David, Lambert A. Motilal, Frances L. Bekele, Séverine Clause, James O. Ariko, Harriet P. Ejang, and Pathmanathan Umaharan. 2019. "Morphological and Genetic Diversity of Cacao (*Theobroma cacao* L.) in Uganda." *Physiology and Molecular Biology of Plants* 25 (2): 361–375. https://doi.org/10.1007/s12298-018-0632-2.

Grassia, Melania, Giancarlo Salvatori, Maria Roberti, Diego Planeta, and Luciano Cinquanta. 2019. "Polyphenols, Methylxanthines, Fatty Acids and Minerals in Cocoa Beans and Cocoa Products." *Journal of Food Measurement and Characterization* 13 (3): 1721–1728. https://doi.org/10.1007/s11694-019-00089-5.

Gu, Liwei, Suzanne E. House, Xianli Wu, Boxin Ou, and Ronald L. Prior. 2006. "Procyanidin and Catechin Contents and Antioxidant Capacity of Cocoa and Chocolate Products." *Journal of Agricultural and Food Chemistry* 54 (11): 4057–4061. https://doi.org/10.1021/jf060360r.

Gutsche, B., and M. Herderich. 1997. "HPLC-MS/MS Profiling of Tryptophan-Derived Alkaloids in Food: Identification of Tetrahydro-β-carbolinedicarboxylic Acids." *Journal of Agricultural and Food Chemistry* 45 (7): 2458–2462. https://doi.org/10.1021/jf960952h.

Hammerstone, John F., Sheryl A. Lazarus, Alyson E. Mitchell, Robert Rucker, and Harold H. Schmitz. 1999. "Identification of Procyanidins in Cocoa (*Theobroma cacao*) and Chocolate Using High-Performance Liquid Chromatography/Mass Spectrometry." *Journal of Agricultural and Food Chemistry* 47 (2): 490–496. https://doi.org/10.1021/jf980760h.

Hatano, Tsutomu, Haruka Miyatake, Midori Natsume, Naomi Osakabe, Toshio Takizawa, Hideyuki Ito, and Takashi Yoshida. 2002. "Proanthocyanidin Glycosides and Related Polyphenols from Cacao Liquor and Their Antioxidant Effects." *Phytochemistry* 59 (7): 749–758. https://doi.org/10.1016/S0031-9422(02)00051-1.

Herraiz, T. 2000. "Tetrahydro-β-Carbolines, Potential Neuroactive Alkaloids, in Chocolate and Cocoa." *Journal of Agricultural and Food Chemistry* 48 (10): 4900–4904. https://doi.org/10.1021/jf000508l.

Ioannone, Francesca, Carla D. Di Mattia, Miriam De Gregorio, Manuel Sergi, Mauro Serafini, and Giampiero Sacchetti. 2015. "Flavanols, Proanthocyanidins and Antioxidant Activity Changes During Cocoa (*Theobroma cacao* L.) Roasting as Affected by Temperature and Time of Processing." *Food Chemistry* 174: 256–262. https://doi.org/10.1016/j.foodchem.2014.11.019.

Jalil, Abbe Maleyki Mhd, and Amin Ismail. 2008. "Polyphenols in Cocoa and Cocoa Products: Is There a Link Between Antioxidant Properties and Health?" *Molecules* 13 (9): 2190–219. https://doi.org/10.3390/molecules13092190.

Jerkovic, Vesna, Meike Bröhan, Elise Monnart, Fanny Nguyen, Sabrina Nizet, and Sonia Collin. 2010. "Stilbenic Profile of Cocoa Liquors from Different Origins Determined by RP-HPLC-APCI(+)-MS/MS. Detection of a New Resveratrol Hexoside." *Journal of Agricultural and Food Chemistry* 58 (11): 7067–7074. https://doi.org/10.1021/jf101114c.

Katz, David L., Kim Doughty, and Ather Ali. 2011. "Cocoa and Chocolate in Human Health and Disease." *Antioxidants & Redox Signaling* 15 (10): 2779–2811. https://doi.org/10.1089/ars.2010.3697.

Kim, Jiyoung, Jaekyoon Kim, Jaesung Shim, Chang Yong Lee, Ki Won Lee, and Hyong Joo Lee. 2014. "Cocoa Phytochemicals: Recent Advances in Molecular Mechanisms on Health." *Critical Reviews in Food Science and Nutrition* 54 (11): 1458–1472. https://doi.org/10.1080/10408398.2011.641041.

Kopustinskiene, Dalia M., Valdas Jakstas, Arunas Savickas, and Jurga Bernatoniene. 2020. "Flavonoids as Anticancer Agents." *Nutrients* 12 (2): 457. https://doi.org/10.3390/nu12020457.

Langer, Swen, Lisa J. Marshall, Andrea J. Day, and Michael R. A. Morgan. 2011. "Flavanols and Methylxanthines in Commercially Available Dark Chocolate: A Study of the Correlation with Nonfat Cocoa Solids." *Journal of Agricultural and Food Chemistry* 59 (15): 8435–8441. https://doi.org/10.1021/jf201398t.

Lechtenberg, Matthias, Katrin Henschel, Ursula Liefländer-Wulf, Bettina Quandt, and Andreas Hensel. 2012. "Fast Determination of N-phenylpropenoyl-l-Amino Acids (NPA) in Cocoa Samples from Different Origins by Ultra-Performance Liquid Chromatography and Capillary Electrophoresis." *Food Chemistry* 135 (3): 1676–1684. https://doi.org/10.1016/j.foodchem.2012.06.006.

Lipp, Markus, and Elke Anklam. 1998. "Review of Cocoa Butter and Alternative Fats for Use in Chocolate—Part A. Compositional Data." *Food Chemistry* 62 (1): 73–97. https://doi.org/10.1016/S0308-8146(97)00160-X.

Lipp, Markus, Catherine Simoneau, Franz Ulberth, Elke Anklam, Colin Crews, Paul Brereton, Wim de Greyt, W. Schwack, and C. Wiedmaier. 2001. "Composition of Genuine Cocoa Butter and Cocoa Butter Equivalents." *Journal of Food Composition and Analysis* 14 (4): 399–408. https://doi.org/10.1006/jfca.2000.0984.

Loureiro, Guilherme A. H. A., Quintino R. Araujo, George A. Sodré, Raúl R. Valle, José O. Souza, Edson M. L. S. Ramos, Nicholas B. Comerford, and Pauline F. Grierson. 2017.

"Cacao Quality: Highlighting Selected Attributes." *Food Reviews International* 33 (4): 382–405. https://doi.org/10.1080/87559129.2016.1175011.

Marais, J. P. J., B. Deavours, R. A. Dixon, and D. Ferreira. 2006. "The Stereochemistry of Flavonoids." In E. Grotewold (Ed.), *The Science of Flavonoids*. New York: Springer. https://doi.org/10.1007/978-0-387-28822-2.

Martínez-Pinilla, Eva, Ainhoa Oñatibia-Astibia, and Rafael Franco. 2015. "The Relevance of Theobromine for the Beneficial Effects of Cocoa Consumption." *Frontiers in Pharmacology* 6: 30. https://doi.org/10.3389/fphar.2015.00030.

Maity, Pradipta, Debasis Adhikari, and Amit Kumar Jana. 2019. "An Overview on Synthetic Entries to Tetrahydro-β-carbolines." *Tetrahedron* 75 (8): 965–1028. https://doi.org/10.1016/j.tet.2019.01.004.

Miller, Kenneth B., W. Jeffrey Hurst, Nancy Flannigan, Boxin Ou, C. Y. Lee, Nancy Smith, and David A. Stuart. 2009. "Survey of Commercially Available Chocolate- and Cocoa-Containing Products in the United States. 2. Comparison of Flavan-3-ol Content with Nonfat Cocoa Solids, Total Polyphenols, and Percent Cacao." *Journal of Agricultural and Food Chemistry* 57 (19): 9169–9180. https://doi.org/10.1021/jf901821x.

Montagna, Maria Teresa, Giusy Diella, Francesco Triggiano, Giusy Rita Caponio, Osvalda De Giglio, Giuseppina Caggiano, Agostino Di Ciaula, and Piero Portincasa. 2019. "Chocolate, 'Food of the Gods': History, Science, and Human Health." *International Journal of Environmental Research and Public Health* 16 (24): 4960. https://doi.org/10.3390/ijerph16244960.

Morze, Jakub, Carolina Schwedhelm, Aleksander Bencic, Georg Hoffmann, Heiner Boeing, Katarzyna Przybylowicz, and Lukas Schwingshackl. 2020. "Chocolate and Risk of Chronic Disease: A Systematic Review and Dose-Response Meta-Analysis." *European Journal of Nutrition* 59 (1): 389–397. https://doi.org/10.1007/s00394-019-01914-9.

Motamayor, Juan C., Philippe Lachenaud, Jay Wallace da Silva e Mota, Rey Loor, David N. Kuhn, J. Steven Brown, and Raymond J. Schnell. 2008. "Geographic and Genetic Population Differentiation of the Amazonian Chocolate Tree (*Theobroma cacao* L)." *PLoS One* 3 (10): e3311. https://doi.org/10.1371/journal.pone.0003311.

Niemenak, Nicolas, Christina Rohsius, Silke Elwers, Denis Omokolo Ndoumou, and Reinhard Lieberei. 2006. "Comparative Study of Different Cocoa (*Theobroma cacao* L.) Clones in Terms of Their Phenolics and Anthocyanins Contents." *Journal of Food Composition and Analysis* 19 (6): 612–619. https://doi.org/10.1016/j.jfca.2005.02.006.

Oracz, Joana, Ewa Nebesny, Dorota Zyzelewicz, Grazyna Budryn, and Boguslawa Luzak. 2020. "Bioavailability and Metabolism of Selected Cocoa Bioactive Compounds: A Comprehensive Review." *Critical Reviews in Food Science and Nutrition* 60 (12): 1947–1985. https://doi.org/10.1080/10408398.2019.1619160.

Oracz, Joanna, Dorota Zyzelewicz, and Ewa Nebesny. 2015. "The Content of Polyphenolic Compounds in Cocoa Beans (*Theobroma cacao* L.), Depending on Variety, Growing Region, and Processing Operations: A Review." *Critical Reviews in Food Science and Nutrition* 55 (9): 1176–1192. https://doi.org/10.1080/10408398.2012.686934.

Ortega, Nàdia, Maria-Paz Romero, Alba Macià, Jordi Reguant, Neus Anglès, José-Ramón Morelló, and Maria-Jose Motilva. 2008. "Obtention and Characterization of Phenolic Extracts from Different Cocoa Sources." *Journal of Agricultural and Food Chemistry* 56 (20): 9621–9627. https://doi.org/10.1021/jf8014415.

Parsons, J. G., and P. G. Keeney. 1969. "Phospholipid Concentration in Cocoa Butter and Its Relationship to Viscosity in Dark Chocolate." *Journal of the American Oil Chemists' Society* 46 (8): 425–427. https://doi.org/10.1007/BF02545628.

Parsons, J. G., P. G. Keeney, and S. Patton. 1969. "Identification and Quantitative Analysis of Phospholipids in Cocoa Beans." *Journal of Food Science* 34 (6): 497–499. https://doi.org/10.1111/j.1365-2621.1969.tb12069.x.

Payne, Mark J., W. Jeffrey Hurst, Kenneth B. Miller, Craig Rank, and David A. Stuart. 2010. "Impact of Fermentation, Drying, Roasting, and Dutch Processing on Epicatechin and Catechin Content of Cacao Beans and Cocoa Ingredients." *Journal of Agricultural and Food Chemistry* 58 (19): 10518–10527. https://doi.org/10.1021/jf102391q.

Pedan, Vasilisa, Norbert Fischer, and Sascha Rohn. 2016. "An online NP-HPLC-DPPH Method for the Determination of the Antioxidant Activity of Condensed Polyphenols in Cocoa." *Food Research International* 89: 890–900. https://doi.org/10.1016/j.foodres.2015.10.030.

Peláez, Pedro P., Inés Bardón, and Pedro Camasca. 2016. "Methylxanthine and Catechin Content of Fresh and Fermented Cocoa Beans, Dried Cocoa Beans, and Cocoa Liquor." *Scientia Agropecuaria* 7 (4): 355–365. http://doi.org/10.17268/sci.agropecu.2016.04.01.

Pereira-Caro, Gema, Gina Borges, Chifumi Nagai, Mel C. Jackson, Takao Yokota, Alan Crozier, and Hiroshi Ashihara. 2013. "Profiles of Phenolic Compounds and Purine Alkaloids During the Development of Seeds of *Theobroma cacao* cv. Trinitario." *Journal of Agricultural and Food Chemistry* 61 (2): 427–434. https://doi.org/10.1021/jf304397m.

Pérez-Jiménez, Jara, Vanessa Neveu, Femke Vos, and Augustin Scalbert. 2010. "Identification of the 100 Richest Dietary Sources of Polyphenols: An Application of the Phenol-Explorer Database." *European Journal of Clinical Nutrition* 64 (3): S112–S120. https://doi.org/10.1038/ejcn.2010.221.

Porter, Lawrence James, Z. Ma, and Bock G. Chan. 1991. "Cacao Procyanidins: Major Flavanoids and Identification of Some Minor Metabolites." *Phytochemistry* 30 (5): 1657–1663. https://doi.org/10.1016/0031-9422(91)84228-K.

Ramiro-Puig, Emma, and Margarida Castell. 2009. "Cocoa: Antioxidant and Immunomodulator." *British Journal of Nutrition* 101 (7): 931–940. https://doi.org/10.1017/s0007114508169896.

Rodríguez-Carrasco, Yelko, Anna Gaspari, Giulia Graziani, Antonello Santini, and Alberto Ritieni. 2018. "Fast Analysis of Polyphenols and Alkaloids in Cocoa-Based Products by Ultra-High Performance Liquid Chromatography and Orbitrap High Resolution Mass Spectrometry (UHPLC-Q-Orbitrap-MS/MS)." *Food Research International* 111: 229–236. https://doi.org/10.1016/j.foodres.2018.05.032.

Rojas, Rosario, Carlos Rodríguez, Candy Ruiz, Rosario Portales, Edgar Neyra, Kirti Patel, Julio Mogrovejo, Gabriela Salazar, and Jasmin Hurtado. 2017. *Cacao Chuncho del Cusco*. Lima: Universidad Peruana Cayetano Heredia. https://issuu.com/jerimo/docs/cacao_chuncho_del_cusco. Accessed November 5, 2020.

Rusconi, Manuel, and Ario Conti. 2010. "*Theobroma cacao* L., the Food of the Gods: A Scientific Approach Beyond Myths and Claims." *Pharmacological Research* 61 (1): 5–13. https://doi.org/10.1016/j.phrs.2009.08.008.

Rýdlová, Ladislava, Jana Prchalová, Tereza Škorpilová, Bo-Anne Rohlík, Helena Čížková, and Aleš Rajchl. 2020. "Evaluation of Cocoa Products Quality and Authenticity by DART/TOF-MS." *International Journal of Mass Spectrometry* 454: 116358. https://doi.org/10.1016/j.ijms.2020.116358.

Sanbongi, Chiaki, Naomi Osakabe, Midori Natsume, Toshio Takizawa, Shuichi Gomi, and Toshihiko Osawa. 1998. "Antioxidative Polyphenols Isolated from *Theobroma cacao*." *Journal of Agricultural and Food Chemistry* 46 (2): 454–457. https://doi.org/10.1021/jf970575o.

Sánchez-Rabaneda, Ferran, Olga Jáuregui, Isidre Casals, Cristina Andrés-Lacueva, Maria Izquierdo-Pulido, and Rosa M. Lamuela-Raventós. 2003. "Liquid Chromatographic/Electrospray Ionization Tandem Mass Spectrometric Study of the Phenolic Composition of Cocoa (*Theobroma cacao*)." *Journal of Mass Spectrometry* 38 (1): 35–42. https://doi.org/10.1002/jms.395.

Schinella, Guillermo, Susana Mosca, Elena Cienfuegos-Jovellanos, María Ángeles Pasamar, Begoña Muguerza, Daniel Ramón, and José Luis Ríos. 2010. "Antioxidant Properties of Polyphenol-Rich Cocoa Products Industrially Processed." *Food Research International* 43 (6): 1614–1623. https://doi.org/10.1016/j.foodres.2010.04.032.

Servent, Adrien, Renaud Boulanger, Fabrice Davrieux, Marie-Neige Pinot, Eric Tardan, Nelly Forestier-Chiron, and Clotilde Hue. 2018. "Assessment of Cocoa (*Theobroma cacao* L.) Butter Content and Composition Throughout Fermentations." *Food Research International* 107: 675–682. https://doi.org/10.1016/j.foodres.2018.02.070.

Sirbu, Diana, Anne Grimbs, Marcello Corno, Matthias S. Ullrich, and Nikolai Kuhnert. 2018. "Variation of Triacylglycerol Profiles in Unfermented and Dried Fermented Cocoa Beans of Different Origins." *Food Research International* 111: 361–370. https://doi.org/10.1016/j.foodres.2018.05.025.

Smit, Hendrik J., Elizabeth A. Gaffan, and Peter J. Rogers. 2004. "Methylxanthines Are the Psycho-Pharmacologically Active Constituents of Chocolate." *Psychopharmacology* 176 (3): 412–419. https://doi.org/10.1007/s00213-004-1898-3.

Stark, Timo, Sabine Bareuther, and Thomas Hofmann. 2005. "Sensory-Guided Decomposition of Roasted Cocoa Nibs (*Theobroma cacao*) and Structure Determination of Taste-Active Polyphenols." *Journal of Agricultural and Food Chemistry* 53 (13): 5407–5418. https://doi.org/10.1021/jf050457y.

Stark, Timo, and Thomas Hofmann. 2005. "Isolation, Structure Determination, Synthesis, and Sensory Activity of N-Phenylpropenoyl-l-amino Acids from Cocoa (*Theobroma cacao*)." *Journal of Agricultural and Food Chemistry* 53 (13): 5419–5428. https://doi.org/10.1021/jf050458q.

Steinberg, Francene M., Monica M. Bearden, and Carl L. Keen. 2003. "Cocoa and Chocolate Flavonoids: Implications for Cardiovascular Health." *Journal of the American Dietetic Association* 103 (2): 215–223. https://doi.org/10.1053/jada.2003.50028.

Tokusoglu, Özlem, and Kemal M. Ünal. 2002. "Optimized Method for Simultaneous Determination of Catechin, Gallic Acid, and Methylxanthine Compounds in

Chocolate Using RP-HPLC." *European Food Research and Technology* 215 (4): 340–346. https://doi.org/10.1007/s00217-002-0565-3.

Tomas-Barberan, Francisco A., Elena Cienfuegos-Jovellanos, Alicia Marín, Begoña Muguerza, Angel Gil-Izquierdo, Begoña Cerda, Pilar Zafrilla, Juana Morillas, Juana Mulero, Alvin Ibarra, Maria A. Pasamar, Daniel Ramón, and Juan Carlos Espín. 2007. "A New Process to Develop a Cocoa Powder with Higher Flavonoid Monomer Content and Enhanced Bioavailability in Healthy Humans." *Journal of Agricultural and Food Chemistry* 55 (10): 3926–3935. https://doi.org/10.1021/jf070121j.

Torres-Moreno, Miriam, Eva Torrescasana, Jordi Salas-Salvadó, and Consol Blanch. 2015. "Nutritional Composition and Fatty Acids Profile in Cocoa Beans and Chocolates with Different Geographical Origin and Processing Conditions." *Food Chemistry* 166: 125–132. https://doi.org/10.1016/j.foodchem.2014.05.141.

Trognitz, Bodo, Emile Cros, Sophie Assemat, Fabrice Davrieux, Nelly Forestier-Chiron, Eusebio Ayestas, Aldo Kuant, Xavier Scheldeman, and Michael Hermann. 2013. "Diversity of Cacao Trees in Waslala, Nicaragua: Associations Between Genotype Spectra, Product Quality and Yield Potential." *PloS One* 8 (1): e54079. https://doi.org/10.1371/journal.pone.0054079.

Tuenter, Emma, Kenn Foubert, and Luc Pieters. 2018. "Mood Components in Cocoa and Chocolate: The Mood Pyramid." *Planta Medica* 84 (12–13): 839–844. https://doi.org/10.1055/a-0588-5534.

Urbańska, Bogumiła, Dorota Derewiaka, Andrzej Lenart, and Jolanta Kowalska. 2019. "Changes in the Composition and Content of Polyphenols in Chocolate Resulting from Pre-Treatment Method of Cocoa Beans and Technological Process." *European Food Research and Technology* 245 (10): 2101–2112. https://doi.org/10.1007/s00217-019-03333-w.

Vázquez-Ovando, Alfredo, Isidro Ovando-Medina, Lourdes Adriano-Anaya, David Betancur-Ancona, and Miguel Salvador-Figueroa. 2016. "Alcaloides y Polifenoles del Cacao, Mecanismos que Regulan su Biosíntesis y sus Implicaciones en el Sabor y Aroma." *Archivos Latinoamericanos de Nutrición* 6 (3): 239–254.

Zarrillo, Sonia, Nilesh Gaikwad, Claire Lanaud, Terry Powis, Christopher Viot, Isabelle Lesur, Olivier Fouet, Xavier Argout, Erwan Guichoux. Franck Salin, Rey Loor Solorzano, Olivier Bouchez, Hélène Vignez, Patrick Severts, Julio Hurtado, Alexandra Yepez, Louis Grivetti, Michael Blake, and Francisco Valdez. 2018. "The Use and Domestication of *Theobroma cacao* During the Mid-Holocene in the Upper Amazon." *Nature Ecology & Evolution* 2 (12): 1879–1888. https://doi.org/10.1038/s41559-018-0697-x.

Zheng, Liyou, Jun Jin, Longkai Shi, Jianhua Huang, Ming Chang, Xingguo Wang, Hui Zhang, and Qingzhe Jin. 2020. "Gamma Tocopherol, Its Dimmers, and Quinones: Past and Future Trends." *Critical Reviews in Food Science and Nutrition*: 1–15. https://doi.org/10.1080/10408398.2020.1711704.

Zięba, Kinga, Magdalena Makarewicz-Wujec, and Małgorzata Kozłowska-Wojciechowska. 2019. "Cardioprotective Mechanisms of Cocoa." *Journal of the American College of Nutrition* 38 (6): 564–575. https://doi.org/10.1080/07315724.2018.1557087.

Zoumas, Barry L., Wesley R. Kreiser, and Roberta Martin. 1980. "Theobromine and Caffeine Content of Chocolate Products." *Journal of Food Science* 45 (2): 314–316. https://doi.org/10.1111/j.1365-2621.1980.tb02603.x.

Żyżelewicz, Dorota, Wiesława Krysiak, Grażyna Budryn, Joanna Oracz, and Ewa Nebesny. 2014. "Tocopherols in Cocoa Butter Obtained from Cocoa Bean Roasted in Different Forms and Under Various Process Parameters." *Food Research International* 63: 390–399. https://doi.org/10.1016/j.foodres.2014.03.027.

Żyżelewicz, Dorota, Wiesława Krysiak, Joanna Oracz, Dorota Sosnowska, Grażyna Budryn, and Ewa Nebesny. 2016. "The Influence of the Roasting Process Conditions on the Polyphenol Content in Cocoa Beans, Nibs and Chocolates." *Food Research International* 89: 918–929. https://doi.org/10.1016/j.foodres.2016.03.026.

Chapter 5

The Impact of Andean Biodiversity on a Healthy Diet and Assessment of the Anti-Inflammatory Potential of the Peruvian Cuisine

Fausto H. Cisneros, Martin J. Talavera, and Luis Cisneros-Zevallos

CONTENTS

DOI: 10.1201/9781003087618-5

5.1 INTRODUCTION

The Andean region, located in the western part of South America, is rich in biodiversity and home to a diversity of crops, some widely known worldwide (e.g., potatoes, quinoa, peanuts, and tomatoes) and many more still relatively unknown. The Andean region comprises not only the Andes mountains but also the coastal area abundant in seafood and the neighboring lush tropical forests of the Amazon.

Advanced pre-Columbian civilizations developed in this part of the world, like the Tiwanaku, Nazca, Moche, Wari, Chimu, and Inca, in part due to this biodiversity and in spite of the difficult geographical terrain. Archeological evidence and depiction of their pottery describe fruits, vegetables, and tubers reportedly consumed as food.

The Spanish conquistadores brought crops and their cooking techniques from the Old World to the Andean region that eventually evolved into the local cuisines of the present-day Andean countries. In the same manner, the Spaniards shipped several Andean crops back to Spain that eventually spread to the rest of Europe and the world. Among these most notably are the potato, Andean corn, and tomato, widely used worldwide nowadays. But the list also includes sweet potato, peanuts, avocado, and other foods.

The African slaves brought by the Spaniards during the colonial period also left their imprint on the cuisine of the Andean countries. The arrival of other immigrants in the latter part of the 19th century and the beginning of the 20th century to Peru, such as the Chinese, Italians, and Japanese, also significantly contributed to shape the fine Peruvian cuisine as it is known today.

In recent years, the Peruvian cuisine experienced unprecedented international recognition due to its unique diversity and blend of flavors; a movement led by renowned chef Gaston Acurio has created a rural and national economic impact, revalued Andean crops, stimulated production and biodiversity, and strengthened social self-esteem and identity. This silent revolution (or "first revolution") has created a platform for continuous growth. Thus, in this chapter we propose a "second revolution" for the Peruvian cuisine that consists of promoting it as a healthy gastronomy in similar fashion as the Mediterranean cuisine famous for preventing inflammation-based chronic diseases compared to the Western diet. In this chapter we will discuss some key elements including a preliminary assessment of the anti-inflammatory potential of Peruvian cuisine, the role of flavor, and a list of selected Andean crops and traditional gastronomic dishes.

5.2 PRELIMINARY ASSESSMENT OF ANTI-INFLAMMATORY POTENTIAL OF PERUVIAN CUISINE

Recently we proposed an integrative model that associates chronic diseases, food, and medicine (Jacobo-Velazquez et al. 2016; Osorio et al. 2016;

Santana-Galvez et al. 2017; Cisneros-Zevallos 2020). The model explains that normal state of health can be challenged based on the diet (e.g., saturated fats, omega-6 fatty acids, sugars), lifestyle, and genetics toward a pre-disease state defined by a low-grade systemic inflammation. If the challenge continues and the inflammation becomes chronic, then the health status will divert to a disease state (Figure 5.1). Conventional medicine puts emphasis on treating a disease state back to a pre-disease state by using pharmaceutical drugs (e.g., metformin for ameliorating symptoms of diabetes/insulin resistance) through intervention/treatment/therapeutic strategies, but not necessarily to a normal state. Usually, conventional medicine is not designed for prevention (exceptions include aspirin). On the other hand, preventive effects against low-grade systemic inflammation by consumption of a healthy diet has been associated to the Mediterranean cuisine known to be a rich source of antioxidants, fiber, and omega-3 fatty acids while low in omega-6 fatty acids, saturated fats, and carbohydrates (Casas et al. 2016). Furthermore, the anti-inflammatory properties of the Mediterranean diet have been associated to a low omega-6:omega-3 ratio of ~1–2:1 compared to the pro-inflammatory Western diet with a ratio value

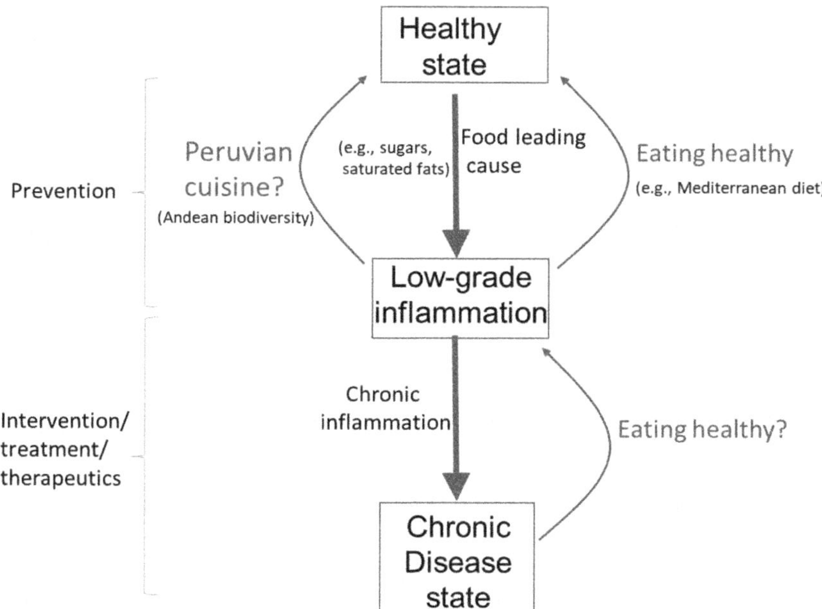

Figure 5.1 Proposed model of the role of Peruvian cuisine as a healthy diet with anti-inflammatory potential for the prevention of low-grade systemic inflammation.

Source: Adapted from Cisneros-Zevallos (2020).

of ~15–16.7:1 (Simopoulos 2004, 2006). In this context, we proposed the question whether the Peruvian cuisine has anti-inflammatory potential (Figure 5.1); however, to the best of our knowledge there is no study in this matter. Thus, in this chapter we propose an assessment of the anti-inflammatory potential of the Peruvian gastronomy by creating an arbitrary scale of the "content of bioactive and chemical groups" present in traditional Peruvian dishes and a "ratio of anti-inflammatory potential" based on anti-inflammatory to pro-inflammatory bioactive and chemical groups.

The arbitrary scale is based on the content of bioactive and chemical groups within each gastronomic dish, with numerical values defined as:

0 = non-available
1 = very low content
3 = low content
6 = moderate content
9 = high content.

The bioactive and chemical groups to assess within each gastronomic dish fall into two main classes including anti-inflammatory groups like antioxidants (AOX), omega-3 fatty acids (ω-3), and fibers (Fb), while pro-inflammatory groups include omega-6 fatty acids (ω-6), saturated fatty acids (SFA), and carbohydrates (Carb). The anti-inflammatory potential of each gastronomic dish uses the arbitrary scale of content of bioactive and chemical groups to establish an anti-inflammatory to pro-inflammatory ratio, defined as:

$$\text{Anti-inflammatory potential} = (\text{AOX} + \omega\text{-3} + \text{Fb})/(\omega\text{-6} + \text{carb} + \text{SFA})$$

This anti-inflammatory potential will give ratio values that will allow to classify the gastronomic dish as neutral (ratio values ≤ 1), low anti-inflammatory potential (ratio values >1–2), moderate anti-inflammatory potential (ratio values >2–3), and high anti-inflammatory potential (ratio values >3).

Table 5.1 shows a list of traditional Peruvian dishes including soups, entrees, sandwiches, main courses, desserts, and beverages; their ingredients; the amounts of bioactive and chemical groups scale values; and the calculated anti-inflammatory potential. To facilitate the analysis, we considered in the assessment that any Peruvian dish with a calculated value above 1 showed anti-inflammatory potential. Accordingly, from the overall 67 selected dishes, 30 show anti-inflammatory potential representing ~44%. Within each category of dishes, there were different results; for instance, in soups (11 dishes), entrees (11 dishes), sandwich (3 dishes), main course (31 dishes), desserts (7 dishes) and beverages (4) the anti-inflammatory potentials were 81%, 27%, 33%, 38%, 14%, and 100%, respectively. Differences among categories of dishes could be associated to the main ingredient quantities utilized. For instance, in soup dishes it is more common to use vegetable ingredients that are an important source of antioxidants, increasing the anti-inflammatory potential of this dish category. Similarly, in the beverage category the presence of grains, fruit, and herbs provide a rich source of antioxidants, increasing their anti-inflammatory potential. On the other hand, in entrees and main course dishes, the use of ingredients rich in carbohydrates overshadow other bioactive and chemical groups present, thus playing a major role in decreasing the anti-inflammatory potential of those dish categories. Similarly, in dessert dishes the presence of ingredients having sugar as major chemical component has an impact by reducing the anti-inflammatory potential of that dish category.

In general, this preliminary assessment of anti-inflammatory potential has major implications for the Peruvian cuisine, which can be used as a guide to aim at creating a healthy cuisine. On the one hand, it may give the opportunity to revisit the ingredient quantities used in main course dishes; for instance, simply reducing the side portion of white rice may confer an anti-inflammatory potential to the selected dish. Alternatively, another strategy would be replacing white rice with quinoa grains that could add antioxidants and fiber and provide an anti-inflammatory potential to the main course dish.

Similarly, it can be an opportunity to revisit ingredient quantities in entrees and desserts to reduce carbohydrate content in the ingredient formulation. In general, the strategy to promote a healthy diet is to increase the anti-inflammatory to pro-inflammatory ratio, thus increasing omega-3 sources through seafood and Andean crops like sacha inchi, and antioxidants and fiber sources through Andean crops like root, tubers, fruits, grains, and herbs will counteract the negative effects of carbohydrates, omega-6, and saturated fatty acid sources.

Another implication of the assessment could be associated to the development of new dishes; for instance, the Novoandina gastronomy, which uses native ingredients to create novel dishes or alternatively revisits traditional ones, can aim at creating a healthy cuisine. Furthermore, the proposed assessment could be used to design processed products as shelf-stable ingredients for the traditional cuisine, aiming at developing novel products with anti-inflammatory potential such as hot pepper sauces, herbal ingredient sauces, healthy drinks, and shelf-stable healthy desserts. However, any effort aiming at health properties has to balance with sensory attributes, as will be discussed in the next section.

Finally, the assessment made for the selected Peruvian dishes is only a preliminary analysis. Further work should focus on refining the arbitrary scale of content of bioactive and chemical groups and the equation for calculating the anti-inflammatory potential. For instance, to improve the arbitrary scale, we suggest finding the specific relationships between food ingredient quantities present in each dish preparation and the corresponding amounts and composition of bioactive and chemical groups associated to each arbitrary scale value. On the other hand, to improve the equation for determining anti-inflammatory potential, there is a need to define a weighting factor of individual bioactive and chemical groups in the formula as an alternative to the linear arithmetic relationship of the present model.

5.3 ROLE OF SENSORY CHARACTERISTICS ON PROMOTING ANDEAN CROPS AND PERUVIAN CUISINE

Food choice is a complex process involving different factors that can be summarized into three unique dimensions: a person's perception, regulated by their psychological and physiological state, their economic and social background, and the food's properties (Shepard 1989). Each of these factors interact with each other and result in the willingness to choose specific foods. More specifically, a food's properties such as sensory attributes (i.e., appearance, flavor, and texture), health information, nutritional content, price, and availability directly affect food choice, and therefore could increase or reduce the intake of specific foods.

Peru is a land of contrasts, where different cultures coexist and melt together, and this arguably can be considered as one of the sources of diversity in regards to how foods are prepared and consumed. Peru is the natural home of potatoes, where a large variety of these tubers have been bred and consumed for hundreds of years because of their flavor and as an energy source. Other foods native of Peru that also have existed for a long time are quinoa, kiwicha, maca, corn, and chili peppers. The Spaniards brought garlic, onions, and rice. At the same time, they brought cows, pigs, and goats that combined with locally bred guinea pigs as a source of protein. Asian immigrants brought ginger and soy sauce, which also have a major influence on Peruvian food preparation. This mixture of cultures and their flavors have unmistakably created a unique environment for new and attractive dishes. The first revolution of Peruvian cuisine can be attributed to the combination of these rich and unique flavors, due in part to this melting pot of cultures and traditions. At the same time, many ingredients used in Peruvian cuisine nowadays are mostly familiar to other societies, which can make them easier to introduce in other markets. Unfamiliar foods can be more difficult to spread to other countries because of food neophobia, which is the "fear" of new or unfamiliar foods. For example, lomo saltado, a flagship dish of Peru, is made of sautéed beef, onions, and tomatoes and served with rice and French fries. This flavorful dish can be more easily commercialized in different markets because it uses mostly mainstream ingredients, keeping its neophobic scores low. It is widely accepted that consumers are programmed since early childhood to prefer familiar foods. Different societies will have different levels of food neophobia, so how easy it is to introduce new flavors to different countries and cultures may vary (Castro and Chambers 2019; Tuorila and Hartmann

2020). Other dishes may also have rich and unique flavors but may be more difficult to introduce in other markets because of less "familiar" ingredients (e.g., anticucho, which is made of beef heart or grilled guinea pig). Another aspect that is very important about Peruvian food is the nutritional aspect. Healthy features of many Peruvian ingredients are evident, but they have not been utilized to their full potential to help improve the popularity of Peruvian cuisine, in addition to palatability. This means that there is a huge potential to further explore the nutritional benefits of several ingredients, and potentially modify existing foods and recipes to improve nutritional content, without affecting the sensory properties of well-known ingredients and dishes. Here is where the science of sensory evaluation can really help.

Sensory science has been defined as a scientific method used to evoke, measure, analyze, and interpret those responses to products as perceived through the senses of sight, smell, touch, taste, and hearing (Stone and Sidel 1992). Descriptive analysis, including the development and establishment of sensory attributes (i.e., sensory lexicons, which are sets of standardized vocabulary for a given product category), is a powerful method to describe the perceived sensory characteristics and intensities of products (Suwonsichon 2019). It is also an integral part of food research. Sensory evaluation helps to get a deeper understanding of specific sensory characteristics of products and their intensities to more easily communicate sensory features among multidisciplinary stakeholders (e.g., growers, manufacturers, product developers, scientists, consumers). For example, in foods, sensory attributes can provide guidance on the appearance, aroma, flavor, texture, and aftertaste of ingredients individually and as a part of a food matrix. This could help understand how ingredients interact with each other to form an overall perception of a product, so we can get an idea if consumers could potentially accept or reject the food. These techniques also help understand how changes on product formulations affect the sensory properties of the final product (e.g., Are the changes perceived or not? If the changes are perceived, is the product more liked or not? What changes can be made?). Sensory lexicons have been developed and actively used in the industry and academia to study the sensory features of products for some time, and several have been published for a variety of products, such as potatoes (Sharma et al. 2020a), green vegetables (Talavera-Bianchi et al. 2010), tomatoes (Hongsoongnern and Chambers 2008), green tea (Lee and Chambers 2007), and coffee (Chambers et al. 2016). Developing lexicons to describe ingredients and complex foods from Peru could help better characterize their properties, discover key quality characteristics, and use as a communication tool among producers and consumers.

A "second revolution" of Peruvian cuisine could be catapulted by the enhancement of health features in Peruvian foods as well as an expansion of research to characterize foods and discover new opportunities. For example, more research should focus on the characterization of the sensory properties of various native Peruvian foods to understand their key characteristics and discover potential white spaces, which could result in opportunities for the development of new products. In addition, it is important to conduct consumer testing to incorporate a hedonic component into the research to further understand drivers of appreciation (i.e., which sensory attributes are most important to drive consumer acceptance). This could help focus development and place emphasis on the most impactful sensory properties of products. Research should also go beyond products and focus on the consumers. This is where segmentation studies could help to characterize the market and discover unique "groups" of consumers. Consumers are heterogeneous, and they want different things. For example, three groups or clusters of potato consumers were identified in the United States. A traditional group focused on price and more conventional potato attributes; a "progressive" group was more focused on natural and nutritional aspects and willingness to experience more novel characteristics; and a group of potato "lovers" who liked most potatoes regardless of their different characteristics (Sharma et al. 2020b). This type of research could help understand the market and potentially result in targeted product development strategy aimed at specific market segments. Finally, if one should focus on features beyond sensory properties (e.g., nutritional aspects), it is important to conduct research to discover what product features are most critical, what information is most relevant to consumers, and what is the best way to reach them (e.g., through marketing and packaging). In this way we can develop products with flavors focusing on key sensory attributes, match specific market segments, and communicate the health benefits of products in the most impactful and effective ways. In the following sections of the chapter, we will list and discuss some traditional Andean crops and Peruvian dishes with a unique balance between flavor and health properties.

5.4 NATIVE ANDEAN CROPS

The following is a list in alphabetical order of selected Andean crops; it is by no means all-inclusive, and some of the species also grow outside the Andean region.

5.4.1 Roots and Tubers

Achira (*Canna edulis*). In the Andes, the rhizomes are boiled and eaten as is; however, in some areas of Southeast Asia it is used as a source of starch for noodle-making due to the unusual properties of its starch (Chansri et al. 2005; Cisneros et al. 2009).

Arracacha (*Arracacia xanthorrhiza*). The smooth-skinned roots that externally resemble a white carrot are frequently added to stews but can also be baked and fried. They have a crisp texture and a delicate flavor (National Research Council 1989).

Maca (*Lepidium meyenii*). The tuberous roots, which are similar to radish, have a tangy taste. They are usually dried and turned into flour. Since they are rich in micronutrients, such as minerals and bioactives, the flour is usually blended with other flours, foods or beverages (e.g., *emoliente*) to increase their nutrient and nutraceutical profile. The worldwide demand for maca has increased significantly in the past few years due to its reported multiple health benefits (da Silva Leitao Peres et al. 2020).

Mashua (*Tropaeolum tuberosum*). The tubers can be eaten raw, but they have a piquant flavor due to the presence of isothiocyanates; they also can be boiled, in which case the flavor becomes mild. In the highlands, it is usually added to soups and stews. Mashua glucosinolate content is high and has a potential medicinal use (Ramallo et al. 2004) and as a source of carotenoids, phenolics, and antioxidant activity (Campos et al. 2006; Chirinos et al. 2007).

Oca (*Oxalis tuberosa*). These are somewhat elongated tuberous roots that come in different colors (white, yellow, red, and purple), and while most varieties are slightly acidic, others are quite sweet (National Research Council 1989). To increase their sweetness, the tubers are exposed to the sun for several days before consumption. It can be prepared in numerous ways, such as baking, boiling, frying, or even added fresh to salads. The tubers can be eaten alone as a snack or added to stews or soups. The vitamin C and iron content are relatively high (León Marroú et al. 2011), which would make oca useful as part of a diet to prevent anemia. High levels of antioxidants for different genotypes have been reported (Campos et al. 2006). Outside the Andean region, it is grown in New Zealand, where it is known as New Zealand yam.

Potatoes. There are several species of native potatoes, including *Solanum tuberosum* (common potato), *S. andigena, S.* x *chaucha, S. stenotomum, S. goniocalyx*, and *S. phureja*. The diversity in size, shape, color, nutrients, and flavor of the tubers is astonishing (National Research Council 1989). These characteristics are taken into account by locals when selecting a potato for a particular dish. Potatoes are present in several Peruvian dishes, including soups (*sancochado, menestrón, aguadito*, several *cremas*), entrees (*papa rellena, causa, papa a la hunacaína, ocopa*), main dishes (in several stews), and desserts (potato starch in *mazamorra morada*). Potatoes are used mashed, sliced, or whole and are considered an important source of antioxidants and textures (Campos et al. 2006; Cisneros et al. 2018).

Ulluco (or **Olluco**; *Ullucus tuberosus*). These tuberous roots exhibit shiny, brilliant colors and a crisp texture when cooked. Due to their high moisture content, they are not baked or fried. Because of their thin skin, they do not need to be peeled (National Research Council 1989). The best known Peruvian dish made with olluco is called *olluquito con charqui* (with dried llama meat) or the variant *olluquito con carne* (with beef). Other dishes are *ajiaco de olluco, chupe de olluco,* and *sopa de olluco.* Olluco is a source of antioxidants that contain betalains and phenolics (Campos et al. 2006).

Yacon (*Smallanthus sonchifolius*). This is a tuberous root that stores carbohydrates in the form of fructooligosaccharides instead of starch, as most roots do (National Research Council 1989). It has a high water content, a crisp texture, and a bland flavor. It can be included in fruit salads for added crispness. A high-fructose syrup is made commercially by concentrating the juice by evaporation, and the syrup is used as a sweetening agent. Due to the presence of fructooligosaccharides, yacon can be considered a prebiotic (Pedreschi et al. 2003; Campos et al. 2012; Choque-Delgado et al. 2013).

5.4.2 Grains and Seeds

Beans. There are several types of beans present in Peruvian dishes that belong to the common bean (*Phaseolus vulgaris*) type, such as *canario, caballero,* and *panamito.* Lima bean, called *pallar* (*P. lunatus*) is another bean used in Peruvian cuisine. *Nuñas* (*P. vulgaris*), also called popping beans, have the interesting characteristic of popping somewhat like popcorn when dry-heated (National Research Council 1989). Many Peruvian dishes carry beans as a main ingredient, for example, *seco de carne con frejoles* and *tacu tacu. Shambar* is a thick soup from the Trujillo area in northern of Peru that includes beans as a main ingredient. There is a dessert made from beans into a sweet paste called *frejol colado.* Beans are high in protein content (Bressani 1983; Sathe et al. 1984), and combined with other sources of protein (e.g., cereals) they could be a good protein replacement for meat products. Experts in sustainable food production systems strongly suggest increasing consumption of beans and at the same time reducing the consumption of food products of animal origin, both for health reasons and for a more environmentally friendly food production process (Roos et al. 2018).

Corn (*Zea mays*). Corn was first domesticated in Mexico but spread thousands of years ago to the rest of the Americas, including the Andean region. This resulted in the development of corn varieties not found outside the Andean region. Some of these varieties are the giant corn (*maíz blanco gigante*), purple corn (*maíz morado*), and several varieties of *maíz cancha* (consumed as a roasted snack). The giant corn is present in several dishes of Peruvian cuisine such as soups (*patasca, sancochado, menestrón, aguadito*), entrees (*tamales, humitas, pastel de choclo, solterito*), stews (*pepián* and *espesado*), and as accompaniment to several dishes such as *ceviche, anticuchos,* and *locro.* An alcoholic beverage called *chicha de jora* is also made from fermented corn. Giant corn can be ground, used as kernels, or simply eaten on the cob. Purple corn is used to make a very typical beverage called *chicha morada* and a dessert called *mazamorra morada*. Anthocyanin pigments present in the corn cob are responsible for the intense purple color of these food products (Cevallos-Casals and Cisneros-Zevallos 2003, 2004). Purple corn is also rich in phenolic compounds, which have been shown to have strong antioxidant, anti-inflammatory, antimutagenic, anticarcinogenic, and anti-angiogenesis properties (Pedreschi and Cisneros-Zevallos 2006, 2007; Lao et al. 2017). *Maíz cancha* is oil-roasted and salted and served as a courtesy appetizer or a snack in practically every Peruvian food restaurant.

Kañiwa (*Chenopodium pallidicaule*). This is a small and dark-colored pseudocereal grain with a high protein content. It is frequently roasted and made into flour. The latter can be added to soups, beverages, and puddings. It can also be mixed with wheat flour for making breads, muffins, and cakes. The high protein content (~16%) is of good quality, with an excellent amino acid profile. Moreover, it has a good protein, carbohydrate, and lipid content balance (Repo-Carrasco et al. 2003).

Kiwicha (*Amaranthus caudatus*). This is a pseudocereal of small, cream-colored grains with a pleasant flavor. When popped by heating, they can be eaten as a snack, used as a cold cereal added to milk, or used as a main component of a granola bar. The grains can also be used as breading for frying meats. They can be milled into flour for use in baking or rolled into flakes and cooked to make porridge. It is a very nutritious grain with a high protein content (13%–18%) and a well-balanced amino acid composition (Venskutonis and Kraujalis 2013).

Quinoa (*Chenopodium quinoa*). The grains of this pseudocereal come in different colors from light to dark. It has received worldwide attention lately, and the demand for the grain has increased significantly due to its nutritive properties, which include a high protein content and a well-balanced amino acid composition (Repo-Carrasco et al. 2003). The protein, carbohydrate, and lipid contents are also well balanced (Nowak et al. 2015). It can be cooked in water, popped by applying dry heat, flaked, or ground into flour, which has multiple uses. It has a mild taste and a firm texture. Some varieties are bitter due to the presence of saponins, which must be removed before consumption.

Sacha inchi (*Plukenetia volubilis*). Also called Inca peanut, its seed has a high oil content, but it is especially prized for its very high omega-3 (α-linolenic acid) oil content (45%–53% of total oil; Hamaker et al. 1992; Bondioli et al. 2006). The cold-pressed oil is bottled and sold in supermarkets and health food stores. The seeds, which also have a high protein content, can be eaten as nuts but must be previously roasted in order to inactivate their antinutritional factors and improve their flavor (Cisneros et al. 2014).

Tarwi (*Lupinus mutabilis*). The bean of this Andean lupin has a very high protein content (~46%) and oil content (~20%) that could turn it into an industrial crop similar to soybeans. However, the presence of toxic and bitter alkaloids in the bean has been an obstacle into achieving this status. The extraction of these alkaloids has proved to be not an easy task, as it requires the use of high volumes of water and a lengthy extraction time of several days (Carvajal-Larenas et al. 2013; Cortés-Avendaño et al. 2020). Tarwi is used in stews, soups, and salads such as *ceviche de tarwi*. It can also be eaten as a snack. The protein presents adequate amounts of the essential amino acids, lysine, and cystine but is deficient in methionine (Schoeneberger et al. 1982).

5.4.3 Vegetables and Herbs

Caigua (*Cyclanthera pedata*). This fruit belongs to the family of the Cucurbitaceae. It is a green and mostly hollow fruit with soft spines. It is used in a few Peruvian dishes, such as salads and in main dishes such as *caigua rellena* (stuffed with seasoned minced meat) and *ajíaco de caigua*. There is interest in caigua for its reportedly anti-inflammatory, antihypercholesterolemic, and antihyperglycemic properties (Rivas et al. 2013).

Peppers (*Capsicum baccatum* 'ají amarillo', *C. pubescens* 'rocoto', *C. chinense* 'ají panca' and 'ají limo', and *C. frutescens* 'ají charapita'). There are several species of peppers (chilis) exhibiting a variety of colors, shapes, sizes, and flavors. They originated in the Americas but today are grown worldwide. A diversity of Andean peppers or chilis (*ajíes*) are used in Peruvian cuisine to add flavor with and without heat or pungency. Peruvian sofrito (*aderezo*) generally consists of red onions, garlic, and one or more ají pastes, such as *ají amarillo* (or *mirasol*) or *ají panca* (or *colorado*). This *aderezo* gives a distinctive flavor to various Peruvian dishes. Usually, the pungent flavor is removed from the peppers used in the *aderezo*. Moreover, the peppers are sometimes dried before grinding them into pastes in order to improve or concentrate the flavor. For example, *ají mirasol* is the dried version of ají amarillo. On the other hand, fresh peppers can be sliced or made into sauces and served as an accompaniment to the main dish if extra pungency is desired. Peppers usually have a high content of carotenoids, which gives them their different colors ranging from yellow to red (although a few peppers are green, violet, and white), as well as their health-promoting functional attributes (Hassan et al. 2019).

Squashes (*Cucurbita maxima* 'zapallo macre'; *C. moschata* 'zapallo loche'). There are two main squashes used in Peruvian cuisine, *zapallo macre* and *zapallo loche*. They are used in soups, stews and desserts. Examples of dishes where squashes are the main ingredient include *crema de zapallo* (soup), *locro de zapallo* (stew), and *picarones* (dessert). *Zapallo loche* is prized in the northern coast of Peru, where dishes such as *arroz con pato* and *cabrito a la chiclayana* owe part of their distinctive flavor to loche. Squashes are highly nutritious, being especially rich in carotenoids and an important source of provitamin A (Azevedo-Meleiro and Rodriguez-Amaya 2007; Jaeger de Carvalho et al. 2012; Kulczyński and Gramza-Michałowska 2019).

Huacatay (*Tagetes minuta*). This aromatic herb has been used for beverage infusions and as a condiment by the indigenous people of certain parts of South America (Soule 1993). There is evidence that the plant has antimicrobial activities (Hethelyi et al. 1986; Hudson 1990). In Peruvian cuisine it is used as a condiment in certain dishes and sauces, such as *salsa ocopa, salsa de ají de pollo a la brasa, locro de zapallo, chupe de camarones*, and *pachamanca*, just to name the best known dishes.

Chincho (*Tagetes elliptica*). This aromatic herb is native to the Peruvian Andes. The leaves are used as a condiment for the meat of the *Pachamanca*, combined with the related herb, *huacatay*. Chincho essential oil has been shown to possess antimicrobial activity (Segovia et al. 2010).

5.4.4 Fruits

Aguaje (*Mauritia flexuosa*). This palm tree that grows in the Amazon basin has fruits with an oily pulp and a color that ranges from yellow to orange, but it has a bland flavor. The pulp is the richest food source of carotene and thus, of provitamin A activity (Pacheco Santos 2005; Candido et al. 2015). In the jungle of Peru, a very popular beverage is made from the pulp, called *aguajina*.

Sauco (*Sambucus peruviana* H.B.K.). This tree grows at high altitudes (2800–3900 meters above sea level). It produces a berry that can be eaten fresh but it usually is made into jam (Verde García y Rengifo Alcántara 2006). The jam is used in some Peruvian pastries as a filling or as a topping.

Camu camu (*Myrciaria dubia*). This plant that grows wild along the banks of the rivers and lakes of the Amazon bears fruits that have a white pulp and a red to purple skin when fully ripe (Camargo Neves et al. 2015). These berries have the highest vitamin C content of any fruit (Rodrigues et al. 2001) and are used by the food industry to increase the vitamin C content of fruit juices and nectars.

Cherimoya (*Annona cherimola*). This fruit of green-colored skin and white pulp has a very sweet and pleasant flavor. Handling must be done with care because the thin skin is very delicate and easily bruised. It is used to flavor diverse Peruvian desserts, including pastries and ice cream. It has health-promoting properties due to its high antioxidant activity (Barreca et al. 2011) and phenolic content (Santos et al. 2016).

Cocona (*Solanum sessiliflorum*). This Amazonian plant bears a fruit that is about the size of an apple whose shape can be spherical or elongated. The skin color is mostly yellow, orange, but some ecotypes are red. The usually cream-colored pulp is acidic and has a seed distribution similar to tomatoes. In Peruvian cuisine, it is used to make hot sauces combined with hot peppers. It is also used to make nectars, usually in combination with other fruits. A recent study has shown cocona pulp has a high vitamin C content (Sereno et al. 2018).

Goldenberry (*Physalis peruviana*). Also called *aguaymanto* in Peru, this plant grows in the Andes and has an orange-colored berry fruit about the size of a grape. The flavor of the fruit is sweet and tangy, similar to that of a sweet tomato. The goldenberry fruit is a great source of carotenoids with provitamin A activity, predominating β-carotene, and lutein (Etzbach et al. 2018), vitamin C and phenolic compounds (Olivares-Tenorio et al. 2016). The iron content of the fruit is higher than that found in most fruits and is equivalent to that encountered in some beans (Rodrigues et al. 2009). Goldenberry can be eaten fresh or used to make nectars and jam, the latter being used in pastries.

Lucuma (*Pouteria lucuma*). Also called the *caramel fruit* because of its flavor, which is reminiscent of caramel. Its skin color ranges from green to orange and the pulp varies from yellow to orange. The pulp owes its color to the high concentration of carotenoids, but it is especially rich in β-cryptoxanthin, a xanthophill with provitamin A activity. In fact, it is one the richest food sources of this type of carotenoid (Nair et al., personal communication, 2021). There is research that also attributes other health-promoting properties to β-cryptoxanthin, such as prevention of certain cancers and osteoporosis. Since lucuma is usually used in Peruvian cuisine to flavor dairy desserts (e.g., ice cream, milkshakes, yogurt), these foods could provide an interesting combination of calcium, vitamin D, and β-cryptoxanthin to potentially promote bone health.

Tumbo (*Passiflora tripartita*). One of the several passion fruit species, tumbo is native to the Andes (National Research Council 1989). It has an ellipsoid shape, and its thick skin has a yellowish green tone. The interior is pleasantly aromatic and is formed by orange arils similar to the common passion fruit. Traditionally, in the world-renowned Peruvian ceviche, fish pieces are marinated in lime juice, however there is a version of ceviche where tumbo is substituted for lime. Tumbo is also used in beverages similar to other passion fruits.

5.5 TRADITIONAL PERUVIAN DISHES

What follows is not a comprehensive list of Peruvian dishes but it represents some of the most traditionally emblematic dishes. We have not included Chinese-Peruvian or Japanese-Peruvian dishes but have selected those based on more traditional and "criollo" dishes. Signature and Novoandina dishes are not included either. The dishes have been categorized into soups, entrées, sandwiches, main courses, desserts, and beverages.

5.5.1 Soups

Aguadito de pollo o choros. A mild cilantro-flavored rice and chicken soup with pieces of vegetables (Andean corn, peas, bell pepper, potato). The sofrito that condiments the soup is made with ají amarillo paste, red onions, cilantro, and cumin.

Chilcano de pescado. A spicy fish broth served with pieces of fish. It is seasoned with red onions, garlic, cilantro, ginger, black pepper, ají limo, and green onions.

Chupe de camarones. A hearty crayfish chowder—from the Arequipa region—containing rice, potatoes and pieces of vegetables (Andean corn, peas, faba bean, huacatay) and soft white cheese (queso fresco) and beaten eggs. The whole crayfish (head and tail) goes into the soup.

Crema de zapallo. This is a creamy squash (zapallo macre or loche may be used) soup. Yellow potatoes (papa amarilla) can be added for a thicker consistency. A sofrito of red onions and garlic adds flavor to the soup. Croutons or Parmesan cheese are optionally added on top.

Inchicapi de gallina con maní. An Amazonian chicken soup thickened and flavored with corn and roasted peanut flour. It is additionally seasoned with red onions, garlic, cilantro, turmeric, and ají charapita and usually served with cassava.

Menestrón. This soup is an adaptation of the Italian Genovese minestrone which has many types of vegetables and is flavored with basil. The Peruvian menestrón also adds New World ingredients such as Andean corn, lima beans, and cassava. It also includes beef and queso fresco that the Genovese minestrone does not have.

Parihuela. This is a hearty and spicy soup made with a diversity of seafood such as crab, fish, mussels, shrimp, and octopus. This soup includes a sofrito of red onions, garlic, ají amarillo, ají mirasol and ají panca paste. It is additionally seasoned with tomatoes, pepper, cumin, oregano, cilantro, parsley, and *chicha de jora* (an alcoholic beverage made by fermenting corn).

Patasca. A hearty soup from the highlands made with *mote* (hominy), beef tripe, and meat. The sofrito is made with red onions, garlic, ají panca, tomatoes, and oregano. It is additionally flavored with bay leaves, hierba buena, and black pepper.

Sancochado. This filling soup includes beef, cabbage, potatoes, cassava, carrots, sweet potato, Andean corn, turnip, and rice. It is additionally seasoned with herbs (cilantro, parsley, hierba buena, and oregano). The vegetables and the beef are removed from the broth and eaten separately. It can replace a main course dish.

Shambar. In this hearty and viscous soup from the Trujillo region, pulses take center stage. It contains an assortment of beans, chickpeas, peas, faba beans, and wheat grains, and is flavored with pork rind and smoked ham. Tradition dictates that it be prepared and eaten on Mondays. This soup can be served as a main course.

Sopa a la minuta. This is a savory but light soup made with ground meat, potatoes, milk (optional), and angel hair. It is seasoned with red onions, tomato paste, garlic, ají panca or mirasol, oregano, and bay leaf.

5.5.2 Entrées

Causa rellena (with tuna, chicken or crab). A cold entree formed by layers of mashed yellow potato acidified with lime juice and flavored with ají amarillo paste and a layer of tuna with mayonnaise; other layers can consist of slices of avocado, eggs, red onions, and/or Peruvian olives. It is often served with lettuce as an accompaniment. Canned tuna can be replaced with chicken or crab.

Ceviche de tarwi. In this salad, the (previously debittered and cooked) tarwi bean (also called *chocho*) is seasoned with lime juice, salt, and pepper and then mixed with julienne-cut red onions, chopped cilantro, garlic, and ají limo.

Humitas. This cooked corn paste (mixed with shortening) is lightly seasoned and securely wrapped in corn leaves before being cooked. There are salty and sweet versions.

Ocopa. This huacatay-flavored sauce from the region of Arequipa is used as a dressing for boiled potatoes. The sauce is made by blending queso fresco, huacatay, vanilla cookies, peanuts (or pecans), evaporated milk, sofrito (garlic, red onion, and ají mirasol), oil, pepper, and salt. A lettuce leaf, a Peruvian olive, and one-half of a boiled egg are usually the accompaniments.

Papa a la huancaína. This creamy sauce is used as a dressing for boiled potatoes. The sauce is made with queso fresco, evaporated milk, ají amarillo, oil, pepper, and salt. To add more consistency, soda crackers can be added. The accompaniment is usually a lettuce leaf, a Peruvian olive, and one-half of a boiled egg.

Papa rellena. This is a large oval-shaped potato croquette. The outside layer is mashed potato turned into a dough and the filling is usually seasoned ground or minced beef, mixed with a slice of hard-boiled egg, slices of Peruvian olive and raisins. The papa rellena is deep fried until a golden-brown crust is formed. It is usually accompanied with *zarza criolla* (julienne-cut red onions in lime juice dressing with chopped parsley and ají limo). Optionally, the potato dough can be mixed with ají amarillo paste to add extra flavor and color.

Pastel de choclo. This is an Andean corn-based baked pie with a filling consisting of seasoned minced or ground beef with raisins. The corn dough is made by blending corn with evaporated milk, shortening or margarine, eggs, sugar, and ají amarillo paste.

Patitas de cerdo en fiambre. Pigs' feet are cooked until tender in flavored water (garlic, red onions, bay leaves, and herbs like hierba buena, oregano, and parsley). They are served covered with a dressing consisting of vinegar, lime juice, julienne-cut red onion, and a sofrito of garlic and ají amarillo, among other condiments that vary across recipes.

Rocoto relleno. This Peruvian version of baked stuffed peppers is originally from the Arequipa region. The hotness or pungency of the rocoto is significantly reduced by deseeding it, removing the veins, and cooking it in hot water. The filling is seasoned ground or minced beef mixed with raisins, slices of hard-boiled eggs, and Peruvian olives. A slice of cheese is placed on top of the rocoto before baking.

Solterito. This salad made with faba beans, corn, fine pieces of red onion, and queso fresco with an oil, lime juice (and/or vinegar) dressing and a touch of rocoto (or any other type of ají) and minced parsley for additional flavor.

Tamal. A cooked dough made with grated corn and lard, to which pork or chicken meat cooked in sofrito (garlic, red onion, ají mirasol) is added. The tamal filling consists of pork (or chicken) meat, a quarter of hard-boiled egg, a Peruvian olive, and a roasted peanut. The dough is wrapped in banana leaves for a final cooking. The tamal is served with zarza criolla.

5.5.3 Sandwiches

The bread usually used in Peruvian sandwiches is called *pan frances* (a round-shaped type of French bread).

Arrebosado de pejerrey. A sandwich made with deep-fried breaded (eggs and wheat flour) pejerrey fish and with zarza criolla.

Chicharrón de cerdo. This consists of cooked and fried pieces of pork meat accompanied with slices of fried sweet potato and zarza criolla.

Butifarra. Made with slices of jamón del país (a type of Peruvian ham), lettuce, zarza criolla, and a sauce made with mayonnaise, mustard, and ají amarillo.

5.5.4 Main Courses

Most of the main dishes are served with rice.

Adobo de cerdo. This dish is from the southern part of Peru and consists of pork meat seasoned with ají panca, garlic, wine vinegar, chicha (a fermented corn alcoholic drink), bay leaves, oregano, hierba buena, black pepper, red onions (minced and slices), cinnamon, and rocoto. It is served with rice and slices of sweet potato.

Ají de pollo. Shredded chicken breast is mixed with a sauce made with bread soaked in evaporated milk (for consistency), sofrito (garlic, red onions, ají Amarillo), ground pecan nuts, and shredded Parmesan cheese. It is served accompanied with rice, a Peruvian olive, and half of a hard-boiled egg.

Ajiaco de olluco. This dish is prepared with olluco, faba beans, and potatoes cooked in a milk-based sauce with queso fresco and sofrito (garlic, red onions, and ají mirasol or ají amarillo).

Anticuchos de corazón. Beef heart kebab is previously marinated in a mixture of ají panca and garlic paste, wine vinegar, cumin, and oregano. It is usually served with half a potato and giant corn.

Arroz a la jardinera. This dish is a mixture of rice, peas (or lima beans), corn, and small pieces of carrots seasoned with sofrito (garlic, red onions, ají amarillo, and turmeric) and cooked in beer and chicken stock. It is usually served with chicken or pork.

Arroz con mariscos. A rice-based dish with shellfish (squid rings, shrimp, scallops, clams) and vegetables (peas, pieces of bell peppers and carrots) cooked in fish stock and white wine. The rice is seasoned with sofrito (garlic, red onions, and ají amarillo), annatto, oregano, bay leaves, cilantro, and Parmesan cheese.

Arroz con pollo o pato. A rice-based dish with chicken (or duck) and vegetables (peas, carrots, and bell pepper) cooked in chicken stock and dark beer. The rice is seasoned with zapallo loche, cilantro, sofrito (garlic, red onions, ají amarillo, and ají panca), pepper, and cumin.

Caiguas rellenas. Caigua is stuffed with seasoned ground or minced beef mixed with raisins, slices of hard-boiled eggs, and Peruvian olives. Ground beef is seasoned with sofrito (garlic, red onion, ají panca), pepper, and cumin.

Carapulcra. A stew made with lightly roasted *papa seca* (small pieces of dried and previously cooked potatoes) and pork cooked in a sauce flavored with sofrito (garlic, red onions, ají panca, ají mirasol, black pepper, and cumin), chicken stock, roasted peanuts, chocolate, cinnamon, and cloves.

Cau cau de mondongo (o pollo). A stew made with tripe (or chicken), potatoes, carrots, peas, and flavored with sofrito (garlic red onions, ají amarillo, cumin, turmeric, and black pepper) and hierba buena.

Ceviche. A dish consisting of raw fish pieces marinated in lime juice mixed with julienne-cut red onions and flavored with ají limo, cilantro, ginger, and black pepper. Corn and half or a slice of sweet potato are usually served as accompaniment.

Espesado de choclo. This dish is from the Chiclayo region. A corn-based sauce is usually cooked with pork and flavored with cilantro, zapallo loche, beef stock, and sofrito (garlic, red onion, and ají amarillo).

Estofado de lengua. A beef tongue stew with potatoes, carrots, tomatoes, and peas. Seasoned with red wine and sofrito (garlic, red onions, and ají panca).

Huatia. From the Andean highlands of Peru and Bolivia. The meat and vegetables are cooked in an oven made with hot stones. Included are native potatoes, oca, faba beans, mashua, corn, sweet potatoes, beef, lamb, llama, or alpaca meat previously marinated in a sauce made with ají panca, chicha de jora, and hierba buena. A sauce made with aromatic herbs like chincho and huacatay can be an accompaniment.

Juane de gallina. A rice-based dish from the jungle region of Peru. The rice is seasoned with sofrito (garlic, red onions, turmeric, oregano, black pepper, and cumin) and chicken stock. Chicken, eggs, Peruvian olives are added before being wrapped in the leaves of bijao, a tropical plant.

Locro de zapallo. A squash-based dish that contains queso fresco, corn, and peas. It is flavored with chicken stock, huacatay, and sofrito (garlic, red onions, ají amarillo, and pepper) and is usually served with a fried egg.

Lomo saltado. In this dish, strips of sirloin are sauteed with onion, tomato, and ají amarillo slices in a sauce consisting of vinegar, soy sauce, minced garlic, and black pepper. Cilantro is sprinkled for additional flavor.

Olluquito con carne (o charqui). A stew of beef and olluco seasoned with sofrito (garlic, red onions, ají panca, and pepper), oregano, and parsley.

Pachamanca. From the Andean highlands of Peru. Potatoes, faba beans, corn, sweet potatoes, together with beef, pork, chicken, or *cuy* meat are cooked in an oven made with hot stones. The meat is previously marinated in a sauce made with ají panca, chincho, and huacatay, among other herbs.

Pepián de choclo. This corn-based sauce is usually cooked with pork and flavored with beef stock and sofrito (garlic, red onion, and ají amarillo).

Pescado a la chorrillana. Fried fish in a sauce containing slices of red onions, tomatoes, and ají amarillo, and seasoned with garlic and ají amarillo paste, oregano, cilantro, and lime juice.

Pescado sudado. Fish and cassava cooked in a spicy sauce containing slices of tomatoes, ají amarillo, bell peppers, and red onions, and seasoned with sofrito (garlic, red onions, ají amarillo, and black pepper), cilantro, chicha de jora, and fish stock.

Picante de mariscos. Seafood chowder consisting of shrimp, squid, scallops, octopus, and sea snails, among other shellfish and mollusks. The sauce consists of dairy cream seasoned with sofrito (butter, red onions, ají panca, and ají amarillo), tomato paste, fish stock, rocoto, cilantro, and Parmesan cheese.

Quinua atamalada. A quinoa dish with milk and queso fresco is cooked to a creamy consistency and flavored with sofrito (garlic, red onions, and ají amarillo) and parsley.

Seco de cabrito con frejoles. This dish is from the northern coast of Peru and consists of goat meat and beans served with rice. The goat meat is seasoned with chicha de jora, zapallo loche, tomato, cilantro, and sofrito (garlic, red onions, ají amarillo, and black pepper).

Seco de res con frijoles. This dish consists of beef in a cilantro sauce, beans, and rice. The cilantro sauce contains carrots and peas and is flavored with cilantro, beef broth, and sofrito (garlic, red onions, ají amarillo, black pepper, and cumin). It is usually served with zarza criolla.

Sopa seca. A dish with chicken and spaghetti in a savory sauce, it is from the Chincha region. The spaghetti sauce is flavored with basil, parsley, chicken broth, bay leaves, tomato dices, and sofrito (garlic, red onions, ají panca, annatto, turmeric, cumin, and pepper). It is frequently served with carapulcra.

Tacacho con cecina. This dish is from the jungle region and consists of small pieces of fried pork mixed with mashed grilled plantain and lard, which is then hand-molded into medium-sized spheres. Cecina (smoked dried pork) is fried and served with tacacho. A usual accompaniment is ají de cocona, a savory hot sauce made with chopped cocona, red onions, cilantro, ají charapita, and lime juice.

Tacu tacu. A fried dough consisting of a mixture of mashed beans and rice. The beans have previously been seasoned with sofrito (garlic, red onions, ají amarillo, oregano, and black pepper). It can be served with zarza criolla, a fried egg, steak, or a fried banana.

Tallarines rojos. Spaghetti in red sauce with chicken. The red sauce is made of ground tomatoes and carrots and flavored with bay leaves, dried mushrooms, and sofrito (garlic, red onions, ají panca, cumin, and black pepper).

Tallarín saltado. This dish consists of strips of sauteed beef steak mixed with spaghetti. The beef is sauteed, along with the red onion and tomato slices, in oil with minced garlic, ginger, and green onion. Soy sauce, vinegar, and oyster sauce are also added for additional flavor.

5.5.5 Desserts and Beverages

Arroz con leche. A rice pudding made with milk, sugar, and raisins and flavored with cinnamon, cloves, vanilla, and lime or orange zest.

Camotillo. An oven dried sweet made with sweet potato, milk, sugar, and flour and flavored with orange and lime zest.

Frijol colado. This sweet, from the Cañete and Chincha region, is made with pureed beans, milk, and sugar. It is flavored with cinnamon, cloves, and sesame seeds.

Mazamorra morada. A purple corn pudding made with purple corn extract, sugar, starch (thickener), lime juice, chopped apples, quince, pineapple, and dried fruits (prunes, apricots). The pudding is flavored with cinnamon, cloves.

Mousse de lucuma. The ingredients are lucuma, condensed milk (milk, sugar), sugar, egg whites, and gelatin.

Picarones. A fried ring-shaped airy or light mass of a yeast-leavened wheat flour, squash, sweet potato, and sugar dough, flavored with aniseed. The picarones are served covered with a chancaca (sugar cane) syrup flavored with cinnamon sticks, cloves, star anise, fig leaves, and orange zest.

Suspiro a la Limeña. A *dulce de leche* custard topped with Port meringue. The custard base is made with condensed, evaporated, egg yolks, and vanilla. The topping is a meringue (sugar and egg whites) flavored with Port wine and sprinkled with cinnamon.

Chicha morada. This traditional Peruvian purple-colored beverage is consumed cold and made by boiling purple corn (grains and cob), pineapple peels, and quince in water. The beverage is additionally flavored with cinnamon, cloves, and lime juice and sweetened with sugar.

Emoliente. This thick beverage can be consumed hot or cold. It is made by boiling roasted barley, flaxseed, *cola de caballo* (an herb), and *uña de gato* (bark) in water. Lime juice and sugar are added for flavor enhancement. Other ingredients such as herbs, grains, and fruits can be added for flavor and/or health-promoting properties.

Te herbales. These hot beverages are prepared by boiling the selected herbs in water and sugar is added for flavor enhancement. Herbal teas are made from lemongrass, muña, and cedroncillo, among other herbs.

Jugos de frutas. These beverages are consumed cold and prepared with the selected fruit in water and sweetened with sugar. These juices are made from native fruits including maracuya (passionfruit), granadilla, tumbo, cherimoya, guava, lucuma, cocona, aguaymanto (goldenberry), camu camu, and sauco, aguaje.

5.6 CONCLUSIONS

In this chapter, we explained that Andean biodiversity and the Peruvian cuisine have the potential to be promoted as a healthy diet, but for this to materialize there is a need of an assessment of their anti-inflammatory potential and to revisit the appropriate quantities of the main ingredients in each dish while keeping the attractiveness of the unique sensory attributes. It is through the sensory attributes that the Peruvian cuisine experienced a "first revolution" internationally recognized and which has been a driving force for the national economy, and in this chapter we proposed a "second revolution" promoting it as a Peruvian healthy diet in similar fashion as the healthy Mediterranean diet. A preliminary assessment of the anti-inflammatory potential of emblematic dishes of Peruvian cuisine shows that there is room for improvement, especially when it comes to readjusting the quantities of some ingredients used in different dish categories. For instance, simply reducing the proportion of carbohydrate sources in main courses like the white rice used as side dish or alternatively replacing white rice with quinoa may improve the anti-inflammatory potential. Similarly, reduction of carbohydrates in entrees, dishes, and desserts will likely promote their anti-inflammatory potential. In general, increasing omega-3 sources through seafood and Andean crops like sacha inchi and antioxidants and fiber sources through Andean crops like root, tubers, fruits, grains, and herbs will counteract the negative effects of carbohydrates, omega-6, and saturated fatty acid sources. There is a unique opportunity to promote the Peruvian cuisine as a healthy diet, and the intent of this chapter is to set the basis as a guide to achieve that goal.

TABLE 5.1 PERUVIAN GASTRONOMIC DISHES (SOUPS, ENTREES, SANDWICHES, MAIN COURSE, DESSERTS AND BEVERAGE CATEGORIES), MAIN INGREDIENTS, CONTENT OF BIOACTIVE AND CHEMICAL GROUPS, AND THEIR CALCULATED ANTI-INFLAMMATORY POTENTIAL

Gastronomic dish (soups)	**Main ingredients**	**Content of bioactive and chemical groups**				**Anti-inflammatory potential*** (AOX + ω-3 + Fb)/(ω-6 + carb + SFA)
		Source of antioxidants (AOX)	**Source of omega-6 or omega-3 fatty acids or saturated fatty acids** (ω-6, ω-3, or SFA)	**Source of fiber** (Fb)	**Source of carbohydrates** (carb)	
Aguadito de pollo or choros	Andean corn, peas, bell pepper, potato, rice, ají amarillo paste, red onions, cilantro, and cumin, chicken, mussels	moderate polyphenols, carotenoids	moderate omega-6 or omega-3	low	low	neutral-low (1–1.6)
Chilcano de pescado	Fish, red onions, garlic, cilantro, ginger, black pepper, ají limo, and green onions	moderate polyphenols, terpenoids, capsaicin	high omega-3	low	low	high (6)
Chupe de camarones	Rice, potatoes and pieces of vegetables (Andean corn, peas, faba bean, huacatay), soft white cheese (queso fresco), milk, and beaten eggs; the whole crayfish (head and tail)	moderate polyphenols, terpenoids	moderate omega-6, omega-3	low	low	low (1.6)

Crema de zapallo	Zapallo macre o loche, yellow potatoes, red onions and garlic, and Parmesan cheese	high carotenoids, polyphenols	N/A	high	moderate	moderate (3)
Inchicapi de gallina con maní	Corn, roasted peanut flour, red onions, garlic, cilantro, turmeric, ají charapita, cassava, and chicken	moderate polyphenols, terpenoids, capsaicin	moderate omega-6	high	moderate	low (1.25)
Menestrón	Many types of vegetables, basil; Andean corn, lima beans and cassava, beef, and cheese	high polyphenols	moderate SFA	moderate	moderate	low (1.25)
Parihuela	Crab, fish, mussels, shrimp, octopus, red onions, garlic, ají amarillo, ají mirasol, and ají panca, with tomatoes, pepper, cumin, oregano, cilantro, parsley, and fermented corn	high polyphenols, carotenoids, capsaicin, terpenoids	high omega-3	low	low	high (7)

(*Continued*)

TABLE 5.1 (CONTINUED)

Gastronomic dish (soups)	**Main ingredients**	**Content of bioactive and chemical groups**				**Anti-inflammatory potential**[*] (AOX + ω-3 + Fb)/(ω-6 + carb + SFA)
		Source of antioxidants (AOX)	**Source of omega-6 or omega-3 fatty acids or saturated fatty acids** (ω-6, ω-3, or SFA)	**Source of fiber** (Fb)	**Source of carbohydrates** (carb)	
Patasca	Mote (hominy), beef tripe, meat, red onions, garlic, ají panca, tomatoes, oregano, bay leaves, hierba buena, and black pepper	moderate polyphenols, capsaicin, terpenoids	moderate SFA	low	moderate	neutral (0.75)
Sancochado	Beef, cabbage, potatoes, cassava, carrots, sweet potato, Andean corn, turnip, rice, cilantro, parsley, hierba buena, and oregano	high polyphenols, terpenoids, carotenoids, glucosinolates	moderate SFA	moderate	moderate	low (1.5)
Shambar	Beans, chickpeas, peas, faba beans, wheat grains, pork rind, and smoked ham	high polyphenols	moderate SFA	high	high	low (1.2)
Sopa a la minuta	Ground meat, potatoes, milk (optional), and angel hair, red onions, tomato paste, garlic, ají panca or mirasol, oregano, and bay leaf	low polyphenols, capsaicin, terpenoids	moderate SFA	low	moderate	neutral (0.5)

Gastronomic dish (entrées)	**Main ingredients**	**Content of bioactive and chemical groups**				**Anti-inflammatory potential*** (AOX + ω-3 + Fb)/(ω-6 + carb + SFA)
		Source of antioxidants (AOX)	**Source of omega-6 or omega-3 fatty acids or saturated fatty acids** (ω-6, ω-3, or SFA)	**Source of fiber** (Fb)	**Source of carbohydrates** (carb)	
Causa rellena	Yellow potato, lime juice, ají Amarillo, avocado, eggs, red onions, olives, lettuce, tuna fish, chicken. or crab	moderate polyphenols, capsaicin	moderate omega-6 or omega-3	low	moderate	neutral–low (0.75–3)
Ceviche de tarwi	Tarwi beans, lime juice, pepper, red onions, cilantro, garlic, and ají limo	high polyphenols, capsaicin	moderate omega-6, low omega-3	high	low	moderate (2.33)
Humitas	Corn	low polyphenols	N/A	moderate	high	neutral (1)
Ocopa	Potatoes, cheese, huacatay, vanilla cookies, peanuts (or pecans), evaporated milk, garlic, red onion, and ají mirasol, oil, pepper, lettuce leaf, Peruvian olive, and boiled egg	moderate polyphenols, capsaicin	moderate omega-6	low	high	neutral (0.6)

(*Continued*)

TABLE 5.1 (CONTINUED)

Gastronomic dish (entrées)	**Main ingredients**	**Content of bioactive and chemical groups**				**Anti-inflammatory potential**[*] (AOX + ω-3 + Fb)/(ω-6 + carb + SFA)
		Source of antioxidants (AOX)	**Source of omega-6 or omega-3 fatty acids or saturated fatty acids** (ω-6, ω-3, or SFA)	**Source of fiber** (Fb)	**Source of carbohydrates** (carb)	
Papa a la huancaína	Potatoes, cheese, evaporated milk, ají amarillo, oil, pepper, soda crackers, lettuce leaf, Peruvian olive, and boiled egg	low polyphenols, capsaicin	moderate omega-6	low	high	neutral (0.4)
Papa rellena	Potato, beef, boiled egg, Peruvian olive, raisins, red onions, lime juice, parsley ají limo, and ají amarillo	moderate polyphenols, capsaicin	moderate SFA	low	high	neutral (0.6)
Pastel de choclo	Andean corn, minced or ground beef, raisins, evaporated milk, shortening (or margarine), eggs, sugar, and ají amarillo	moderate polyphenols, capsaicin	moderate omega-6, SFA	low	high	neutral (0.42)

Patitas de cerdo en fiambre	Pig's feet, garlic, red onions, bay leaves, herbs like hierba buena, oregano and parsley, vinegar, lime juice, red onion, garlic, and ají amarillo	low polyphenols, terpenoids, capsaicin	high SFA	low	low	neutral (0.5)
Rocoto relleno	Rocoto peppers, ground or minced beef, raisins, boiled eggs, Peruvian olives, garlic, red onion, ají panca, pepper, cumin, and cheese	moderate polyphenols, carotenoids, capsaicin	moderate SFA, low omega-6	moderate	low	neutral (1)
Solterito	Faba beans, corn, red onion, cheese, oil, lime juice/or vinegar, rocoto, pepper, and parsley	high polyphenols, capsaicin	low omega-6	high	low	moderate (3)
Tamal	Corn, lard, pork or chicken meat, garlic, red onion, ají mirasol, boiled egg, a Peruvian olive, roasted peanut, and zarza criolla (onions, ají, lime)	low polyphenols, capsaicin	high omega-6, SFA	moderate	high	neutral (0.33)

(*Continued*)

TABLE 5.1 (CONTINUED)

Gastronomic dish (sandwich)	**Main ingredients**	**Content of bioactive and chemical groups**				**Anti-inflammatory potential**[*] (AOX + ω-3 + Fb)/(ω-6 + carb + SFA)
		Source of antioxidants (AOX)	**Source of omega-6 or omega-3 fatty acids or saturated fatty acids** (ω-6, ω-3, or SFA)	**Source of fiber** (Fb)	**Source of carbohydrates** (carb)	
Arrebosado de pejerrey	Bread, pejerrey fish, eggs, wheat flour, with zarza criolla (onions, ají, and lemon)	low polyphenols, capsaicin	high omega-3	very low	moderate	moderate (2.16)
Chicharrón de cerdo	Bread, pork, sweet potato, zarza criolla (onions, ají, and lemon)	low polyphenols, capsaicin	high SFA	very low	moderate	neutral (0.26)
Butifarra	Bread, jamón del país (a Peruvian ham), lettuce, zarza criolla mayonnaise, mustard, and ají amarillo	Low polyphenols, capsaicin	moderate SFA	very low	moderate	neutral (0.33)

Gastronomic dish (main course)	**Main ingredients**	**Content of bioactive and chemical groups**				**Anti-inflammatory potential*** (AOX + ω-3 + Fb)/(ω-6 + carb)
		Source of antioxidants (AOX)	**Source of omega-6 or omega-3 fatty acids or saturated fatty acids** (ω-6, ω-3, or SFA)	**Source of fiber** (Fb)	**Source of carbohydrates** (carb)	
Adobo de cerdo	Pork, ají panca, garlic, vinegar, chicha (a fermented corn alcoholic drink), bay leaves, oregano, hierba buena, black pepper, red onions, cinnamon, and rocoto; served with rice and slices of sweet potato	moderate polyphenols, capsaicin	moderate SFA	low	moderate	neutral (0.75)
Ají de pollo	Chicken, bread, evaporated milk, garlic, red onions, ají amarillo, pecan nuts, and Parmesan cheese; served with rice, Peruvian olive, and boiled egg	low polyphenols, capsaicin	moderate omega-6,	low	moderate	neutral (0.5)
Ajiaco de olluco	Olluco, faba beans, potatoes, milk, cheese, garlic, red onions, and ají mirasol or amarillo	high polyphenols, betalains, capsaicin	moderate omega-6	high	low	low (2)

(*Continued*)

TABLE 5.1 (CONTINUED)

Gastronomic dish (main course)	**Main ingredients**	**Content of bioactive and chemical groups**				**Anti-inflammatory potential*** (AOX + ω-3 + Fb)/(ω-6 + carb)
		Source of antioxidants (AOX)	**Source of omega-6 or omega-3 fatty acids or saturated fatty acids** (ω-6, ω-3, or SFA)	**Source of fiber** (Fb)	**Source of carbohydrates** (carb)	
Anticuchos de corazón	Beef heart, ají panca, garlic, vinegar, cumin, oregano, potato, and corn	low polyphenols, capsaicin	high SFA	low	low	neutral (0.66)
Arroz a la jardinera	Rice, peas (or lima beans), corn, carrots, garlic, red onions, ají amarillo, turmeric; usually served with chicken or pork	moderate polyphenols, capsaicin	moderate omega-6 or SFA	moderate	high	neutral (0.8)
Arroz con mariscos	Rice, shellfish (squid rings, shrimp, scallops, clams), peas, bell peppers, carrots, garlic, red onions, ají amarillo, annatto, oregano, bay leaves, cilantro, and Parmesan cheese	moderate polyphenols, capsaicin	high omega-3	moderate	high	moderate (2.33)

Arroz con pollo o pato	Rice, chicken (or duck), peas, carrots, bell pepper, dark beer, zapallo loche, cilantro, garlic, red onions, ají amarillo, ají panca, pepper, and cumin	moderate polyphenols, capsaicin, terpenoids	moderate omega-6	low	high	neutral (0.6)
Caiguas rellenas	Caigua, ground or minced beef, raisins, boiled eggs, Peruvian olives, garlic, red onion, ají panca, pepper, and cumin; served with rice	moderate polyphenols, capsaicin	moderate SFA	high	moderate	low (1.25)
Carapulcra	Potatoes, pork, garlic, red onions, ají panca, ají mirasol, black pepper, cumin, chicken stock, roasted peanuts, cinnamon, and cloves; served with rice	moderate polyphenols, capsaicin	moderate SFA	low	high	neutral (0.6)
Cau cau de mondongo (o pollo)	Tripe (or chicken), potatoes, carrots, peas, garlic red onions, ají Amarillo, cumin, turmeric, black pepper, and hierba buena; served with rice	moderate polyphenols, capsaicin	moderate SFA or omega-6	low	high	neutral (0.6)

(Continued)

TABLE 5.1 (CONTINUED)

Gastronomic dish (main course)	**Main ingredients**	**Content of bioactive and chemical groups**				**Anti-inflammatory potential*** (AOX + ω-3 + Fb)/(ω-6 + carb)
		Source of antioxidants (AOX)	**Source of omega-6 or omega-3 fatty acids or saturated fatty acids** (ω-6, ω-3, or SFA)	**Source of fiber** (Fb)	**Source of carbohydrates** (carb)	
Ceviche	Raw fish, lime juice, red onions, ají limo, cilantro, ginger, black pepper, corn, and sweet potatoes	high polyphenols, capsaicin	high omega-3	low	low	high (7)
Espesado de choclo	Corn, pork, cilantro, zapallo loche, beef stock, garlic, red onion, and ají amarillo	moderate polyphenols, capsaicin	moderate SFA	low	moderate	neutral (0.75)
Estofado de lengua	Beef tongue, potatoes, carrots, tomatoes, peas, red wine, garlic, red onions, and ají panca; served with rice	moderate polyphenols, capsaicin	moderate SFA	low	moderate	neutral (0.75)
Huatia	Native potatoes, oca, faba beans, mashua, corn, sweet potatoes, beef, lamb, llama, or alpaca meat, ají panca, chicha de jora, hierba buena, aromatic herbs like chincho and huacatay	high polyphenols, capsaicin, terpenoids	moderate SFA	high	moderate	low (1.5)

Juane de gallina	Rice, garlic, red onions, turmeric, oregano, black pepper, cumin, chicken stock, chicken, eggs, and Peruvian olives	moderate polyphenols	moderate omega-6	N/A	high	neutral (0.4)
Locro de zapallo	Squash, cheese, corn, peas, chicken stock, huacatay, garlic, red onions, ají amarillo, pepper, and fried egg; served with rice	moderate polyphenols, carotenoids, capsaicin, terpenoids	low omega-6	high	moderate	low (1.66)
Lomo saltado	Sirloin, onion, tomato, ají amarillo, vinegar, soy sauce, garlic, black pepper, cilantro; served with rice and fried potatoes	moderate polyphenols, capsaicin	moderate SFA	low	moderate	neutral (0.75)
Olluquito con carne (o charqui)	Beef, olluco, garlic, red onions, ají panca, pepper, oregano, and parsley; served with rice	moderate polyphenols, capsaicin, terpenoids	moderate SFA	low	moderate	neutral (0.75)
Pachamanca	Potatoes, faba beans, corn, sweet potatoes, beef, pork, chicken, cuy meat, ají panca, chincho, and huacatay, among other herbs	high polyphenols, capsaicin, terpenoids	moderate omega-6, SFA	high	moderate	neutral (1)
Pepián de choclo	Corn, pork, beef stock, garlic, red onion, and ají amarillo; served with rice	low polyphenols, capsaicin	moderate SFA	low	moderate	neutral (0.5)

(*Continued*)

TABLE 5.1 (CONTINUED)

Gastronomic dish (main course)	**Main ingredients**	**Content of bioactive and chemical groups**				**Anti-inflammatory potential**[*] (AOX + ω-3 + Fb)/(ω-6 + carb)
		Source of antioxidants (AOX)	**Source of omega-6 or omega-3 fatty acids or saturated fatty acids** (ω-6, ω-3, or SFA)	**Source of fiber** (Fb)	**Source of carbohydrates** (carb)	
Pescado a la chorrillana	Fried fish, red onions, tomatoes, and ají amarillo, garlic, oregano, cilantro, and lime juice; served with rice	moderate polyphenols, capsaicin, terpenoids	high omega-3	low	moderate	moderate (3)
Pescado sudado	Fish, cassava, tomatoes, ají amarillo, bell peppers, red onions, garlic, black pepper, cilantro, chicha de jora, and fish stock	moderate polyphenols, capsaicin	high omega-3	low	low	high (6)
Picante de mariscos	Seafood (shrimp, squid, scallops, octopus, sea snails, among other shellfish and mollusks), rice, dairy cream, butter, red onions, ají panca, ají amarillo, tomato paste, fish stock, rocoto, cilantro, and Parmesan cheese	low polyphenols, capsaicin	high omega-3, moderate omega-6	low	moderate	low (1.25)

Quinua atamalada	Quinoa, milk, cheese, garlic, red onions, ají amarillo, and parsley	moderate polyphenols, capsaicin, terpenoids	moderate omega-6	high	low	low (1.66)
Seco de cabrito con frejoles	Goat meat, beans, chicha de jora, zapallo loche, tomato, cilantro, garlic, red onions, ají amarillo, and black pepper; served with rice	moderate polyphenols, capsaicin, terpenoids	moderate SFA	high	moderate	low (1.25)
Seco de res con frijoles	Beef, cilantro, beans, carrots, peas, garlic, red onions, ají amarillo, black pepper, and cumin; served with zarza criolla and rice	moderate polyphenols, capsaicin, terpenoids	moderate SFA	high	moderate	low (1.25)
Sopa seca	Chicken, spaghetti, basil, parsley, chicken broth, bay leaves, tomato, garlic, red onions, ají panca, annatto, turmeric, cumin, and pepper; served with carapulcra	low polyphenols, capsaicin, terpenoids	moderate omega-6	low	high	neutral (0.4)
Tacacho con cecina	Pork, plantain, lard, cecina (smoked dried pork), cocona, red onions, cilantro, ají charapita, and lime juice	low polyphenols, capsaicin	moderate SFA	N/A	moderate	neutral (0.25)

(*Continued*)

TABLE 5.1 (CONTINUED)

Gastronomic dish (main course)	**Main ingredients**	**Content of bioactive and chemical groups**				**Anti-inflammatory potential*** (AOX + ω-3 + Fb)/(ω-6 + carb)
		Source of antioxidants (AOX)	**Source of omega-6 or omega-3 fatty acids or saturated fatty acids** (ω-6, ω-3, or SFA)	**Source of fiber** (Fb)	**Source of carbohydrates** (carb)	
Tacu tacu	Beans, rice, garlic, red onions, ají amarillo, oregano, black pepper; served with zarza criolla, fried egg, a steak, or a fried banana.	moderate polyphenols, capsaicin	moderate omega-6 or SFA	moderate	moderate	neutral (1)
Tallarines rojos	Spaghetti, chicken, tomatoes, carrots, bay leaves, dried mushrooms, garlic, red onions, ají panca, cumin, and black pepper	moderate carotenoids, polyphenols, capsaicin	moderate omega-6	N/A	high	neutral (0.4)
Tallarín saltado	beef steak, spaghetti, red onion, tomato, oil, garlic, ginger, green onion, soy sauce, vinegar, and oyster sauce	moderate polyphenols, carotenoids	moderate SFA	N/A	high	neutral (0.4)

Gastronomic dish (desserts and beverages)	**Main ingredients**	**Content of bioactive and chemical groups**				**Anti-inflammatory potential*** (AOX + ω-3 + Fb)/(ω-6 + carb + SFA)
		Source of antioxidants (AOX)	**Source of omega-6 or omega-3 fatty acids or saturated fatty acids** (ω-6, ω-3, or SFA)	**Source of fiber** (Fb)	**Source of carbohydrates** (carb)	
Arroz con leche	Rice, milk, sugar, raisins, cinnamon, cloves, vanilla, and lime or orange zest	low polyphenols	moderate omega-6	N/A	high	neutral (0.2)
Camotillo	Sweet potato, milk, sugar, flour, and orange and lime zest	moderate polyphenols, carotenoids	moderate omega-6	moderate	high	neutral (0.8)
Frijol colado	Beans, milk, sugar, cinnamon, cloves, and sesame seeds	moderate polyphenols	moderate omega-6	moderate	high	neutral (0.8)
Mazamorra morada	Purple corn extract, sugar, starch (thickener), lime juice, apples, quince, pineapple, and dried fruits (prunes, apricots, cinnamon, and cloves)	high polyphenols	N/A	moderate	high	low (1.66)
Mousse de lucuma	Lucuma, condensed milk, sugar, egg whites, and gelatin	moderate carotenoids	moderate omega-6	low	high	neutral (0.6)

(*Continued*)

TABLE 5.1 (CONTINUED)

Gastronomic dish (desserts and beverages)	**Main ingredients**	**Content of bioactive and chemical groups**				**Anti-inflammatory potential*** (AOX + ω-3 + Fb)/(ω-6 + carb + SFA)
		Source of antioxidants (AOX)	**Source of omega-6 or omega-3 fatty acids or saturated fatty acids** (ω-6, ω-3, or SFA)	**Source of fiber** (Fb)	**Source of carbohydrates** (carb)	
Picarones	Wheat flour, squash, sweet potato, sugar aniseed, chancaca (sugar cane syrup), cinnamon, cloves, star anise, fig leaves. and orange zest	moderate carotenoids	N/A	low	high	neutral (1)
Suspiro a la limeña	Condensed milk, evaporated, egg yolks, vanilla, sugar, egg whites, Port wine, and cinnamon	N/A	moderate omega-6	N/A	high	neutral (0)
Chicha morada	Purple corn, quince, pineapple peels, additionally flavored with cinnamon, cloves, and lime juice, and sweetened with sugar	high polyphenols	N/A	N/A	moderate	low (1.5)

Emoliente	Barley, flaxseed, cola de caballo (herb), uña de gato (bark), lime juice, sugar, and other ingredients such as herbs, grains, and fruits	high polyphenols, terpenoids	N/A	moderate	low	high (5)
Te herbales	Herbal teas based on lemongrass, muña, and cedroncillo, among other herbs	high polyphenols, terpenoids	N/A	N/A	low	moderate (3)
Jugos de frutas	Juices based on native fruits including maracuya (passion fruit), granadilla, tumbo, cherimoya, guava, lucuma, cocona, aguaymanto (goldenberry), camu camu, sauco, and aguaje, among others	high polyphenols, vitamin C, carotenoids	N/A	N/A	moderate	low (1.5)

* Reported values of anti-inflammatory potential were determined based on the arbitrary scale values assigned in the four columns of content of bioactive and chemical groups in Table 5.1, following the arbitrary scale described in section 5.2. In this assessment, fruit, vegetables, grains, and herbs are a source of antioxidants (e.g., polyphenols, carotenoids, vitamin C, betalains, terpenoids, and glucosinolates); seafood and grains a source of omega-3; dairy and chicken a source of omega-6; grains and vegetables a source of fiber; grains, vegetables, and fruit a source of carbohydrates; and red meats a source of saturated fatty acids.

Abbreviation: N/A (not available).

REFERENCES

Azevedo-Meleiro, Cristiane H., and Delia B. Rodriguez-Amaya. 2007. "Qualitative and Quantitative Differences in Carotenoid Composition Among *Cucurbita Moschata*, *Cucurbita Maxima*, and *Cucurbita pepo*." *Journal of Agricultural and Food Chemistry* 55 (10): 4027–4033. https://doi.org/10.1021/jf063413d.

Barreca, Davide, Giuseppina Lagana, Silvana Ficarra, Ester Tellone, Ugo Leuzzi, Antonio Galtieri, and Ersilia Bellocco. 2011. "Evaluation of the Antioxidant and Cytoprotective Properties of the Exotic Fruit *Annona cherimola* Mill. (Annonaceae)." *Food Research International* 44 (7): 2302–2310. https://doi.org/10.1016/j.foodres.2011.02.031.

Bondioli, Paolo, Laura Della Bella, and Petra Rettke. 2006. "Alpha Linolenic Acid Rich Oils: Composition of *Plukenetia volubilis* (Sacha Inchi) Oil from Peru." *La Rivista Italiana Delle Sostanze Grasse* 83: 120–123.

Bressani, Ricardo. 1983. "Research Needs to Up-Grade the Nutritional Quality of Common Beans (Phaseolus vulgaris)." *Plant Foods for Human Nutrition* 32: 101–110. https://doi-org.srv-proxy2.library.tamu.edu/10.1007/BF01091330.

Camargo Neves, Leandro, Vanuza Xavier da Silva, Edvan Alves Chagas, Christinny Giselly Barcelar Lima, and Sergio Ruffo Roberto. 2015. "Determining the Harvest Time of Camu-Camu (*Myrciaria dubia* (HBK) Mc Vaugh) Using Measured Pre-arvest Attributes." *Scientia Horticulturae* 186: 15–23. https://doi.org/10.1016/j.scienta.2015.02.006.

Campos, David, Giuliana Noratto, Rosana Chirinos, Carlos Arbizu, William Roca, and Luis Cisneros-Zevallos. 2006. "Antioxidant Capacity and Secondary Metabolites in Four Species of Andean Tuber Crops: Native Potato (Solanum sp.), Mashua (*Tropaeolum tuberosum* Ruiz & Pavón), Oca (*Oxalis tuberosa* Molina) and ulluco (*Ullucus tuberosus* Caldas)." *Journal of the Science of Food and Agriculture* 86: 1481–1488. https://doi.org/10.1002/jsfa.2529.

Campos, David, Indira Betalleluz-Pallardel, Rosana Chirinos, Ana Aguilar-Galvez, and Giuliana Noratto. 2012. "Prebiotic Effects of Yacon (*Smallanthus Sonchifolius* Poepp. & Endl), a Source of Fructooligosaccharides and Phenolic Compounds With Antioxidant Activity." *Food Chemistry* 135: 1592–1599. https://doi.org/10.1016/j.foodchem.2012.05.088.

Candido, Thalita Lin Netto, Mara Reis Silva, and Tania Da Silveira Agostini-Costa. 2015. "Bioactive Compounds and Antioxidant Capacity of Buriti (*Mauritia flexuosa* L.f.) from the Cerrado and Amazon Biomes." *Food Chemistry* 177: 313–319. https://doi.org/10.1016/j.foodchem.2015.01.041.

Carvajal-Larenas, Francisco, M. J. Rob Nout, Martinus A. J. S. Tiny van Boekel, M. Koziol, and Anita R. Linnemann. 2013. "Modelling of the Aqueous Debittering Process of *Lupinus mutabilis* Sweet." *LWT—Food Science and Technology* 53 (2): 507–516. https://doi.org/10.1016/j.lwt.2013.03.017.

Casas, Rosa, Emilio Sacanella, and Ramon Estruch. 2016. "The Immune Protective Effect of the Mediterranean Diet Against Chronic Low-Grade Inflammatory Diseases." *Endocrine, Metabolic and Immune Disorders Drug Targets* 14: 245–254. https://doi.org/10.2174/1871530314666140922153350.

Castro, Mauricio, and Edgar Chambers. 2019. "Willingness to Eat an Insect-Based Product and Impact on Brand Equity: A Global Perspective." *Journal of Sensory Studies* 34 (2): e12486. https://doi.org/10.1111/joss.12486.

Cevallos-Casals, Bolivar, and Luis Cisneros-Zevallos. 2003. "Stoichiometric and Kinetic Studies of Phenolic Antioxidants from Andean Purple Corn and Red-Fleshed Sweet Potato." *Journal of Agricultural and Food Chemistry* 51: 3313–3319. https://doi.org/10.1021/jf034109c.

Cevallos-Casals, Bolivar, and Luis Cisneros-Zevallos. 2004. "Stability of Anthocyanin-Based Aqueous Extracts of Andean Purple Corn and Red-Fleshed Sweet Potato Compared to Synthetic and Natural Colorants." *Food Chemistry* 86: 69–77. https://doi.org/10.1016/j.foodchem.2003.08.011.

Chambers, Edgar, Karolina Sanchez, Uyen X. T. Phan, Rhonda Miller, Gail V. Civille, and Brizio Di Donfrancesco. 2016. "Development of a 'Living' Lexicon for Descriptive Sensory Analysis of Brewed Coffee." *Journal of Sensory Studies* 31 (6): 465–480. https://doi.org/10.1111/joss.12237.

Chansri, Ruethaipak, Chureerat Puttanlek, Vilai Rungsadthogy, and Dudsadee Uttapap. 2005. "Characteristics of Clear Noodles Prepared from Edible Canna Starches." *Journal of Food Science* 70 (5): S337–S342. https://doi.org/10.1111/j.1365-2621.2005.tb09988.x.

Chirinos, Rosana, David Campos, Carlos Arbizu, Herve Rogez, Jean-Francois Rees, Yvan Larondelle, Giuliana Noratto, and Luis Cisneros-Zevallos. 2007. "Effect of Genotype, Maturity Stage and Post-Harvest Storage on Phenolic Compounds, Carotenoid Content and Antioxidant Capacity, of Andean Mashua Tubers (*Tropaeolum tuberosum* Ruiz & Pavón)." *Journal of the Science of Food and Agriculture* 87: 437–446. https://doi.org/10.1002/jsfa.2719.

Choque-Delgado, Grethel Teresa, Wirla Maria da Silva-Cunha-Tamashiro, Mario Roberto Maróstica-Junior, and Glaucia Maria Pastore. 2013. "Yacon (*Smallanthus sonchifolius*): A Functional Food." *Plant Foods for Human Nutrition* 68: 222–228. https://doi.org/10.1007/s11130-013-0362-0.

Cisneros, Fausto H., Daniel Paredes, Adrian Arana, and Luis Cisneros-Zevallos. 2014. "Chemical Composition, Oxidative Stability and Antioxidant Capacity of Oil Extracted from Roasted Seeds of Sacha-Inchi (*Plukenetia volubilis* L.)." *Journal of Agricultural and Food Chemistry* 62 (22): 5191–5197. https://doi.org/10.1021/jf500936j.

Cisneros, Fausto H., Roberto Zevillanos, and Luis Cisneros-Zevallos. 2009. "Characterization of Starch from Two Ecotypes of Andean Achira Roots (*Canna edulis*)." *Journal of Agricultural and Food Chemistry* 57: 7363–7368. https://doi.org/10.1021/jf9004687.

Cisneros, Fausto H., Roberto Zevillanos, Mariella Figueroa, Gabriel Gonzalez, and Luis Cisneros-Zevallos. 2018. Characterization of Starch from Two Andean Potatoes: Ccompis (Solanum tuberosum spp. andigena) and Huayro (Solanum x chaucha). Starch - Stärke, 70: 1700134. https://doi.org/10.1002/star.201700134.

Cisneros-Zevallos, Luis. 2020. "The Power of Plants: How Fruit and Vegetables Work as Source of Nutraceuticals and Supplements." *International Journal of Food Science and Nutrition* 1–5. https://doi.org/10.1080/09637486.2020.1852194.

Cortés-Avendaño, Paola, Marko Tarvainen, Jukka-Pekka Suomela, Patricia Glorio-Paulet, Baoru Yang, and Ritva Repo-Carrasco-Valencia. 2020. "Profile and Content of Residual Alkaloids in Ten Ecotypes of *Lupinus Mutabilis* Sweet After Aqueous Debittering Process." *Plant Foods for Human Nutrition* 75: 184–191. https://doi.org/10.1007/s11130-020-00799-y.

da Silva Leitao Peres, Natalia, Leticia Cabrera Parra Bortoluzzi, Leila Larisa Medeiros Marques, Maysa Formigoni, Renata Hernandez Barros Fuchs, Adriana Aparecida Droval, and Flavia Aparecida Reitz Cardoso. 2020. "Medicinal Effects of Peruvian Maca (*Lepidium meyenii*): A Review." *Food and Function* 11 (1): 83–92. https://doi.org/10.1039/C9FO02732G.

Etzbach, Lara, Anne Pfeiffer, Fabian Weber, and Andreas Schieber. 2018. "Characterization of Carotenoid Profiles in Goldenberry (*Physalis peruviana* L.) Fruits at Various Ripening Stages and in Different Plant Tissues by HPLC-DAD-APCI-MS^n." *Food Chemistry* 245: 508–517. https://doi.org/10.1016/j.foodchem.2017.10.120.

Hamaker, Bruce, C. Valles, R. Gilman, R. Hardmeier, D. Clark, H. Garcia, A. Gonzales, I. Kohlstad, M. Castro, R. Valdivia, T. Rodriguez, and M. Lescano. 1992. "Amino Acid and Fatty Acid Profiles of the Inca Peanut (*Plukenetia volubilis*)." *Cereal Chemistry* 69: 461–463.

Hassan, Norazian Mohd, Nurul Asyiqin Yusof, Amirah Fareeza Yahaya, Nurul Nasyitah Rozali, and Rashidi Othman. 2019. "Carotenoids of Capsicum Fruits: Pigment Profile and Health-Promoting Functional Attributes—Review." *Antioxidants* 8: 469–494. https://doi.org/10.3390/antiox8100469.

Hethelyi, Eva, Bela Danos, Peter Tetenyi, and Istvan Koczka. 1986. "GC-MS Analysis of the Essential Oils of Four Tagetes Species and the Anti-Microbial Activity of *Tagetes minuta*." *Flavour and Fragrance Journal* 1: 169–173. https://doi.org/10.1002/ffj.2730010408.

Hongsoongnern, Pairin, and Edgar Chambers IV. 2008. "A Lexicon for Texture and Flavor Characteristics of Fresh and Processed Tomatoes." *Journal of Sensory Studies* 23 (5): 583–599. https://doi.org/10.1111/j.1745-459X.2008.00174.x.

Hudson, James B. 1990. *Antiviral Compounds from Plants*. Boca Raton, FL: CRC Press.

Jacobo-Velazquez DA, Ortega-Hernandez E, CisnerosZevallos L. 2016. Book chapter: vegetable containing juices: carrot, kale and sprout juices for prevention and therapeutics. In: Shahidi F and Alasalvar C, editors. *Handbook offunctional beverages and human health*. Boca Rato. https://doi.org/10.1201/b19490.

Jaeger de Carvalho, Lucia Maria, Patricia Barros Gomes, Ronoel Luiz de Oliveira Godoy, Sidney Pacheco, Pedro Henrique Fernandes do Monte, et al. 2012. "Total Carotenoid Content, A-Carotene and B-Carotene, of Landrace Pumpkins (*Cucurbita moschata* Duch): A Preliminary Study." *Food Research International* 47: 337–340. https://doi.org/10.1016/j.foodres.2011.07.040.

Kulczyński, Bartosz, and Anna Gramza-Michałowska. 2019. "The Profile of Secondary Metabolites and Other Bioactive Compounds in *Cucurbita pepo* L. and *Cucurbita moschata* Pumpkin Cultivars." *Molecules* 24: 2945–2967. https://doi.org/10.3390/molecules24162945.

Lao, Fei, Gregory T. Sigurdson, and M. Monica Giusti. 2017. "Health Benefits of Purple Corn (*Zea mays* L.) Phenolic Compounds." *Comprehensive Reviews in Food Science and Food Safety* 16: 234–246. https://doi.org/10.1111/1541-4337.12249.

Lee, Jeehyun, and Delores H. Chambers. 2007. "A Lexicon for Flavor Descriptive Analysis of Green Tea." *Journal of Sensory Studies* 22 (3): 256–272. https://doi.org/10.1111/j.1745-459X.2007.00105.x.

León Marroú, Maria E., Misael Y. Villacorta González, and Sandra E. Pagador Flores. 2011. "Composición Química de Oca (*Oxalis tuberosa*), Arracacha (*Arracacia xanthorriza*) y Tarwi (*Lupinus mutabilis*): Formulación de una Mezcla Base para Productos Alimenticios." *Revista Venezolana de Ciencia y Tecnología de Alimentos* 2 (2): 239–252.

Nair, Vimal, Liz A. Salvatierra, Renato A. Siu, Fausto H. Cisneros, and Luis Cisneros-Zevallos. 2021. *Phenolic, Carotenoid and Phytosterol Characterization of Lucuma (Pouteria Lucuma) Fruit by HPLC-ESI/APCI-Ms^n and GC-MS, and Its Associated Antihyperglycemic Activities.* College Station, TX: Department of Horticultural Sciences, Texas A&M University, Department of Agroindustrial Engineering and Agribusiness, Universidad San ignacio de Loyala. *Personal communication.*

National Research Council. 1989. *Lost Crops of the Incas: Little-Known Plants of the Andes with Promise for Worldwide Cultivation.* Washington, DC: National Academy Press.

Nowak, Verena, Juan Du, and U. Ruth Charrondiere. 2016. "Assessment of the Nutritional Composition of Quinoa (*Chenopodium quinoa* Willd.)." *Food Chemistry* 193: 47–54. https://doi.org/10.1016/j.foodchem.2015.02.111.

Olivares-Tenorio, Mary Luz, Matthijs Dekker, R. Ruud Verkerk, and Martinus A. J. S. van Boekel. 2016. "Health-Promoting Compounds in Cape Gooseberry (*Physalis peruviana* L.): Review from a Supply Chain Perspective." *Trends in Food Science and Technology* 57 (Part A): 83–92. https://doi.org/10.1016/j.tifs.2016.09.009.

Osorio C, Schreckinger E, Bhargava P, Bang WY, JacoboVelazquez DA, Cisneros-Zevallos L. 2016. Book chapter: golden berry and selected tropical (Acai, Aguaymanto, Acerola and Maqui) Juices. In: Shahidi F and Alasalvar C, Editors. *Handbook of functional beverages and human health.* Boca Raton: CRC Press Taylor & Francis Group. pp. 251–269. https://doi.org/10.1201/b19490.

Pacheco Santos, Leonor Maria. 2005. "Nutritional and Ecological Aspects of Buriti or Aguaje (*Mauritia flexuosa* Linnaeus filius): A Carotene-Rich Palm Fruit from Latin America." *Ecology of Food and Nutrition* 44 (5): 345–358. https://doi.org/10.1080/03670240500253369.

Pedreschi, Romina, David Campos, Giuliana Noratto, Rosana Chirinos, and Luis Cisneros-Zevallos. 2003. "Andean Yacon Root (*Smallanthus sonchifolius* Poepp. Endl) Fructooligosaccharides as a Potential Novel Source of Prebiotics." *Journal of Agricultural and Food Chemistry* 51: 5278–5284. https://doi.org/10.1021/jf0344744.

Pedreschi, Romina, and Luis Cisneros-Zevallos. 2006. "Antimutagenic and Antioxidant Properties of Phenolic Fractions from Andean Purple Corn (*Zea mays* L.)." *Journal of Agricultural and Food Chemistry* 54: 4557–4567. https://doi.org/10.1021/jf0531050.

Pedreschi, Romina, and Luis Cisneros-Zevallos. 2007. "Phenolic Profiles of Andean Purple Corn (*Zea mays* L.)." *Food Chemistry* 100: 956–963. https://doi.org/10.1016/j.foodchem.2005.11.004.

Ramallo, Rodrigo, Jean Paul Wathelet, Eric Le Boulenge, Elizabeth Torres, Michel Marlier, Jean Francois Ledent, Augusto Guidi, and Yvan Larondelle. 2004. "Glucosinolates in Isaño (*Tropaeolum tuberosum*) Tubers: Qualitative and Quantitative Content and Changes After Maturity." *Journal of the Science of Food and Agriculture* 84: 701–706. https://doi.org/10.1002/jsfa.1691.

Repo-Carrasco, Ritva, Clara Espinoza, and Sven-Erik Jacobsen. 2003. "Nutritional Value and Use of the Andean Crops Quinoa (*Chenopodium quinoa*) and Kañiwa (*Chenopodium pallidicaule*)." *Food Reviews International* 19 (1): 179–189. https://doi.org/10.1081/FRI-120018884.

Rivas, Marisa, Dora Vignale, Roxana M. Ordoñez, Iris C. Zampini, Maria R. Alberto, Jorge E. Sayago, and Maria I. Isla. 2013. "Nutritional, Antioxidant and Anti-Inflammatory Properties of *Cyclanthera pedata*, an Andean Fruit and Products

Derived from Them." *Food and Nutrition Sciences* 4: 55–61. http://doi.org/10.4236/fns.2013.48A007.

Rodrigues, Roberta B., Hilary C. Menezes, Lourdes M. C. Cabral, Manuel Dornier, and Max Reynes. 2001. "An Amazonian Fruit with a High Potential as a Natural Source of Vitamin C: The Camu-Camu (*Myrciaria dubia*)." *Fruits* 56: 345–354. https://doi.org/10.1051/fruits:2001135.

Rodrigues, Eliseu, Ismael I. Rockenbach, Ciriele Cataneo, Luciano V. Gonzaga, Eduardo S. Chaves, and Roseane Fett. 2009. "Minerals and Essential Fatty Acids of the Exotic Fruit *Physalis peruviana* L." *Ciencia e Tecnología de Alimentos, Campinas* 29 (3): 642–645. http://doi.org/10.1590/S0101-20612009000300029.

Roos, Elin, Georg Carlsson, Ferawati Ferawati, Mohammed Hefni, Andreas Stephan, Pernilla Tidaker, and Cornelia Witthoft. 2018. "Less Meat, More Legumes: Prospects and Challenges in the Transition toward Sustainable Diets in Sweden." *Renewable Agriculture and Food Systems*: 1–14. https://doi.org/10.1017/S1742170518000443.

Santana-Galvez J, Cisneros-Zevallos L, Jacobo-Velazquez DA. 2017. Chlorogenic acid: recent advances on its dual role as a food additive and a nutraceutical against metabolic syndrome. *Molecules.* 22(3):358. https://doi.org/10.3390/molecules22030358.

Santos, Sonia A., Carla Vilela, Joao F. Camacho, Nereida Cordeiro, Manuel Gouveia, Carmen S. Freire, and Armando J. Silvestre. 2016. "Profiling of Lipophilic and Phenolic Phytochemicals of Four Cultivars from Cherimoya (*Annona cherimola* Mill.)." *Food Chemistry* 211: 845–852. https://doi.org/10.1016/j.foodchem.2016.05.123.

Sathe, Shridhar K., Sanyukta Deshpande, D. K. Chip Salunkhe, and Joseph J. Rackis. 1984. "Dry Beans of Phaseolus. A Review. Part 1. Chemical Composition: Proteins." *Critical Reviews in Food Science and Nutrition* 20 (1): 1–46. https://doi.org/10.1080/10408398409527382.

Schoeneberger, Hans, Rainer Gross, Hans D. Cremer, and Ibrahim Elmadfa. 1982. "Composition and Protein Quality of *Lupinus mutabilis*." *Journal of Nutrition* 112 (1): 70–76. https://doi.org/10.1093/jn/112.1.70.

Segovia, Ingrid K. B., Lucybel L. Suárez de la Cruz, Américo J. L. Castro, Silvia C. Suárez, and Julio R. Q. Ruiz. 2010. "Composición Química del Aceite Esencial De *Tagetes elliptica* Smith 'Chincho' y Actividades Antioxidante, Antibacteriana y Antifúngica." *Ciencia e investigación* 13 (2): 82–87. https://doi.org/10.15381/ci.v13i2.3231.

Sereno, Aiane Benevide, Marlene Bampi, Isabela Eloise dos Santos, Sila M.R. Ferreira, Renata Labronici Bertin, and Claudia C. H. Krüger. 2018. "Mineral Profile, Carotenoids and Composition of Cocona (*Solanum Sessiliflorum* Dunal), a Wild Brazilian Fruit." *Journal of Food Composition and Analysis* 72: 32–38. https://doi.org/10.1016/j.jfca.2018.06.001.

Sharma, Chetan, Edgar Chambers IV, Sastry S. Jayanty, Vidyasagar Sathuvalli Rajakalyan, David G. Holm, and Martin J. Talavera. 2020a. "Development of a Lexicon to Describe the Sensory Characteristics of a Wide Variety of Potato Cultivars." *Journal of Sensory Studies* 35 (4): e12577. https://doi.org/10.1111/joss.12577.

Sharma, Chetan, Sastry S. Jayanty, Edgar Chambers IV, and Martin J. Talavera. 2020b. "Segmentation of Potato Consumers Based on Sensory and Attitudinal Aspects." *Foods* 9: 161. https://doi:10.3390/foods9020161.

Simopoulos, Artemis P. 2004. "Omega-6/Omega-3 Essential Fatty Acid Ratio and Chronic Diseases." *Food Reviews International* 20: 77–90. https://doi.org/10.1081/FRI-120028831.

Simopoulos, Artemis P. 2006. "Evolutionary Aspects of Diet, the Omega-6/Omega-3 Ratio and Genetic Variation: Nutritional Implications for Chronic Diseases." *Biomedicine and Pharmacotherapy* 60: 502–507. https://doi.org/10.1016/j.biopha.2006.07.080.

Soule, Jacqueline A. 1993. "*Tagetes Minuta*: A Potential New Herb from South America." In J. Janick and J. E. Simon (Eds.), *New Crops*. New York: Wiley, 649–654.

Stone, Herbert, and Joel Sidel. 1992. *Sensory Evaluation Practices, Food and Science Technology Series*. Cambridge: Academic Press, 2nd edition.

Suwonsichon, Suntaree. 2019. "The Importance of Sensory Lexicons for Research and Development of Food Products." *Foods* 8 (1): 1–16, 27. https://doi.org/10.3390/foods8010027.

Talavera-Bianchi, Martin, Edgar Chambers IV, and Delores H. Chambers. 2010. "Lexicon to Describe Flavor of Fresh Leafy Vegetables." *Journal of Sensory Studies* 25 (2): 163–183. https://doi.org/10.1111/j.1745-459X.2009.00249.x.

Tuorila, Hely, and Christina Hartmann. 2020. "Consumer Responses to Novel and Unfamiliar Foods." *Current Opinion in Food Science* 33: 1–8. https://doi.org/10.1016/j.cofs.2019.09.004.

Venskutonis, Petras R., and Paulius Kraujalis. 2013. "Nutritional Components of Amaranth Seeds and Vegetables: A Review on Composition, Properties, and Uses." *Comprehensive Reviews in Food Science and Food Safety* 12: 381–412. https://doi.org/10.1111/1541-4337.12021.

Verde García, Peter T., and Eduardo P. Rengifo Alcántara. 2006. "Transformación Integral e Industrialización del Sauco (*Sambucus peruviana*) para el Desarrollo Microrregional Sostenible." Tesis, Universidad Nacional de Trujillo, Trujillo.

Chapter 6

Nutritional Attributes and Effect of Processing on Peruvian Chili Peppers

Eduardo Morales-Soriano and Roberto Ugás

CONTENTS

6.1 PERUVIAN CHILI PEPPERS: GENERAL INFORMATION

Archeological records show that chili peppers have been used by human beings since at least 10,000 years, as evidenced by remains found in northern Peru (Dillehay et al. 2017). Although it is generally accepted that Bolivia was the area of origin of the genus *Capsicum* in South America, or rather a continuous belt from southeastern Brazil to the Andes, recent research postulated that the ancestors of *Capsicum* may have come into existence in the present region of Peru, Ecuador,

DOI: 10.1201/9781003087618-6

and Colombia, with later expansion in a clockwise direction around the Amazon basin, toward central and southeastern Brazil, then back to western South America, and finally northwards to Central America (Carrizo García et al. 2016). There are five cultivated species in the genus *Capsicum* (Jarret et al. 2019), the most abundant worldwide being *C. annuum*, which originated in Mexico (Kraft et al. 2014), while the other four cultivated species originated in South America and in the Caribbean region. Bolivia has the highest diversity of wild *Capsicum*, while the highest diversity of cultivated *Capsicum* is found in Peru (van Zonneveld et al. 2015). Today all five cultivated species of *Capsicum* can be found in Peruvian markets, the most abundant being *C. chinense*, *C. baccatum*, and *C. pubescens* (Ugás 2012). In the Spanish spoken in most South American countries, chili peppers are called *ají*, except *C. pubescens*, which is called *rocoto* in Peru.

Capsicum fruits are generally classified in vegetable types (sweet pepper) and spices (chili peppers and sweet paprika). Nowadays, chili peppers are a globally consumed species, as fresh fruits, as dried and ground fruits and as a variety of hot sauces, with China, India, and Mexico responsible of about 50% of fresh chili pepper world production (FAO 2018). In pre-Columbian times in Latin America, chili peppers were used in various ways: as a tribute, currency, in rituals, as medicine, or as punishment, in addition to food (Bosland and Votava 2012). Nowadays, some of these uses are still present in many rural communities, but mainly due to the pungency attribute and their characteristic aroma and flavor, chili peppers are widely used in traditional and modern gastronomy. Therefore, *Capsicum* consumption is increasing and represents an important source of food diversity because a large range of sizes, shapes, colors, and other attributes are present in these materials. This diversity is also reflected in the nutritional composition, which is determined by the species, cultivar, growing condition, and fruit maturity (Bosland and Votava 2012).

Ugás (2012) presented a first approach to the classification of traditional Peruvian chili peppers, based on the conservation and evaluation work carried out by the Vegetable Crops Research Program of UNALM, with its germplasm collection being the most representative of the Peruvian diversity of *Capsicum*. The proposed classification is shown in Table 6.1 and a group of representative chili peppers is reported in Table 6.2, which were used in the studies on pungency profiling and aroma fingerprinting reported later in this chapter.

From a commercial point of view *C. baccatum* is by large the most important chili pepper in Peru, grown throughout the country but mostly in medium-sized and large fields in the irrigated coastal area in order to supply national markets as fresh fruit, as dried fruit, and as an input for processing into creams and sauces. In spite of its commercial importance, there are no registered cultivars yet, let alone hybrids, and several landraces of the Escabeche or Amarillo type can be found along the Coast of Peru. Escabeche is the quintessential chili pepper of Peruvian gastronomy, highly aromatic and of medium-low pungency;

TABLE 6.1 CLASSIFICATION OF MAIN VARIETAL GROUPS OF PERUVIAN CHILI PEPPERS

Main growing region	Varietal group	Scientific name	Location	Diversity level
Chili peppers from the Northern Coast	Cerezos	*C. annuum*	Lambayeque	Low
	Cacho de cabra	*C. baccatum*	Lambayeque	Low
	Verde o Largo	*C. baccatum*	Tumbes, Piura	Low
	Limo	*C. chinense*	Northern Coast	High
	Mochero	*C. chinense*	La Libertad	Low
	Arnaucho	*C. chinense*	North of Lima	Medium
Chili peppers from the Coast with intensive production	Escabeche, Amarillo, Mirasol	*C. baccatum*	Coast	Medium
	Pacae	*C. baccatum*	Arequipa, Moquegua, Tacna	Low
	Panca, Especial, Negro, Rojo	*C. chinense*	Coast	Low
Chili peppers from the Amazon	Ayuyo	*C. baccatum*	Jungle (Ucayali and San Martin)	Medium
	Challuaruro	*C. baccatum*	Jungle	Medium
	Charapita	*C. chinense*	Jungle	High
	Dulce	*C. chinense*	Jungle	High
	Pucunucho	*C. chinense*	Jungle	Medium
	Malagueta	*C. frutescens*	Jungle	Low
Chili peppers from the Andes	Rocoto	*C. pubescens*	Low Andes and medium altitude	Medium
	Rocoto de huerta	*C. pubescens*	Low and medium altitude	Medium
	Rocoto de la selva central	*C. pubescens*	Mountainous rainforest	Low

Source: Based on Ugás (2012).

very often it is home-processed to remove the pungency, later used in bases for stews and soups. Other types of *C. baccatum* are small-fruited landraces popular in the Amazon region. *C. chinense* is the most diverse species in the country, as explained by the large number of types and landraces, the most popular being Limo, originally grown by smallholders in the northern coast but now

TABLE 6.2 REPRESENTATIVE PERUVIAN CHILI PEPPERS

Name	Photo	Species	Region of origin in Peru
Cerezo triangular		*C. annuum*	Lambayeque
Chico		*C. baccatum*	Huánuco
Cacho de cabra amarillo		*C. baccatum*	La Libertad
Escabeche		*C. baccatum*	Lima
Amarillo de Chachapoyas		*C. baccatum*	Amazonas
Mochero		*C. chinense*	La Libertad

Name	Photo	Species	Region of origin in Peru
Arnaucho		*C. chinense*	Lima
Miscucho		*C. chinense*	La Libertad
Picante		*C. chinense*	San Martín
Bola		*C. chinense*	Piura
Panca		*C. chinense*	La Libertad

found in larger commercial fields in other parts of the country, mainly in the forested slopes found where the Andes slowly gives place to the Amazon lowlands. Limo is the typical chili pepper for ceviche, considered the most popular dish in Peruvian gastronomy, where raw fish briefly marinated in lemon juice is served with onion, chili pepper, corn, and—to reduce the heat sensation—a piece of sweet potato. *Panca* is the most important commercial *C. chinense*, grown in larger fields in the northern and southern coasts and is always marketed in dry form to be used in stews and sauces, very often when meat or fish are prepared. *C. pubescens* is grown throughout the mountain areas, in warmer valleys and is the *Capsicum* that can be grown at higher altitudes, where it can be kept as perennial in home gardens; this Rocoto is pungent and tasty, the larger fruits often stuffed with meat and cheese. *C. annuum* is a rare chili pepper in Peru, basically restricted to the *Cerezo* types of Northern Peru, while *C. frutescens* is restricted to the Amazonian type Malagueta. While all chili peppers are expected to be pungent or spicy, there is one exception: Ají Dulce (sweet chili pepper) is non-pungent and of the same characteristics of other non-pungent *C. chinense* found throughout the tropics in Latin America.

Capsicum components also have nutritional and medical importance. The *capsicum* fruits are an excellent source of natural antioxidants (vitamins C, E, and carotenoids), which appear to be critically important in preventing or reducing chronic and age-related diseases. Carotenoids of chili peppers provide remarkable provitamin A activity. Moreover, chili peppers with red color can be also considered as a noticeable source of capsanthin, the most powerful antioxidant compound among pepper carotenoids (Rodríguez-Burruezo et al. 2010a).

Chili peppers are historically employed in traditional medicine and are currently being used in modern herbology and conventional medicines. Capsaicin, the predominant compound in pungent types of *Capsicum*, induces depletion of neuropeptides from sensory nerve terminals. Moderate consumption of capsaicinoids per day has beneficial health effects, such as decreasing glucose level, LDL cholesterol, and C-reactive protein (Kenig et al. 2018). Chronic pain associated with post-herpetic neuralgia, diabetic neuropathy, and other pain syndromes receives treatment with different capsaicin creams in dermatologic therapy (Palevitch and Cracker 1996). Additionally, pepper sprays (irritant repellent products) contain up to 10% *Capsicum* oleoresin as well as other capsaicinoid compounds (Barceloux 2009). Even though it is best characterized in the field of nociception and pain, several experimental and clinical studies also demonstrate the role of capsaicin in other important pathological states such as cancer, obesity, skin disorders, cardiovascular diseases, and urinary disorders (Hernández-Pérez et al. 2020); additionally, it has also been implicated in other activities including treatment of the upper respiratory reflexes, the prevention of adipogenesis, boosting metabolic rate, and regulation of innate and adaptive immune responses (Basith et al. 2016).

6.2 NUTRITIONAL AND NUTRACEUTICAL COMPONENTS OF PERUVIAN CHILI PEPPERS

In general, chili peppers are raw materials that are characterized by having unique nutritional compounds such as capsaicinoids, in addition to having

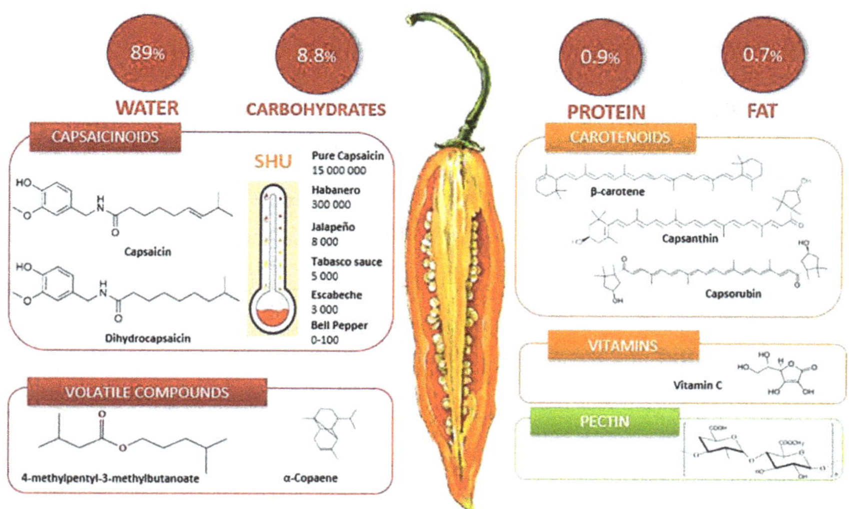

Figure 6.1 Main components of chili peppers.

Sources: Based on Arora et al. (2011), Zewdie and Bosland (2001), Kollmannsberger et al. (2011), National Center for Biotechnology Information (2019), Valderrama and Ugás (2009).

complex carotenoid profiles, the presence of vitamins such as C, and fiber, among the most important. Chili pepper compounds are schematically represented in Figure 6.1.

6.2.1 Capsaicinoids

Pungency is probably the most important attribute of *Capsicum* fruits. This characteristic depends on capsaicinoids that bind to the vanilloid receptors in the mouth. This burning sensation is stimulated by heat or physical abrasion (Kollmannsberger et al. 2011). Glands located in the epidermal cells of the fruit placenta are responsible for the synthesis and accumulation of capsaicinoids (Guzman et al. 2011). Moreover, capsaicinoid profiles prove that capsaicin and dihydrocapsaicin are the most important components determining pungency (accounting for more than 90% of the total pungency), usually expressed in Scoville Heat Units (SHU; Zewdie and Bosland 2001). Capsaicinoids are insoluble in water but soluble in hot water and solvents such as ethanol, ether, benzene, chloroform, and acetonitrile. Chemically, capsaicin ($C_{18}H_{27}NO_3$) is a phenolic compound including three regions: an aromatic ring (vanillyl amine), an amide bond, and a hydrophobic side chain (fatty acids of 8–13 carbon atoms). The substituents in positions 3 and 4, and the phenol 4-OH group in the aromatic ring, determine the extent of the pungent activity (Arora et al. 2011; Barbero et al. 2010). Figure 6.1 shows the molecule of capsaicin and dihydrocapsaicin.

The length of the lateral chain is important for the capsaicinoid bioactivity, related with lipophilia, which indicates the capacity of a compound to cross cell membranes and the absorption of bioactive compounds (Barbero et al. 2010). Capsaicinoids have been reported with healthy benefits, such as antioxidant, anticancer activity, pain reduction, effects on cardiovascular system, and antiobesity properties (Backonja et al. 2010; Harada and Okajima 2009; Kawada et al. 1986; Lu et al. 2017; Reyes-Escogido et al. 2011; Tundis et al. 2013). Capsaicinoids bind to specific receptors (transient receptor potential vanilloid) and create a kind of cation-channel receptors, increasing intracellular cation content (especially Ca^{2+}). This increment provokes the release of inflammatory neuropeptides with analgesic effects, for example. Additionally, accumulation of Ca^{2+} within cancer cells induces cellular damage (Barbero et al. 2010; Reyes-Escogido et al. 2011).

Information about capsaicinoids is available on common pepper characterization (Dubey et al. 2015; Duelund and Mouritsen 2017; Giuffrida et al. 2013; Pino et al. 2007; Rodríguez-Burruezo et al. 2010b; Topuz and Ozdemir 2007), evolution during ripening (Menichini et al. 2009; Barbero et al. 2014), effects of location (Antonious et al. 2009) and processing (Loizzo et al. 2015; Moreno-Escamilla et al. 2015; Ornelas-Paz et al. 2010; Topuz et al. 2011), and bioaccessibility (Victoria-Campos et al. 2015).

In case of Peruvian chili peppers, some studies are available. In the research of Kollmannsberger et al. (2011), two typical Andean species were analyzed (*C. baccatum* var. *pendulum* and *C. pubescens* R. & P.). These two species were mildly pungent compared with a *C. chinense* sample (13–352 vs. 1605 mg kg^{-1} total capsaicinoids). In the study of Meckelmann et al. (2013), 147 accessions were selected to represent the biodiversity of Peruvian chili peppers (*C. annuum, C. baccatum, C. chinense*, and *C. frutescens*). The highest capsaicinoid concentration was found in a *C. frutescens* accession, with 1560.1 mg/100 g of total capsaicinoids and a pattern of 68.5% capsaicin, 29.5% dihydrocapsaicin, and 1.7% nordihydrocapsaicin (250 000 SHU). In some samples, no capsaicinoids at all could be detected.

In another investigation (Meckelmann et al. 2015), 23 Peruvian chili pepper accessions from three different locations (two from the coast and one from the jungle) were analyzed. The concentrations of capsaicinoids ranged from 1.0 mg/100 g to 1515.5 mg/100 g. The results show that the highest values of SHU are not the same in each locality, but they do belong to the *C. chinense* species in the three locations. The maximum value of pungency was in a sample from a coastal locality, with a pungency level like a Jalapeño pepper (Arora et al. 2011).

Finally, a capsaicinoid analysis was conducted on 20 representative Peruvian chili peppers (*C. annuum, C. baccatum*, and *C. chinense)* (Morales Soriano et al. 2018). In all cases, capsaicin concentrations were higher than dihydrocapsaicin amounts, with values between 13 and 200 mg capsaicin per 100 g of sample (dry weight). Concerning pungency, a *C. baccatum* landrace (common name *Chico*) shows the highest SHU value, although other authors reported *C. chinense* to be the species with highest pungency levels (De Witt and Bosland 1996; Nuez et al. 1996). Probably, this result is mainly due to its unusually high dihydrocapsaicin level. All the results show a large variability in capsaicin, dihydrocapsaicin and SHU values among the different species, without a correlation between pungency levels and species classifications (Zewdie and Bosland 2001).

6.2.2 Carotenoids

One important characteristic in *Capsicum* peppers is their diversity in fruit color, which depends mainly on the carotenoid content and profile (Bosland and Votava 2012; Rodríguez-Burruezo et al. 2010a). Carotenoids are mainly C_{40} tetraterpenoids and can be classified in two groups: (1) carotenoids containing only hydrogen and carbon, named carotenes, and (2) carotenoids also containing oxygen, called xanthophylls (Bosland and Votava 2012; Rodriguez-Amaya 2001). Figure 6.1 shows the structure of β-carotene, and the xanthophylls capsanthin and capsorubin, red carotenoids only present in chili peppers (Minguez-Mosquera and Hornero-Mendez 1993; Schweiggert et al. 2005a).

These natural colorants are well known to have beneficial effects on human health, such as an important source of provitamin A (Bosland and Votava 2012), reducing the risk of chronic diseases (like cardiovascular), and certain types of cancer (Krinsky and Johnson 2005; Rao and Rao 2007). Antioxidant properties are also reported in chili pepper carotenoids, capsanthin being the most important (Rodríguez-Burruezo et al. 2010a). Moreover, a study demonstrates the capsanthin promotion on lipid metabolism in mice (Kim et al. 2017).

Literature contains information about carotenoid profiles in fresh chili peppers (Giuffrida et al. 2013; Kim et al. 2016; Rodríguez-Burruezo et al. 2010a); carotene and xanthophyll evolution during ripening of chili peppers (Hornero-Méndez et al. 2000; Hornero-Méndez and Mínguez-Mosquera 2000); the shelf life of red chili peppers in powder (Giuffrida et al. 2014); and the effect of thermal processing (Cervantes-Paz et al. 2014; Guerra-Vargas et al. 2001).

However, the investigations of the profile and abundance of carotenoids, as well as their changes during ripening, processing, and storage, are limited in the case of Peruvian chili peppers. Rodríguez-Burruezo et al. (2010a) studied 12 Bolivian chili peppers (*C. baccatum* and *C. pubescens*) in terms of carotenoid profile. These authors identified 16 different carotenoids, being capsanthin the main carotenoid in the red fruits, followed by capsanthin 5,6 epoxide, antheraxanthin, and violaxanthin. In the case of yellow- and orange-colored fruits, violaxanthin was the major carotenoid, followed by cis-violaxanthin, antheraxanthin, and lutein.

Total content and individual carotenoid profile were investigated in 20 representative Peruvian chili peppers (Morales-Soriano et al. 2019). A large variability of total carotenoid content was found (330 to 28 418 μg equivalent β-carotene/100 g fresh sample), being the highest value for *Panca* (*C. chinense*; brown color), followed by the red samples *Cerezo redondo* (*C. annuum*) and *Bola* (*C. chinense*). Similarly, a wide range of carotenoid profiles was found, especially in red and orange chili peppers. In the case of yellow samples, the number of carotenoids and their content was low. β-carotene was the most abundant in almost all the samples and capsanthin was present only in red landraces. High retinol activity equivalent levels were found in red peppers, due to provitamin A activity of β-carotene, α-carotene, and β-cryptoxanthin (Ordóñez-Santos et al. 2017; Trumbo et al. 2001).

6.2.3 Vitamin C

Fresh chili peppers are also a good source of vitamins, especially vitamin C (Bosland and Votava 2012; Meckelmann et al. 2013). In fact, 100 g of green sweet peppers provide almost the 100% of recommended daily intake of this vitamin. The vitamin C content increases with the ripening of the fruit (Xavier and Pérez-Gálvez 2015).

In the case of Peruvian chili peppers, 147 accessions were analyzed in ascorbic acid (vitamin C) content (Meckelmann et al. 2013). Fresh chili peppers contain up to 250 mg/100 g fresh weight, and this content is influenced by the degree of ripeness. The maximum of ascorbic acid was found in a *C. chinense* sample with 295 mg/100 g after a drying process (samples were dried for stability of other compounds and for the transfer to the analysis laboratory). Besides the antioxidant and vitamin function of ascorbic acid, high concentrations help to protect other components like carotenoids, maintaining the color intensity during subsequent processes.

6.2.4 Other Compounds

Fresh chili peppers have been recognized as an important source of phenolic compounds, metabolites with antioxidant activity (Silva et al. 2013; Hernández-Pérez et al. 2020). It has been reported that phytochemical changes occurring during ripening affect the antioxidant activity and composition

(Hernández-Pérez et al. 2020). Regarding to the species, Rodríguez-Burruezo et al. (2009) found that *C. pubescens* showed lower total phenolic levels than *C. baccatum* and *C. annuum*, although differences were not as remarkable as for ascorbic acid. Total phenolic levels in *C. baccatum* were similar to those found in *C. annuum* or even higher in some accessions, suggesting that some *C. baccatum* fruits have a very high nutritional value.

The contribution of the raw material cell wall polysaccharides (e.g., pectin) and dietary fiber to the textural properties of chili pepper-based pastes/sauces plays a crucial role (Christiaens et al. 2012b). Moreover, the contribution of chili peppers to the consumption of dietary fiber can be important from the nutritional and health point of view (Martínez et al. 2014). Capsicuman, a pectic polysaccharide found in peppers—which comprises galacturonic acid (74.0%), rhamnose (1.6%), arabinose (2.6%) and galactose (2.4%) residues—was characterized; the importance of this polysaccharide was the immunomodulatory activity demonstrated in mice (Popov et al. 2011).

Twenty Peruvian chili pepper landraces were characterized (*C. annum, C. baccatum*, and *C. chinense* accessions) in terms of structural properties, including the Bostwick consistency index of chili purées, dry matter, dietary fiber, and pectin content (Morales-Soriano et al. 2019). A wide range of stability (presence or absence of syneresis) and consistency were found, partially explained by the dry matter, dietary fiber, and pectin content. *C. annuum* samples displayed the highest dry matter and fiber content. The degree of syneresis (related with stability) and the consistency are usually attributed to the characteristic of both the dispersed (particles) and continuous (serum) phases (Christiaens et al. 2012a).

Capsinoids (different from capsaicinoids) are other compounds found in almost all peppers. They are structurally and with biological properties like capsaicinoids, except that they are not hot or spicy. These compounds have several notable biological properties: they are powerful antioxidants, promote energy consumption and decrease fat accumulation, increase body temperature and the consumption of oxygen, and present anti-inflammatory activity (Lim 2013; Barbero et al. 2010; Coutinho et al. 2015; Baenas et al. 2019).

6.2.5 Toxicity of Components

Toxicological studies of *C. annum* in mice diet apparently do not affect general health and body weight at any level of pepper, concluding that red chili was relatively non-toxic at the doses tested (Lim 2013). However, references of heartburn, problems with ingestion or skin problems have been reported in some cases after eating chili peppers (Idrees et al. 2020).

6.3 EFFECT OF PROCESSING ON CHILI PEPPERS

Peru is one of the origins of chili peppers, comprising more than 200 landraces (Jäger et al. 2013; Valderrama and Ugás 2009; Ugás 2012). Only a handful of chili pepper varieties and landraces are commercialized or transformed in different products, such as sauces or fermented or dehydrated products. The importance of processing is to give stability to this material, since chili peppers are an important source of nutritional compounds. In other countries like Mexico, consumption of chili sauces is considerable in their population. These chili pepper sauces are considered a source of bioactive compounds in combination with other ingredients (tomatoes, onions, garlic, and different kinds of leaves and seeds). Effects of processing were also analyzed (Cárdenas-Castro et al. 2019).

Usually, chili pepper sauces production considered at least four stages: the mix and blending of the ingredients, the cooking, the cooling step, and the packing in the final container. In the industry, the rheological performance is an important characteristic (Gamonpilas et al. 2011).

Presence of enzymes is also an important consideration during the chili pepper processing. In addition to lipoxygenase (LOX), peroxidase (POD), and poliphenoloxydase (PPO), other enzymes such as polygalacturonase (PG), pectin methyl esterase (PME), cellulase and β-galactosidase have been studied in *Capsicum* samples (Schweiggert et al. 2005b; Rao et al. 2011). Blanching is the main operation to denature these enzymes, and the evaluation of residual activity of POD, PPO, and LOX was studied in paprika and chili powder (Schweiggert et al. 2005b); residual activity of POD was studied in fresh chili pepper using high-humidity hot air impingement blanching (HHAIB; Wang et al. 2020).

6.3.1 Mechanical Disintegration

One of the first operations applied to reduce the particle size is blending, in addition to modifying the food texture or rheological characteristics. This particle size reduction causes color changes in chili purées or sauces due to the increasing of surface area (Ahmed et al. 2000). This situation also provokes exposition of different compounds, and significant nutritional losses can occur, such as degradation of carotenoids, vitamin C, polyphenols, and others.

High-pressure homogenization is also applied in chili pepper sauces to reduce and make the particle size uniform, but in this case the fluid is forced through a narrow opening at high speed with a pressure around 100 MPa. This operation is also used to produce emulsions. With regard to the nutritional aspect, the influence of the particle size reduction induced by high pressure homogenization results in an improvement of lipid digestion and β-carotene bioaccessibility (Salvia-Trujillo et al. 2013). Additionally, if oil is present, high-pressure homogenization can increase the transfer of carotenoids to the oil phase, thus facilitating carotenoid micellization and bioaccessibility, and eventually improving the nutritional quality of the final product (Mutsokoti et al. 2015).

6.3.2 Thermal Processing

Thermal process is applied to obtain microbiologically safe and stable products with a good sensorial quality. Pasteurization is the most used at the industrial level, and the required temperature/time combination depends on the pH of the product.

Minimal changes in sensorial and nutritional characteristics are expected using moderate temperatures. Furthermore, thermal processing can induce the disruption of natural cell barriers, increasing the extractability and bioavailability of compounds such as carotenoids or chlorophyll (Palmero et al. 2014). Carotenoid and capsaicinoid bioaccessibility of chili peppers has been studied considering thermal treatment and other factors such as initial concentration or ripening stage of the fruit (Victoria-Campos et al. 2013, 2015). Bioaccessibility of carotenoids can increase in thermal treatments, when oil is added due to the transfer of lipophilic nutrients to the oil (Mutsokoti et al. 2016). However, a decrease of carotenoid content was also noticed when thermal treatment was applied in chili pepper products (Pugliese et al. 2013). Discoloration and a decrease in apparent viscosity were also reported after thermal processing application, depending on the intensity (Ismail and Revathi 2006; Ramos-Aguilar et al. 2015).

6.3.3 Other Processes

Drying is another process applied to chili peppers, solar dehydration being the most popular for some species, since the consumption of *Amarillo* (*C. baccatum*) or *Panca* (*C. chinense*) is very typical in their dry state. Carotenoids are the most affected compound during a drying process, and their bioaccessibility was studied on yellow bell pepper when hot air drying (HAD) was applied. The HAD process resulted in cell wall disruption and induced an anticipated enhancement of carotenoid release. In this study, the carotenoids in fresh pepper were dissolved in globular lipid fraction, meaning a high level of bioaccessibility in comparison with other raw materials, like carrot. Dried pepper had a relatively high bioaccessibility of all its carotenoids (Zhang et al. 2018).

Different drying methods were also studied on paprika elaboration form Jalapeno. A novel contact drying method, Influence of Refractance Window Drying (RWD), was investigated in comparison with freeze-drying (FD), oven drying (OD), and natural convective drying (NCD) methods, and their effect on main components. Carotenoids and capsaicinoids were significantly decreased by all the methods except NCD. Due to ongoing synthesis, the NCD method resulted in higher capsaicinoids and carotenoids except violaxanthin and mutatoxanthin (Topuz et al. 2011).

High-pressure processing (HPP) is a non-thermal method employing elevated pressures as the main lethal agent for pathogen reduction. Pressure application is in the range from 400 to 600 MPa for a few minutes. HPP results in minimal changes in the nutritional and sensory properties of the final product (Balasubramaniam et al. 2016; Oey et al. 2008). Some studies have proved this process to be an alternative means of pasteurization of chili pepper pieces (Hernández-Carrión et al. 2014). HPP can cause enzymatic and non-enzymatic conversions. In some cases, including pepper samples, HPP can promote pectinmethylesterase activity, obstruct polygalacturonase action, and retard β-elimination which causes texture degradation. Furthermore, minimal reduction on soluble protein and ascorbic acid content was noticed in HPP treated peppers in comparison with thermal processing (Castro et al. 2008).

6.4 CONCLUSION AND FUTURE PERSPECTIVES

In general, Peruvian chili peppers have unique attributes when each landrace is analyzed. Capsaicinoids, carotenoids, vitamin C, and fiber are the most important compounds, in addition to polyphenols, other antioxidants, minerals, and volatile compounds. The analysis of these compounds shows the large biodiversity of chili peppers and the complex relation with sensorial, nutritional, and texture attributes on raw material and final product after processing. Furthermore, several studies demonstrate chili pepper has bioactive compounds with positive effect on health.

Analysis of different process must be studied, especially their effect on the chili pepper compounds. An integrated approach considering enzyme activity and cell structures which contains bioactive compounds must be addressed. Operations such as blending or high-pressure homogenization involve different particle size reduction modifying the consistency and viscosity of final products. Then, application of stabilization techniques must be considered in function of the final functional attributes in order to preserve the more important healthy compounds or minimize undesirable changes. Bioaccessibility of carotenoids or capsaicinoids must be considered by selecting the appropriate technology.

Specific studies on Peruvian chili peppers must be promoted. New product development using chili pepper landraces must be considered with the appropriate technology and scientific characterization, showing their main potential attributes. Valorization of the raw material and the by-products generated, such as seeds or veins as a source of capsaicinoids or fiber, for example, must be increased in order to increase the consumption and added value of Peruvian chili peppers.

Capsicum chili peppers can be used as model crops when understanding the evolution, diversity, and market potential of biodiversity. In spite of having the world's largest diversity of chili peppers, most Peruvians rarely consume more than 4 or 5 varieties or landraces, a trend augmented by the fact that 80% of the population lives in urban areas. In the last two decades, chili peppers have been regarded as the DNA of Peruvian cuisine, and the gastronomic movement has shown its potential to promote social change, a process where good eating habits are promoted as well as stronger connections of the markets and kitchens with smallholder agriculture, which is responsible for over 70% of the food consumed in the country, but historically underestimated in terms of policies, investment, and public services. If Peru can feel proud of this agricultural biodiversity it is because smallholder families have been involved in the conservation of traditional landraces and the knowledge associated with it. Improved value chains for biodiversity need to prove useful for these families and their livelihoods.

REFERENCES

Ahmed, Jasim, U. S. Shivhare, and G. S. V. Raghavan. 2000. "Rheological Characteristics and Kinetics of Colour Degradation of Green Chilli Puree." *Journal of Food Engineering* 44 (4): 239–244. https://doi.org/10.1016/S0260-8774(00)00034-0.

Antonious, George F., Terry Berke, and Robert L. Jarret. 2009. "Pungency in *Capsicum chinense*: Variation among Countries of Origin." *Journal of Environmental Science and Health Part B* 44 (2): 179–184. https://doi.org/10.1080/03601230802599118.

Arora, Rajendra, N. S. Gill, Gaurav Chauhan, and A. C. Rana. 2011. "An Overview About Versatile Molecule Capsaicin." *International Journal of Pharmaceutical Sciences and Drug Research* 3 (4): 280–286. http://ijpsdr.com/index.php/ijpsdr/article/view/447.

Backonja, Misha Miroslav, T. Philip Malan, Geertrui F. Vanhove, and Jeffrey K. Tobias. 2010. "NGX-4010, a High-Concentration Capsaicin Patch, for the Treatment of Postherpetic Neuralgia: A Randomized, Double-Blind, Controlled Study with an Open-Label Extension." *Pain Medicine*: 600–608. https://doi.org/10.1111/j.1526-4637.2009.00793.x.

Baenas, Nieves, Miona Belović, Nebojša Ilic, Diego A. Moreno, and Cristina García-Viguera. 2019. "Industrial Use of Pepper (*Capsicum annum* L.) Derived Products: Technological Benefits and Biological Advantages." *Food Chemistry* 274: 872–885. https://doi.org/10.1016/j.foodchem.2018.09.047.

Balasubramaniam, V. M. Bala, Gustavo V Barbosa-Cánovas, and Huub Lelieveld. 2016. "High Pressure Processing of Food." In V. M. Balasubramaniam, Gustavo V.

Barbosa-Cánovas, and Huub L. M. Lelieveld (Eds.), *Food Engineering Series*. New York: Springer. https://doi.org/10.1007/978-1-4939-3234-4.

Barbero, Gerardo F., José M. G. Molinillo, Rosa Varela, Miguel Palma, Francisco Macias, and Carmelo G. Barroso. 2010. "Application of Hansch's Model to Capsaicinoids and Capsinoids: A Study Using the Quantitative Structure—Activity Relationship: A Novel Method for the Synthesis of Capsinoids." *Journal of Agricultural and Food Chemistry* 58 (6): 3342–3349. https://doi.org/10.1021/jf9035029.

Barbero, Gerardo F., Aurora G. Ruiz, Ali Liazid, Miguel Palma, Jesús C. Vera, and Carmelo G. Barroso. 2014. "Evolution of Total and Individual Capsaicinoids in Peppers During Ripening of the Cayenne Pepper Plant (*Capsicum annuum* L.)." *Food Chemistry* 153: 200–206. https://doi.org/10.1016/j.foodchem.2013.12.068.

Barceloux, Donald G. 2009. "Pepper and Capsaicin (*Capsicum* and Piper Species)." *Disease-a-Month* 55 (6): 380–390. https://doi.org/10.1016/j.disamonth.2009.03.008.

Basith, Shaherin, Minghua Cui, Sunhye Hong, and Sun Choi. 2016. "Harnessing the Therapeutic Potential of Capsaicin and Its Analogues in Pain and Other Diseases." *Molecules* 21 (8): 966. https://doi.org/10.3390/molecules21080966.

Bosland, Paul W., and Eric Votava. 2012. *Peppers: Vegetable and Spice Capsicums*. Wallingford: CABI, 2nd edition.

Cárdenas-Castro, Alicia Paulina, Guadalupe del Carmen Perales-Vázquez, Laura A. De la Rosa, Víctor Manuel Zamora-Gasga, Víctor Manuel Ruiz-Valdiviezo, Emilio Alvarez-Parrilla, and Sonia Guadalupe Sáyago-Ayerdi. 2019. "Sauces: An Undiscovered Healthy Complement in Mexican Cuisine." *International Journal of Gastronomy and Food Science* 17: 100154. https://doi.org/10.1016/j.ijgfs.2019.100154.

Carrizo García, Carolina, Michael H. J. Barfuss, Eva M. Sehr, Gloria E. Barboza, Rosabelle Samuel, Eduardo A. Moscone, and Friedrich Ehrendorfer. 2016. "Phylogenetic Relationships, Diversification and Expansion of Chili Peppers (*Capsicum, Solanaceae*)." *Annals of Botany* 118: 35–51. https://doi:10.1093/aob/mcw079.

Castro, Sónia M., Jorge A. Saraiva, José A. Lopes-da-Silva, Ivonne Delgadillo, Ann Van Loey, Chantal Smout, and Marc Hendrickx. 2008. "Effect of Thermal Blanching and of High Pressure Treatments on Sweet Green and Red Bell Pepper Fruits (*Capsicum annuum* L.)." *Food Chemistry* 107 (4): 1436–1449. https://doi.org/10.1016/j.foodchem.2007.09.074.

Cervantes-Paz, Braulio, Elhadi M. Yahia, José De Jesús Ornelas-Paz, Claudia I. Victoria-Campos, Vrani Ibarra-Junquera, Jaime David Pérez-Martínez, and Pilar Escalante-Minakata. 2014. "Antioxidant Activity and Content of Chlorophylls and Carotenoids in Raw and Heat-Processed Jalapeño Peppers at Intermediate Stages of Ripening." *Food Chemistry* 146: 188–196. https://doi.org/10.1016/j.foodchem.2013.09.060.

Christiaens, Stefanie, Sandy Van Buggenhout, Davis Chaula, Katlijn Moelants, Charlotte C. David, Johan Hofkens, Ann M. Van Loey, and Marc E. Hendrickx. 2012a. "In Situ Pectin Engineering as a Tool to Tailor the Consistency and Syneresis of Carrot Purée." *Food Chemistry* 133 (1): 146–155. https://doi.org/10.1016/j.foodchem.2012.01.009.

Christiaens, Stefanie, Sandy Van Buggenhout, Ken Houben, Davis Chaula, Ann M. Van Loey, and Marc E. Hendrickx. 2012b. "Unravelling Process-Induced Pectin

Changes in the Tomato Cell Wall: An Integrated Approach." *Food Chemistry* 132 (3): 1534–1543. https://doi.org/10.1016/j.foodchem.2011.11.148.

Coutinho, Janclei P., Gerardo F. Barbero, Oreto F. Avellán, A. Garcés-Claver, Helena T. Godoy, Miguel Palma, and Carmelo G. Barroso. 2015. "Use of Multivariate Statistical Techniques to Optimize the Separation of 17 Capsinoids by Ultra Performance Liquid Chromatography Using Different Columns." *Talanta* 134: 256–263. https://doi.org/10.1016/j.talanta.2014.11.004.

De Witt, David, and Paul W. Bosland. 1996. *Peppers of the World: An Identification Guide.* Berkeley: Ten Speed Press.

Dillehay, Tom, Steve Goodbred, Mario Pino, Víctor Vásquez Sánchez, Teresa Rosales, James Adovasio, Michael Collins, Patricia Netherly, Christine Hastorf, Katherine Chiou, Dolores Piperno, Isabel Rey and Nancy Velchoff. 2017. "Simple Technologies and Diverse Food Strategies of the Late Pleistocene and Early Holocene at Huaca Prieta, Coastal Peru." *Sciences Advances* 3 (5). https://doi: 10.1126/sciadv.1602778.

Dubey, Rakesh K., Vikas Singh, Garima Upadhyay, A. K. Pandey, and Dhan Prakash. 2015. "Assessment of Phytochemical Composition and Antioxidant Potential in Some Indigenous Chilli Genotypes from North East India." *Food Chemistry* 188: 119–125. https://doi.org/10.1016/j.foodchem.2015.04.088.

Duelund, Lars, and Ole G. Mouritsen. 2017. "Contents of Capsaicinoids in Chillies Grown in Denmark." *Food Chemistry* 221: 913–918. https://doi.org/10.1016/j.foodchem.2016.11.074.

FAO. 2018. "FAOSTAT." www.fao.org/faostat/en/#data/QC.

Gamonpilas, Chaiwut, W. Pongjaruvat, Asira Fuongfuchat, P. Methacanon, N. Seetapan, and N. Thamjedsada. 2011. "Physicochemical and Rheological Characteristics of Commercial Chili Sauces as Thickened by Modified Starch or Modified Starch/Xanthan Mixture." *Journal of Food Engineering* 105 (2): 233–240. https://doi.org/10.1016/j.jfoodeng.2011.02.024.

Giuffrida, Daniele, Paola Dugo, Germana Torre, Chiara Bignardi, Antonella Cavazza, Claudio Corradini, and Giacomo Dugo. 2013. "Characterization of 12 *Capsicum* Varieties by Evaluation of Their Carotenoid Profile and Pungency Determination." *Food Chemistry* 140 (4): 794–802. https://doi.org/10.1016/j.foodchem.2012.09.060.

Giuffrida, Daniele, Paola Dugo, Germana Torre, Chiara Bignardi, Antonella Cavazza, Claudio Corradini, and Giacomo Dugo. 2014. "Evaluation of Carotenoid and Capsaicinoid Contents in Powder of Red Chili Peppers During One Year of Storage." *Food Research International* 65 (PB): 163–170. https://doi.org/10.1016/j.foodres.2014.06.019.

Guerra-Vargas, María, María Eugenia Jaramillo-Flores, L. Dorantes-Alvarez, and Humberto Hernández-Sánchez. 2001. "Carotenoid Retention in Canned Pickled Jalapeno Peppers and Carrots as Affected by Sodium Chloride, Acetic Acid, and Pasteurization." *Journal of Food Science* 66 (4): 620–626. https://doi.org/10.1111/j.1365-2621.2001.tb04611.x.

Guzman, Ivette, Paul W. Bosland, and Mary A. O'Connell. 2011. "Heat, Color, and Flavor Compounds in *Capsicum* Fruit." In David R. Gang (Ed.), *The Biological Activity of Phytochemicals.* New York: Springer, 109–126. https://doi.org/10.1007/978-1-4419-7299-6.

Harada, Naoaki, and Kenji Okajima. 2009. "Effects of Capsaicin and Isoflavone on Blood Pressure and Serum Levels of Insulin-Like Growth Factor-I in Normotensive and Hypertensive Volunteers with Alopecia." *Bioscience, Biotechnology, and Biochemistry* 73 (6): 1456–1459. https://doi.org/10.1271/bbb.80883.

Hernández-Carrión, María, Isabel Hernando, and Amparo Quiles. 2014. "High Hydrostatic Pressure Treatment as an Alternative to Pasteurization to Maintain Bioactive Compound Content and Texture in Red Sweet Pepper." *Innovative Food Science and Emerging Technologies* 26: 76–85. https://doi.org/10.1016/j.ifset.2014.06.004.

Hernández-Pérez, Talía, Gómez-García, María del Rocío, Valverde, María Elena, Paredes-López, Octavio. 2020. "*Capsicum annuum* (Hot Pepper): An Ancient Latin-American Crop with Outstanding Bioactive Compounds and Nutraceutical Potential: A Review." *Comprehensive Reviews in Food Science and Food Safety* 19 (6): 2972–2993. https://doi.org/10.1111/1541-4337.12634.

Hornero-Méndez, Dámaso, Ricardo Gómez-Ladrón de Guevara, and María Isabel Mínguez-Mosquera. 2000. "Carotenoid Biosynthesis Changes in Five Red Pepper (*Capsicum annuum* L.) Cultivars During Ripening. Cultivar Selection for Breeding." *Journal of Agricultural and Food Chemistry* 48 (9): 3857–3864. https://doi.org/10.1021/jf991020r.

Hornero-Méndez, Dámaso, and M. Isabel Mínguez-Mosquera. 2000. "Xanthophyll Esterification Accompanying Carotenoid Overaccumulation in Chromoplast of *Capsicum annuum* Ripening Fruits Is a Constitutive Process and Useful for Ripeness Index." *Journal of Agricultural and Food Chemistry* 48 (5): 1617–1622. https://doi.org/10.1021/jf9912046.

Idrees, Saba, Muhammad Asif Hanif, Muhammad Adnan Ayub, Asma Hanif, and Tariq Mahmood Ansari. 2020. "Chili Pepper." In Muhammad Asif Hanif, Haq Nawaz, Muhammad Mumtaz Khan, and Hugh Byrne (Eds.), *Medicinal Plants of South Asia*. Amsterdam: Elsevier, 113–124. https://doi.org/10.1016/B978-0-08-102659-5.00009-4.

Ismail, Noryati, and R. Revathi. 2006. "Studies on the Effects of Blanching Time, Evaporation Time, Temperature and Hydrocolloid on Physical Properties of Chili (*Capsicum annuum* var. Kulai) Puree." *LWT—Food Science and Technology* 39 (1): 91–97. https://doi.org/10.1016/j.lwt.2004.12.003.

Jäger, Matthias, Alejandra Jiménez, and Karen Amaya. 2013. *Las Cadenas de Valor de Los Ajíes Nativos de Perú. Compilación de Los Estudios Realizados Dentro Del Marco Del Proyecto 'Rescate y Promoción de Ajíes Nativos En Su Centro de Origen' Para Perú*. Cali: Bioversity International. www.bioversityinternational.org/e-library/publications/detail/las-cadenas-de-valor-de-los-ajies-nativos-de-peru/.

Jarret, Robert L., Gloria Estela Barboza, Fabiane Rabelo da Costa Batista, Terry Berke, and Yu-Yu Chou. 2019. "*Capsicum*—An Abbreviated Compendium." *Journal of the American Society for Horticultural Science* 144 (1): 3–22. https://doi.org/10.21273/JASHS04446-18.

Kawada, Teruo, Koh-Ichiro Hagihara, and Kazuo Iwai. 1986. "Effects of Capsaicin on Lipid Metabolism in Rats Fed a High Fat Diet." *Journal of Nutrition* 116 (7): 1272–1278. https://doi.org/10.1093/jn/116.7.1272.

Kenig, Sasa, Alenka Baruca-Arbeiter, Nina Mohorko, Mojca Stubelj, Masa Cernelic-Bizjak, Dunca Bandelj, Zala Jenko-Praznikar, and Ana Petelin. 2018. "Moderate but Not High Daily Intake of Chili Pepper Sauce Improves Serum Glucose and Cholesterol Levels." *Journal of Functional Foods* 44: 209–217. https://doi.org/10.1016/j.jff.2018.03.014.

Kim, Ji Sun, Chul Geon An, Jong Suk Park, Yong Pyo Lim, and Suna Kim. 2016. "Carotenoid Profiling from 27 Types of Paprika (*Capsicum annuum* L.) with Different Colors, Shapes, and Cultivation Methods." *Food Chemistry* 201: 64–71. https://doi.org/10.1016/j.foodchem.2016.01.041.

Kim, Ji Sun, Tae Youl Ha, Suna Kim, Sung Joon Lee, and Jiyun Ahn. 2017. "Red Paprika (*Capsicum annuum* L.) and Its Main Carotenoid Capsanthin Ameliorate Impaired Lipid Metabolism in the Liver and Adipose Tissue of High-Fat Diet-Induced Obese Mice." *Journal of Functional Foods* 31: 131–140. https://doi.org/10.1016/j.jff.2017.01.044.

Kollmannsberger, Hubert, Adrián Rodríguez-Burruezo, Siegfried Nitz, and Fernando Nuez. 2011. "Volatile and Capsaicinoid Composition of Ají (*Capsicum baccatum*) and Rocoto (*Capsicum pubescens*), Two Andean Species of Chile Peppers." *Journal of the Science of Food and Agriculture* 91 (9): 1598–1611. https://doi.org/10.1002/jsfa.4354.

Kraft, Kraig H., Cecil H. Brown, Gary P. Nabhan, Eike Luedeling, José de Jesús Luna Ruiz, Geo Coppens d'Eeckenbrugge, Robert J. Hijmans, and Paul Gepts. 2014. "Multiple Lines of Evidence for the Origin of Domesticated Chili Pepper, *Capsicum annuum*, in Mexico." *Proceedings of the National Academy of Sciences of the United States of America* 111 (17): 6165–6170. https://doi.org/10.1073/pnas.1308933111.

Krinsky, Norman I., and Elizabeth J. Johnson. 2005. "Carotenoid Actions and Their Relation to Health and Disease." *Molecular Aspects of Medicine* 26 (6): 459–516. https://doi.org/10.1016/j.mam.2005.10.001.

Lim, Tae-Kang. 2013. "*Capsicum annuum*." In *Edible Medicinal and Non-Medicinal Plants*. Dordrecht: Springer, vol. 6: 161–196. https://doi.org/10.1007/978-94-007-5628-1_28.

Loizzo, Monica R., Alessandro Pugliese, Marco Bonesi, Francesco Menichini, and Rosa Tundis. 2015. "Evaluation of Chemical Profile and Antioxidant Activity of Twenty Cultivars from *Capsicum* Annuum, *Capsicum* Baccatum, *Capsicum* Chacoense and *Capsicum* Chinense: A Comparison Between Fresh and Processed Peppers." *LWT—Food Science and Technology* 64 (2): 623–631. https://doi.org/10.1016/j.lwt.2015.06.042.

Lu, Muwen, Chi Tang Ho, and Qingrong Huang. 2017. "Extraction, Bioavailability, and Bioefficacy of Capsaicinoids." *Journal of Food and Drug Analysis* 25 (1): 27–36. https://doi.org/10.1016/j.jfda.2016.10.023.

Martínez, Mario M., Álvaro Díaz, and Manuel Gómez. 2014. "Effect of Different Microstructural Features of Soluble and Insoluble Fibres on Gluten-Free Dough Rheology and Bread-Making." *Journal of Food Engineering* 142: 49–56. https://doi.org/10.1016/j.jfoodeng.2014.06.020.

Meckelmann, Sven W., Dieter W. Riegel, Maarten J. van Zonneveld, Llermé Ríos, Karla Peña, Roberto Ugás, Lourdes Quinonez, Erika Mueller-Seitz, and Michael Petz. 2013. "Compositional Characterization of Native Peruvian Chili Peppers (*Capsicum* Spp.)." *Journal of Agricultural and Food Chemistry* 61 (10): 2530–2537. https://doi.org/10.1021/jf304986q.

Meckelmann, Sven W., Dieter W. Riegel, Maarten van Zonneveld, Llermé Ríos, Karla Peña, Erika Mueller-Seitz, and Michael Petz. 2015. "Capsaicinoids, Flavonoids,

Tocopherols, Antioxidant Capacity and Color Attributes in 23 Native Peruvian Chili Peppers (*Capsicum* Spp.) Grown in Three Different Locations." *European Food Research and Technology* 240 (2): 273–283. https://doi.org/10.1007/s00217-014-2325-6.

Menichini, F., R. Tundis, M. Bonesi, M. Loizzo, F. Conforti, G. Statti, B. Decindio, P. Houghton, and F. Menichini. 2009. "The Influence of Fruit Ripening on the Phytochemical Content and Biological Activity of *Capsicum* Chinense Jacq. Cv Habanero." *Food Chemistry* 114 (2): 553–560. https://doi.org/10.1016/j.foodchem.2008.09.086.

Minguez-Mosquera, M. Isabel, and Damaso Hornero-Mendez. 1993. "Separation and Quantification of the Carotenoid Pigments in Red Peppers (*Capsicum annuum* L.), Paprika, and Oleoresin by Reversed-Phase HPLC." *Journal of Agricultural and Food Chemistry* 41 (10): 1616–1620. https://doi.org/10.1021/jf00034a018.

Morales-Soriano, Eduardo, Biniam Kebede, Roberto Ugás, Tara Grauwet, Ann Van Loey, and Marc Hendrickx. 2018. "Flavor Characterization of Native Peruvian Chili Peppers through Integrated Aroma Fingerprinting and Pungency Profiling." *Food Research International* 109: 250–259, February. https://doi.org/https://doi.org/10.1016/j.foodres.2018.04.030.

Morales-Soriano, Eduardo, Agnese Panozzo, Roberto Ugás, Tara Grauwet, Ann Van Loey, and Marc Hendrickx. 2019. "Carotenoid Profile and Basic Structural Indicators of Native Peruvian Chili Peppers." *European Food Research and Technology* 245 (3): 717–732. https://doi.org/10.1007/s00217-018-3193-2.

Moreno-Escamilla, Jesús Omar, Laura A. de la Rosa, José Alberto López-Díaz, Joaquín Rodrigo-García, José Alberto Núñez-Gastélum, and Emilio Alvarez-Parrilla. 2015. "Effect of the Smoking Process and Firewood Type in the Phytochemical Content and Antioxidant Capacity of Red Jalapeño Pepper During Its Transformation to Chipotle Pepper." *Food Research International* 76: 654–660. https://doi.org/10.1016/j.foodres.2015.07.031.

Mutsokoti, Leonard, Agnese Panozzo, Edwin T. Musabe, Ann Van Loey, and Marc Hendrickx. 2015. "Carotenoid Transfer to Oil upon High Pressure Homogenisation of Tomato and Carrot Based Matrices." *Journal of Functional Foods* 19: 775–785. https://doi.org/10.1016/j.jff.2015.10.017.

Mutsokoti, Leonard, Agnese Panozzo, Ann Van Loey, and Marc Hendrickx. 2016. "Carotenoid Transfer to Oil During Thermal Processing of Low Fat Carrot and Tomato Particle Based Suspensions." *Food Research International* 86: 64–73. https://doi.org/10.1016/j.foodres.2016.05.019.

National Center for Biotechnology Information. 2019. "PubChem Database." https://pubchem.ncbi.nlm.nih.gov/compound/.

Nuez, Fernando, Ramiro Gil, and Joaquín Costa. 1996. *El Cultivo de Pimientos, Chiles y Ajíes*. Madrid: Mundi Prensa Libros S.A.

Oey, Indrawati, Martina Lille, Ann Van Loey, and Marc Hendrickx. 2008. "Effect of High-Pressure Processing on Colour, Texture and Flavour of Fruit- and Vegetable-Based Food Products: A Review." *Trends in Food Science and Technology* 19 (6): 320–328. https://doi.org/10.1016/j.tifs.2008.04.001.

Ordóñez-Santos, Luis Eduardo, Jader Martínez-Girón, and Maria Enith Arias-Jaramillo. 2017. "Effect of Ultrasound Treatment on Visual Color, Vitamin C, Total Phenols, and Carotenoids Content in Cape Gooseberry Juice." *Food Chemistry* 233: 96–100. https://doi.org/10.1016/j.foodchem.2017.04.114.

Ornelas-Paz, José de Jesús, J. Manuel Martínez-Burrola, Saúl Ruiz-Cruz, Víctor Santana-Rodríguez, Vrani Ibarra-Junquera, Guadalupe I. Olivas, and J. David Pérez-Martínez. 2010. "Effect of Cooking on the Capsaicinoids and Phenolics Contents of Mexican Peppers." *Food Chemistry* 119 (4): 1619–1625. https://doi.org/10.1016/j.foodchem.2009.09.054.

Palevitch, Dan, and Lyle E. Cracker. 1996. "Nutritional and Medical Importance of Red Pepper (*Capsicum Spp.*)." *Journal of Herbs, Spices & Medicinal Plants* 3 (2): 55–83. https://doi.org/10.1300/J044v03n02_08.

Palmero, Paola, Lien Lemmens, Marc Hendrickx, and Ann Van Loey. 2014. "Role of Carotenoid Type on the Effect of Thermal Processing on Bioaccessibility." *Food Chemistry* 157: 275–282. https://doi.org/10.1016/j.foodchem.2014.02.055.

Pino, Jorge, Marilú González, Liena Ceballos, Alma Rosa Centurión-Yah, Jorge Trujillo-Aguirre, Luis Latournerie-Moreno, and Enrique Sauri-Duch. 2007. "Characterization of Total Capsaicinoids, Colour and Volatile Compounds of Habanero Chilli Pepper (*Capsicum Chinense* Jack.) Cultivars Grown in Yucatan." *Food Chemistry* 104 (4): 1682–1686. https://doi.org/10.1016/j.foodchem.2006.12.067.

Popov, Sergey V., Raisa G. Ovodova, Victoria V. Golovchenko, Galina Yu Popova, Feodor V. Viatyasev, Alexandre S. Shashkov, and Yury S. Ovodov. 2011. "Chemical Composition and Anti-Inflammatory Activity of a Pectic Polysaccharide Isolated from Sweet Pepper Using a Simulated Gastric Medium." *Food Chemistry* 124 (1): 309–315. https://doi.org/10.1016/j.foodchem.2010.06.038.

Pugliese, Alessandro, Monica Rosa Loizzo, Rosa Tundis, Yvonne O'Callaghan, Karen Galvin, Francesco Menichini, and Nora O'Brien. 2013. "The Effect of Domestic Processing on the Content and Bioaccessibility of Carotenoids from Chili Peppers (*Capsicum* Species)." *Food Chemistry* 141 (3): 2606–2613. https://doi.org/10.1016/j.foodchem.2013.05.046.

Ramos-Aguilar, Olivia P., José De Jesús Ornelas-Paz, Saul Ruiz-Cruz, Paul B. Zamudio-Flores, Braulio Cervantes-Paz, Alfonso A. Gardea-Béjar, Jaime D. Pérez-Martínez, Vrani Ibarra-Junquera, and Jaime Reyes-Hernández. 2015. "Effect of Ripening and Heat Processing on the Physicochemical and Rheological Properties of Pepper Pectins." *Carbohydrate Polymers* 115: 112–121. https://doi.org/10.1016/j.carbpol.2014.08.062.

Rao, Anjali, and Leticia Rao. 2007. "Carotenoids and Human Health." *Pharmacological Research* 55 (3): 207–216. https://doi.org/10.1016/j.phrs.2007.01.012.

Rao, Tadapaneni V. Ramana, Neeta B. Gol, and Khilana K. Shah. 2011. "Effect of Postharvest Treatments and Storage Temperatures on the Quality and Shelf Life of Sweet Pepper (*Capsicum annuum* L.)." *Scientia Horticulturae* 132 (1): 18–26. https://doi.org/10.1016/j.scienta.2011.09.032.

Reyes-Escogido, Maria, Edith G. Gonzalez-Mondragon, and Erika Vazquez-Tzompantzi. 2011. "Chemical and Pharmacological Aspects of Capsaicin." *Molecules* 16 (12): 1253–1270. https://doi.org/10.3390/molecules16021253.

Rodriguez-Amaya, Delia B. 2001. *A Guide to Carotenoid Analysis in Foods: Life Sciences.* Washington, DC: ILSI Press.

Rodríguez-Burruezo, Adrián, Maria del Carmen González-Mas, and Fernando Nuez. 2010a. "Carotenoid Composition and Vitamin A Value in Ají (*Capsicum baccatum* L.) and Rocoto (C. *pubescens* R. & P.), 2 Pepper Species from the Andean Region." *Journal of Food Science* 75 (8): S446–S453. https://doi.org/10.1111/j.1750-3841.2010.01795.x.

Rodríguez-Burruezo, Adrián, H. Kollmannsberger, J. Prohens, S. Nitz, and A. Fita. 2010b. "Comparative Analysis of Pungency and Pungency Active Compounds in Chile Peppers (*Capsicum Ssp.*)." *Bulletin UASVM Horticulture* 67 (1): 270–273. https://journals.usamvcluj.ro/index.php/horticulture/article/viewFile/4972/4627.

Rodríguez-Burruezo, Adrián, Jaime Prohens, María D. Raigón, and Fernando Nuez. 2009. "Variation for Bioactive Compounds in Ají' (*Capsicum baccatum* L.) and Rocoto (*C. pubescens* R. & P.) and Implications for Breeding." *Euphytica* 170 (1): 169–181. https://doi.org/10.1007/s10681-009-9916-5.

Salvia-Trujillo, Laura, C. Qian, Olga Martín-Belloso, and David J. McClements. 2013. "Influence of Particle Size on Lipid Digestion and β-Carotene Bioaccessibility in Emulsions and Nanoemulsions." *Food Chemistry* 141 (2): 1475–1480. https://doi.org/10.1016/j.foodchem.2013.03.050.

Schweiggert, Ute, Dietmar R. Kammerer, Reinhold Carle, and Andreas Schieber. 2005a. "Characterization of Carotenoids and Carotenoid Esters in Red Pepper Pods (*Capsicum annuum* L.) by High-Performance Liquid Chromatography/Atmospheric Pressure Chemical Ionization Mass Spectrometry." *Rapid Communications in Mass Spectrometry* 19 (18): 2617–2628. https://doi.org/10.1002/rcm.2104.

Schweiggert, Ute, Andreas Schieber, and Reinhold Carle. 2005b. "Inactivation of Peroxidase, Polyphenoloxidase, and Lipoxygenase in Paprika and Chili Powder after Immediate Thermal Treatment of the Plant Material." *Innovative Food Science and Emerging Technologies* 6 (4): 403–411. https://doi.org/10.1016/j.ifset.2005.05.001.

Shepherd, Richard. 1989. "Factors Influencing Food Preferences and Choice." In R. Shepherd (Ed.), *Handbook of the Psychophysiology of Human Eating*. Chichester: Wiley, 3–24.

Silva, Luis, Jessica Azevedo, María J Pereira, Patricia Valentão, and Paula B. Andrade. 2013. "Chemical Assessment and Antioxidant Capacity of Pepper (*Capsicum annuum* L.) Seeds." *Food and Chemical Toxicology* 53 (1): 240–248. https://doi.org/10.1016/j.fct.2012.11.036.

Topuz, Ayhan, Cuneyt Dincer, Kubra Sultan Özdemir, Hao Feng, and Mosbah Kushad. 2011. "Influence of Different Drying Methods on Carotenoids and Capsaicinoids of Paprika (Cv.; Jalapeno)." *Food Chemistry* 129 (3): 860–865. https://doi.org/10.1016/j.foodchem.2011.05.035.

Topuz, Ayhan, and Feramuz Ozdemir. 2007. "Assessment of Carotenoids, Capsaicinoids and Ascorbic Acid Composition of Some Selected Pepper Cultivars (*Capsicum annuum* L.) Grown in Turkey." *Journal of Food Composition and Analysis* 20 (7): 596–602. https://doi.org/10.1016/j.jfca.2007.03.007.

Trumbo, Paula, Allison A. Yates, Sandra Schlicker, and Mary Poos. 2001. "Dietary Reference Intakes." *Journal of the American Dietetic Association* 101 (3): 294–301. https://doi.org/10.1016/S0002-8223(01)00078-5.

Tundis, Rosa, Federica Menichini, Marco Bonesi, Filomena Conforti, Giancarlo Statti, Francesco Menichini, and Monica R. Loizzo. 2013. "Antioxidant and Hypoglycaemic Activities and Their Relationship to Phytochemicals in *Capsicum Annuum* Cultivars During Fruit Development." *LWT—Food Science and Technology* 53 (1): 370–377. https://doi.org/10.1016/j.lwt.2013.02.013.

Ugás, Roberto. 2012. "Clasificación de Los Ajíes Del Perú." In Roberto Ugás and Víctor Mendoza (Eds.), *El Punto de Ají*. Lima: Programa de Hortalizas, UNALM & Programa VLIR-UNALM, 8–17. www.lamolina.edu.pe/hortalizas/webdocs/PUNTODEAJI.pdf.

Valderrama, Mariano, and Roberto Ugás. 2009. *Ajíes Peruanos—Sazón Para El Mundo*. Edited by Sociedad Peruana de Gastronomía. Lima: Empresa Editora El Comercio S.A., 1st edition.

van Zonneveld Maarten, Marleni Ramirez, David E Williams, Michael Petz, Sven W Meckelmann, Teresa Avila, Carlos Bejarano, Llermé Ríos, Karla Peña, Matthias Jäger, Dimary Libreros, Karen Amaya, and Xavier Scheldeman. 2015. "Screening Genetic Resources of *Capsicum* Peppers in Their Primary Center of Diversity in Bolivia and Peru." *PLoS One* 10 (9): e0134663. https://doi.org/10.1371/journal.pone.0134663.

Victoria-Campos, Claudia I., José De Jesús Ornelas-Paz, Olivia P. Ramos-Aguilar, Mark L. Failla, Chureeporn Chitchumroonchokchai, Vrani Ibarra-Junquera, and Jaime D. Pérez-Martínez. 2015. "The Effect of Ripening, Heat Processing and Frozen Storage on the in Vitro Bioaccessibility of Capsaicin and Dihydrocapsaicin from Jalapeño Peppers in Absence and Presence of Two Dietary Fat Types." *Food Chemistry* 181: 325–332. https://doi.org/10.1016/j.foodchem.2015.02.119.

Victoria-Campos, Claudia I., José de Jesús Ornelas-Paz, Elhadi M. Yahia, and Mark L. Failla. 2013. "Effect of the Interaction of Heat-Processing Style and Fat Type on the Micellarization of Lipid-Soluble Pigments from Green and Red Pungent Peppers (*Capsicum annuum*)." *Journal of Agricultural and Food Chemistry* 61 (15): 3642–3653. https://doi.org/10.1021/jf3054559.

Wang, Hui, Qian Zhang, Arun S. Mujumdar, Xiao Ming Fang, Jun Wang, Yu Peng Pei, Wei Wu, Magdalena Zielinska, and Hong Wei Xiao. 2020. "High-Humidity Hot Air Impingement Blanching (HHAIB) Efficiently Inactivates Enzymes, Enhances Extraction of Phytochemicals and Mitigates Brown Actions of Chili Pepper." *Food Control* 111: 107050. https://doi.org/10.1016/j.foodcont.2019.107050.

Xavier, Ana A. O., and Antonio Pérez-Gálvez. 2015. "Peppers and Chilies." *Encyclopedia of Food and Health*: 301–306. https://doi.org/10.1016/B978-0-12-384947-2.00533-X.

Zewdie, Yayeh, and Paul W. Bosland. 2001. "Capsaicinoid Profiles Are Not Good Chemotaxonomic Indicators for *Capsicum* Species." *Biochemical Systematics and Ecology* 29 (2): 161–169. https://doi.org/10.1016/S0305-1978(00)00041-7.

Zhang, Zhongyuan, Qiuyu Wei, Meimei Nie, Ning Jiang, Chunju Liu, Chunquan Liu, Dajing Li, and Lang Xu. 2018. "Microstructure and Bioaccessibility of Different Carotenoid Species as Affected by Hot Air Drying: Study on Carrot, Sweet Potato, Yellow Bell Pepper and Broccoli." *LWT—Food Science and Technology* 96: 357–363. https://doi.org/10.1016/j.lwt.2018.05.061.

Chapter 7

Characterization and Preservation of the Bioactive Compounds of Sacha Inchi (*Plukenetia volubilis* and *P. huayllabambana*) Oils

Nancy Chasquibol Silva, Rafael Alarcón Rivera, Raquel B. Gómez-Coca, Wenceslao Moreda, and M. Carmen Pérez-Camino

CONTENTS

7.1 INTRODUCTION

People have long enjoyed consuming what have now come to be known as superfoods. Superfoods are mainly seeds and fruits, many of them unknown to a vast majority of consumers because they belong to a specific and restricted area of the planet. Nevertheless, today these foods are promoted, added to salads,

DOI: 10.1201/9781003087618-7

yogurts, or drunk as infusions to maintain a healthy lifestyle. Consumers can find them in gourmet stores, specialized dietary stores, seed shops, and on specialty shelves in large supermarkets.

Among the superfoods, there is an important variety of native Andean crops: legumes and grains, such as kiwicha, quinoa, kañiwa, and giant corn; roots, such as maca and yacon; and vegetables, such as the family of the *opuntia* or the *muña*. They all fall within the denomination of superfoods, as they have a high protein content with an adequate balance of amino acids. They are also important crops for their high resistance and adaptability to environmental conditions. Most of them have high nutritional value and are present in the daily diet of Andean populations, who cultivate appropriate ecotypes from 100 to 4300 m above sea level. Superfoods also have strong social implications, supporting small and medium-sized farms and promoting the sustainable management of natural resources.

There are also some species with oleo potential, which could be included within the superfood group such as sacha inchi seeds. *Sacha inchi* is a Quechua term that means "wild peanut." Sacha inchi seeds belong to the Euphorbiaceae family and to the *Plukenetia* genus. It is a climbing plant first identified in 1753 by the naturalist Linnaeus, and 12 species have since been described to date in South and Central America (Bussmann et al. 2009). Special mention should be made of the taxa *P. brachybotrya, P. loretensis, P. volubilis, P. polyandenia*, and *P. huayllabambana*. Among them, *P. volubilis* is the most widespread.

P. huayllabambana is not as common as *P. volubilis*. It grows in the wild in the Rodriguez de Mendoza, department of Amazonas-Peru, at about 1295 m above sea level. It has larger seeds than *P. volubilis* (Rodríguez et al. 2010) and needs live tutors to grow, as shown in Figure 7.1. It has a dehiscent capsular

Figure 7.1 Sacha inchi (*Plukenetia huayllabambana*) plant, grains, almonds, oil, and protein.

fruit that usually consists of 2, 4, 5, and up to 7 lobes where the seeds are found. It is green and turns brown when harvested.

7.2 PRODUCTION AND USES OF SACHA INCHI OILS

In native communities, sacha inchi seeds have traditionally been used for human consumption. The seeds are lightly roasted and consumed as such. At present, outside these communities, sacha inchi seeds have become popular as snacks. It is possible to find these Amazonian seeds in large supermarket chains specializing in health food. They are presented as peanuts (natural, with sea salt, caramelized, dark-chocolate coated, etc.), although there is a significant difference: their higher content in amino acids (AA). Sacha inchi is unusually high in tryptophan (44 mg/g of protein), an amino acid essential to the production of serotonin, a chemical in the nervous system involved in regulating the appetite (Sathe et al. 2002). It is also rich in cysteine, tyrosine, and threonine. Because of its special AA content, it has been suggested that it could complement or even replace the protein intake derived from meat products.

The oil obtained from the extraction process, usually by cold pressure of the mature seed, is, at present, the principal way of consuming sacha inchi. This oil is the main component of the seed, with around 35–60 wt%, depending on the taxa, and it is normally eaten raw (virgin). Virgin oil from the *P. volubilis* species, the one with the highest production, has a yellow to amber color, and is the most commonly bottled and marketed.

Additional uses of the extracted oil other than in food have been reported. In the pharmaceutical industry, it is dispensed as capsules, vitamin supplements, ointment, and cosmetics. Sacha inchi oils restructure and protect the skin, hair, and nails, limiting dehydration and strengthening and rebuilding their natural barriers. For all those reasons, it is highly appreciated in cosmetics (Mosquera et al. 2012). This broadens the market niche for farmers, since natural products in the cosmetic industry are becoming more relevant and of higher importance for consumers. The defatted cake, or solid residue that remains after obtaining the oil, has a high protein content (>50%) and is mainly used for flour but also a nutritional supplement.

The interest in sacha inchi oils lies mainly in the fact that it is one of the few vegetable oils with a considerably high content in α-linolenic (Ln) and linoleic acids (L), which are essential fatty acids for humans. They cannot be synthesized in the body; therefore, they must be obtained directly from food. In addition to its content in essential fatty acids, sacha inchi oils are highly valued for their taste. Unlike other similar polyunsaturated seed oils such as chia or linseed oils and most edible vegetable oils coming from seeds, sacha inchi oils, with their appreciated flavor, are considered a gourmet oil and are consumed virgin, with no other processing than that of extraction by cold pressing.

7.3 SAFETY AND BENEFITS OF SACHA INCHI OIL CONSUMPTION

The Amazon basin, in Peru, Ecuador, and Colombia, has centuries-old evidence that supports the nutritional and therapeutic properties of daily life products from its biodiversity, particularly of the sacha inchi, where it is native. However, those health benefits have not been sufficiently proven in the rest of the world, where the consumption of sacha inchi oils is relatively recent. An exception is China, where sacha inchi seeds were introduced in 2006 and are now growing in the Xishuangbanna area. Countries such as Laos and Thailand also have begun the cultivation of sacha inchi plants (Liu et al. 2014).

Opportunities for sacha inchi oil in the international markets are driven by the health and wellness trend. Sacha inchi oils are highly valued for their omega fatty acid composition (ω-3 and ω-6), which meet the expectations of consumers searching for healthy oils. They help to counteract irritable colon and fatty liver problems, favoring the reduction in low-density lipoproteins (LDL), or "bad" cholesterol, in the blood; stimulating the increase in high-density lipoproteins, or (HDL) "good" cholesterol; and regulating blood pressure, preventing myocardial infarction and arterial thrombosis (Mozaffarian and Wu 2011). In people with hypertriglyceridemia, they have been shown to induce reductions of up to 30.9% (Kastelein et al. 2014).

Some of the first investigations on the health benefits of sacha inchi oils were carried out by Huamán et al. (2008), who conducted a postprandial study on young adults from 18 to 25, who ingested 50 grams of sacha inchi *P. volubilis* oil. This work demonstrated that the consumption of this oil reduces the levels of triglycerides, total cholesterol, and LDL cholesterol, and increases HDL cholesterol levels compared to an ingestion of olive oil.

Gorriti et al. (2010), working with rats which were administered sacha inchi oil for 60 days in amounts of 0.5 mL per kg, concluded that there is no toxicity in the animals when the oil is included in their diet. The biochemical results concur with those achieved by Huamán et al. (2008), in humans. In patients with hiperlipoproteinemia, it was determined that sacha inchi oil produces a decrease in the average values of total cholesterol and non-esterified fatty acids with an elevation of HDL cholesterol in groups that, for four months, consumed 5 and 10 mL per day of a suspension containing 2 g/5 mL of sacha inchi oil (Garmendia et al. 2011). Many other studies were focused on the investigation of the acceptability, safety, and health-promoting effects of oral administration of sacha inchi oils (Gonzales and Gonzales 2014). Thus, Alayón et al. (2016) updated the literature published over a 15-year period (2000–2015) on the composition and dietary consumption of sacha inchi seeds and oil, reflecting on the advantages of sacha inchi intake. Wang et al. (2018) reviewed the chemical composition and biological activity of different parts, seeds, and leaves of the sacha inchi *P. volubilis* plant.

The polyunsaturated fatty acids EPA (eicosapentaenoic fatty acid, C20:5) and DHA (docosahexaenoic acid, C22:6) from fish oils, have clearly shown to improve lipid profiles in humans, enhancing cardiovascular health (Nasiff-Hadad and Meriño-Ibarra 2003). However, their flavor hinders the acceptance of these oils into a regular diet. There is continuous research for a substitute vegetable oil rich in ω-3 fatty acids, as is the case of oils rich in α-linolenic acid (ALA). However, the conversion of ALA to EPA and EPA to DHA in humans is limited (0.2%–8% of ALA is converted to EPA and 0%–4% of ALA to DHA). Nevertheless, there are studies that support that ALA by itself benefits cardiovascular activity. Sacha inchi oil is offered as a possible substitute for fish oils and there are studies recommending these oils as good ω-3-rich supplements for populations with no access to marine sources (Mozaffarian and Wu 2011).

7.4 CHARACTERIZATION OF THE SACHA INCHI OILS

Regardless of the taxa, sacha inchi oil is one of the most unsaturated edible oils, with a content of polyunsaturated fatty acids (PUFAs) of >75%. The PUFA fraction comprises high levels of α-linolenic and linoleic essential fatty acids. Around 50% of the oil is α-linolenic acid (C18:3, ω-3, ALA, Ln), and about 40% is linoleic acid (C18:2, ω-6, L). The monounsaturated (MUFA) fraction is mainly formed of 8.4%–10.7% of oleic acid (C18:1, ω-9, O). The saturated fatty acids are composed of palmitic (C16:0, 4.7%–5.7%) and stearic acid (C18:0, 3.0%–3.7%; Guillén et al. 2003; Bondioli et al. 2006). Table 7.1 summarizes the main fatty acids in sacha inchi *P. volubilis* and *P. huayllabambana* oils compared with, chia, linseed, sunflower, safflower, and olive oils. As observed, sacha inchi oils are similar to chia and linseed oils regarding α-linolenic acid, but not to oleic or linoleic acids.

Exhaustive studies on the composition of the oils from the *P. volubilis* have been published, which show its physicochemical characteristics, quality, and purity parameters (Chirinos et al. 2015). The oils obtained by the cold extraction process have been characterized by both the glyceridic and non-glyceridic fractions of *P. volubilis* and *P. huayllabambana*. The oils were obtained from seeds of Rodriguez de Mendoza, department of Amazonas-Peru, during consecutive harvests. The results proved the narrow variability of the data obtained for each analytical parameter from one harvest to the next (Chasquibol et al. 2014).

Regarding the major sacha inchi oil components, the molecular species of triacylglycerols (TG) were studied in addition to the fatty acid composition for its characterization. The TG were analyzed by high performance liquid

TABLE 7.1 MAIN FATTY ACIDS (PERCENTAGE OVER TOTAL) OF THE SACHA INCHI (*P. volubilis* AND *P. huayllabambana*) OILS COMPARED TO OTHER VEGETABLE OILS

Vegetable oils	Fatty acid composition (%)					Reference
	Palmitic C16:0	Stearic C18:0	Oleic C18:1	Linoleic C18:2	Linolenic C18:3	
Sacha inchi (*P. volubilis*)	3.9	3.9	9.9	34.4	47.9	Chasquibol et al. (2014)
Sacha inchi (*P. huayllabambana*)	1.3–1.7	4.3–5.7	7.6–9.9	24.5–27.3	55.5–60.4	
Chia oil	7.0–7.1	2.8–3.2	7.3–10.5	18.9–20.4	59.7–63.8	Kulczyński et al. (2019)
Linseed oil	5.7–6.8	4.5–5.0	20.5–24.0	14.0–18.2	46.6–54.2	Berto et al. (2020) Herchi et al. (2014)
Sunflower oil	6.2–7.5	2.9–3.9	23.2–36.4	52.3–64.3	Nd-0.2	Uncu et al. (2019)
Safflower oil	6.7–6.9	2.5–3.0	11.9–16.4	75.7–77.1	0.16–0.17	
Olive oil	7.5–20.0	0.5–5.0	55.0–83.0	3.5–21.0	<1	IOC (2019)

chromatography coupled to a refractive index detector (HPLC-RI; Table 7.2). It was determined that the molecular species of triacylglycerols with even equivalent carbon numbers (ECN) between 36 and 42 were the main ones, with those formed by two Ln and one L acid molecules, namely dilinolenoyl-linoleoylglycerol (LnLnL; ECN = 38) as the majority (22.6%–27.7%), followed by the groups which conform to the ECN = 40 and 42 (Chasquibol et al. 2014). Fanali et al. 2011) reported similar results for *P. volubilis* using a mass spectrometry detector (HPLC-MS). Although sacha inchi oils contain α-linolenic as the main fatty acid, its trilinolenin (LnLnLn, ECN = 36; 17.7%–23.0% and 12.3% for *P. huayllabambana and P. volubilis*, respectively) is not one of the major triacylglycerols as occurs in chia or linseed oils (Ixtaina et al. 2011; Herchi et al. 2014).

In addition to their special triacylglycerol profile, sacha inchi oils contain a fraction of unsaponifiable components of high biological interest. Generally, the unsaponifiable matter of edible vegetable edible oils (2%–3%) comprises a group of compounds present in extremely low quantities. Despite being minor compounds, it is especially important since some of the major components are the phytosterol and the tocopherol families, known to display significant health benefits.

As in all vegetable oils, phytosterols and tocopherols are also the main components of the unsaponifiable matter in sacha inchi oils and their occurrence is of great qualitative and quantitative importance. Together with a large amount of total sterols (>1500 mg/kg; Table 7.2), sacha inchi oils have a particular sterol profile, analyzed by split-gas chromatography coupled to a flame ionization detector (GC-FID; Chirinos et al. 2013; Chasquibol et al. 2016).

Unlike other edible vegetable oils, the main sterol after β-sitosterol is stigmasterol instead of campesterol (Chasquibol et al. 2014). The typical gas-chromatogram profile of the sterols corresponding to a genuine virgin sacha inchi oil includes the two main peaks of the above-mentioned compounds, β-sitosterol (~60%) and stigmasterol (~30%) and a series of small peaks corresponding to cholesterol (~0.3%), campesterol (~5%), Δ7-campesterol (~0.5%), clerosterol (~0.8%), sitostanol (~0.6%), Δ5-avenasterol (~1.5%), Δ5,24-stigmastadienol (~0.3%), Δ7-stigmastenol (~0.6%), and Δ7-avenasterol (~0.3%).

Tocopherols are important members of the family of the non-glyceridic minor compounds that belong to the unsaponifiable matter of sacha inchi oils. Variations found in their quantities could correspond to oil extraction technology, crop cultivation conditions, or genetic factors, and even the analysis procedure (Kodahl 2020). The most widespread procedure for their determination is by direct injection of a hexane oil solution in a liquid chromatograph provided with a fluorescence detector, setting the wavelengths at 290 and 330 nm for excitation and emission, respectively. The major molecular species of

TABLE 7.2 TRIACYLGLYCEROL PROFILE AND UNSAPONIFIABLE COMPONENTS OF GENUINE AND COMMERCIAL SACHA INCHI (UNSPECIFIED VARIETY) OILS

Determination	Genuine Sacha inchi oils		Commercial Sacha inchi oils	References
	P. huayllabambana	*P. volubilis*		
Triacylglycerols (%)				
ECN 36 (LnLnLn)	17.7–23.0	12.3	1.4–17.7	Fanali et al. (2011), Chasquibol et al. (2016), Ramos-Escudero et al. (2019)
ECN 38 (LnLnL)	22.6–27.7	22.2	2.8–23.5	
ECN 40 (LnLL + LnLnP + LnLnO)	19.2–23.1	28.7	5.5–23.8	
ECN 42 (LLL + OLLn + PLLn + SLnLn)	19.0–24.0	21.1	13.5–27.1	
ECN 44 (LLO + OOLn + LLP + POLn + SLLn)	6.0–10.5	11.7	8.6–27.1	
ECN 46 (OOL + LLS + LnOS + PLO)	1.4–5.0	3.5	3.6–27.3	
ECN 48 (OOO + SLO + POO)	0.1–2.7	0.5	0.8–25.4	
ECN 50 (SSO)	0–1.0	–	0.3–3.1	
Sterols (%)				
Cholesterol	0.2–0.4	0.3	0.1–0.4	Bondioli et al. (2006), Chirinos et al. (2015)
Campesterol	4.4–5.6	7.1–8.8	5.1–18.9	
Stigmasterol	26.5–32.2	21.2–27.9	10.4–26.5	
β-Sitosterol	57.9–63.1	45.2–56.5	50.6–66.2	
Total ($mg \cdot kg^{-1}$)	1653–2236	2472	2001–2950	

(Continued)

TABLE 7.2 (CONTINUED)

Tocopherols (%)				
α	0.2	<0.1–1.3	0.1–15.8	Chasquibol et al. (2014), Chririnos et al. (2015), Chasquibol et al. (2016)
β	<0.1	<0.1–0.9	nd-1.3	
γ	53.6–70.0	49.0–81.4	49.4–67.8	
δ	30.0–46.4	12.6–54.6	24.7–51.6	
Total (mg · kg^{-1})	1410–2978	2112–2218	624–3332	
Aliphatic hydrocarbons (%)				
H21:0	10.2–14.0	–	7.1–11.6	Chasquibol et al. (2014), Chasquibol et al. (2016)
H22:0	10.0–12.6	–	6.9–14.0	
H23:0	36.4–41.8	–	18.0–40.5	
H24:0	17.9–19.7	–	9.2–22.5	
H25:0	13.7–16.3	–	9.0–19.1	
Total (mg · kg^{-1})	114.8–170.7	–	21.7–131.3	
Alcohols (%)				
Lineals A22:0+A24:0+A26:0	–	–	0.32–7.62	Ramos-Escudero et al. (2019)
Diterpenic (phytol)	–	–	0.10–43.51	
Triterpenics	–	–	50.43–96.8	
Total (mg · kg^{-1})	–	–	881.9–3987.0	

tocopherols found in sacha inchi oils are γ-tocopherol followed by δ-tocopherol, with percentages of 53.6%–70.0% and 49.0%–81.4% for γ-tocopherol and 30.0%–46.4% and 12.6%–54.6% for δ-tocopherol of the total tocopherol contents in *P. huayllabambana* and *P. volubilis* oils, respectively (Table 7.2; Chirinos et al. 2015; Chasquibol et al. 2016).

Aliphatic hydrocarbons conform a remarkable group of minor unsaponifiable compounds present in edible fats and oils and are considered as intermediate metabolites in the formation of other minor compounds. They are determined by on-column GC-FID and have a specific profile for each edible oil and, in the particular case of sacha inchi, are formed by the saturated series of odd and even carbon atom numbers that range from 21:0 to 29:0, with 23:0 hydrocarbon as the main one (36.4%–41.8% for *P. huayllabambana* oils; Chasquibol et al. 2014).

Ramos-Escudero et al. (2019) provided data on the composition of alcohols corresponding to 27 commercial sacha inchi oils (Table 7.2). The alcoholic fraction is determined by GC-FID and is divided into the aliphatic alcohols with 22:0, 24:0, 26:0 and 28:0 carbon atoms, the diterpenic alcohols phytol and geranilgeraniol, and the triterpenics or 4,4-dimethylsterols formed by β-amirine, butirospermol, cicloartenol, and 24-methylencicloartenol.

7.5 QUALITY AND REGULATION

As a result of all the research work performed, sacha inchi oil is present in new markets far from its origin. In the United States, it is now a GRAS food (generally recognized as safe). In Europe, it is available in specialized stores, and it is at present in the European Union's list of "novel foods." Up to now, the Codex Alimentarius does not include sacha inchi among edible oils, but the Food Safety Authority of Ireland recognized it as substantially equivalent to linseed oil, which is a common vegetable oil in Europe.

Currently, sacha inchi oils have a regulation (NTP 2018) based on the one published in 2009 (NTP 151,400 2009), which is the Peruvian Technical Standards Regulation. It contains the physicochemical parameters that the oil extracted from the *Plukenetia* genus seeds must meet when intended for human consumption. This is the only standard pertaining to sacha inchi oils in the world to date and is mandatory only in Peru. It includes all the analytical quality and purity parameters the oils from *P. volubilis* and *P. huayllabambana* species must specifically meet to be classified as "extra virgin" or "virgin" oil. Table 7.3 summarizes the parameters, the established values, and the testing method to be used, which are usually ISO (International Organization for Standardization) or AOCS (American Oil Chemists' Society) approaches.

TABLE 7.3 PHYSICOCHEMICAL QUALITY AND PURITY REQUIREMENTS FOR SACHA INCHI (*P. volubilis* AND *P. huayllabambana*) OILS

Requirement	Minimum		Maximum		Testing Method
	P. Volubilis	*P. huayllabambana*	*P. volubilis*	*P. huayllabambana*	
Density at 20 °C	0.926	0.926	0.931	0.931	ISO 6883
Iodine Index (HANUS solution)	183	183	199	199	AOAC 920.158
Saponification index	192	192	196	196	AOAC 920.160
Refractive index at 20 °C	1.478	1.478	1.481	1.481	AOCS Cc 7–25
Unsaponifiable matter (%)	–	–	0.36	0.36	ISO 3596
Free acidity (% oleic) Extra virgin oil Virgin oil	–	–	1.0 2.0	1.0 2.0	ISO 660 AOCS Ca 5a-40
Peroxide Value (mEqO_2/Kg oil)	–	–	10.0	10.0	ISO 3960 AOCS Cd 8b-90
Humidity and Volatile matter (%)	–	–	0.14	0.14	ISO 662 AOCS Ca 2b-38
Impurities (%)	–	–	0.02	0.02	ISO 663 AOCS Ca 3a-46
Fatty acid composition (%) Oleic acid Linoleic acid Linolenic acid	8.9 32.1 44.7	7.9 24.0 55.0	–	–	–
Tocopherols	1900	1900	–	–	

7.6 GENUINENESS OF SACHA INCHI OILS

Sacha inchi seeds grow in very few areas in the world and the oils are usually extracted by careful cool pressing, and consequently they are relatively expensive. Because of their high cost and valuable properties, their quality and authenticity must be certified. It is especially important to guarantee that oils marketed are authentic sacha inchi oils. Studies conducted with genuine oils, obtained directly from seeds collected and processed, have shown that their compositions comply with the regulatory provisions, regardless of the harvest.

When the quality and authenticity of commercialized oils labeled as "extra virgin sacha inchi oil" were studied, it was demonstrated that some of those oils did not fulfill the basic requirement established in the regulations (NTP 151,400 2009, 2018; Chasquibol et al. 2016). Even more, a substantial amount of sacha inchi oils sold in the market contained other cheaper oils. This was determined through various analytical tests, including fatty acid composition and sterol profile. The first and more basic parameter to be determined is the α-linolenic acid content, which must be higher than 44.7% and 55.0% in the cases of *P. volubilis* and *P. huayllabambana*, respectively. The profile of the fatty acid methyl esters (FAMEs) of commercialized oils revealed that some samples did not comply with the percentages required for ω-3, ω-6, and ω-9 fatty acids. Even more, in some commercial oils, *trans* isomers of fatty acids were present in quantities much higher than those expected for crude or virgin oils, which indicates the use of heat in some of the steps of the extraction process, which is not allowed when obtaining virgin oils. Recently, Ramos Escudero et al. (2019), when analyzing a considerable number of commercialized sacha inchi oils, verified that a good deal of these oils failed to meet NTP 2018 requirements regarding their fatty acid compositions.

Sterols, determined by GC, also help to detect adulteration since in genuine sacha inchi oils the stigmasterol/campesterol ratio is higher than that obtained for most other edible vegetable oils. The study of this ratio from a huge number of oils led Chasquibol et al. (2016), to stablish that for authentic sacha inchi oils, of any of the species, the mentioned value must be higher than 4 (Bondioli et al. 2006; Chirinos et al. 2013, 2015).

Delta 5,23-stigmastadienol is a sterol which by GC elutes before and almost overlaps with clerosterol, closed to β-sitosterol. Although absent in crude or virgin oils, it is only formed during the refining process. Both Δ5,23-stigmastadienol and clerosterol are present in quantities <1% and most of the studies do not present the data.

The major animal sterol cholesterol is also present in vegetable oils, including sacha inchi oils, but in small quantities, ~0.5% of total sterols. There are a few edible vegetable oils which contain higher quantities, such as palm oils

(2%–7%). Genuine sacha inchi oils are among the edible vegetable oils with low percentages of this animal sterol.

In most cases, there is no certainty regarding the type of oil used in such adulteration, probably because of not going deep into this subject, but it is suspected that it is sunflower oil or even soybean oil because they are the cheapest oils for this matter. In some of the commercialized oils the presence of very characteristic compounds such as the flavons, sesamin, and sesamolin clearly indicates the presence of sesame oils. In other cases, as reported by Ramos-Escudero et al. (2019), the presence of brasicasterol in a few of their studied samples indicates that they were adulterated with rapeseed oil.

The regulation for sacha inchi oils also includes indications in relation to the total amount of tocopherols that these oils must contain, establishing a maximum of 1900 mg · kg^{-1}. There are no indications concerning the percentages of each one individually or even if some of them, α-, β-, γ-, or δ-tocopherol, are absent in the genuine oils (NTP 2009, 2018). The published studies on the individual composition of tocopherols indicate that the molecular species α- and β-tocopherol are not usually present and therefore, their occurrence in an important percentage is indicative of a mixture with another vegetable oil as can be observed for some commercial samples (Chasquibol et al. 2016).

The traditional testing methods combined with chemometric techniques are improving the characterization and authentication of sacha inchi oils and have been shown to present a good alternative. Thus, Ramos-Escudero et al. (2019) found significant differences between clusters for palmitic acid, oleic acid, γ-tocopherol, δ-tocopherol, campesterol, and stigmasterol and propose the study of these compounds as markers of authenticity in commercial sacha inchi oils.

In the same way, Maurer et al. 2012, proposed fast and robust methods such as spectroscopy. The authors used a temperature-controlled ZnSe ATR-MIR portable spectrometer and studied the results by a precise mathematical approach to characterize and authenticate sacha inchi oils.

7.7 PRESERVATION OF SACHA INCHI OILS

As a good source of essential fatty acids, (ω-3Ln; ω-6L), antioxidants and phytosterols, sacha inchi oils are ideal for preparing foods for people with special nutritional requirements (athletes, malnourished children, or people with cardiac disease). There is sufficient evidence about the benefits of ω-3 fatty acids to human health, (Alayón et al. 2016; Garmendia et al. 2011; Kastelein et al. 2014; Mozaffarian and Wu 2011) and the content of ω-3 fatty acids in sacha inchi oils is among the highest. Sacha inchi oils are normally consumed raw in salads, but for a better intake, the oils could be included in energy drinks, infant formulations, snacks, and bakery products, and so on, and some of the most recent research is headed in that direction. However, the special fatty acid composition of sacha inchi oils makes them vulnerable to oxidation. They easily develop an unpleasant smell, which affects the organoleptic characteristics of the foods containing them, as well as their commercialization period. Their oxidative stability index, determined by Rancimat at 100 °C was 1.6–3.4 hours (Chasquibol et al. 2014; Rodríguez et al. 2015). Their estimated shelf life at usual storage temperatures of 30 °C, 25 °C, and 20 °C was 0.79, 1.79, and 3.29 years, respectively (Rodríguez et al. 2015).

In the literature, some processes are proposed to increase oil stability, protecting their bioactive components and increasing their resistance during conservation and food processes, such as pasteurization, sterilization, baking, or frying, with minimal alterations. In the specific case of sacha inchi, the main proposed process is the roasting of the seeds before extraction, the addition of antioxidant extracts and microencapsulation of the oils.

7.7.1 The Roasting Process of Sacha Inchi Seeds

To minimize the negative effect of the polyunsaturated fatty acids on the oxidative stability of sacha inchi oils, it was proposed to roast the seeds before obtaining the oil. The roasting process consists of subjecting the seeds to high temperatures (<100 °C) during a short period of time (<10 min). Cisneros et al. (2014), studied the effect of roasting sacha inchi (*P. volubilis* L.) seeds on the stability and the composition of the oils. The authors established 75 °C and 9 minutes as the mildest conditions and minimum requirement to eliminate the astringent taste of the sacha inchi seeds. The authors concluded that the partial roasting of the seeds increased the resistance of the oil to oxidation, without affecting the fatty acid profile. Furthermore, their studies show that when oil is stored, there is a slower rate of oxidation from roasted seeds, compared to the unroasted ones and a higher tocopherol content remains in the oils from roasted seeds compared to those from unroasted seeds. Additionally, an increase in the antioxidant capacity of the sacha inchi oils coming from roasted seeds was observed when measured by the DPPH (2,2-diphenyl-1-picrylhydrazyl) assay, probably due to the formation of Maillard reaction products. Roasting has been reportedly responsible for the development of Maillard-derived compounds of a polar nature, which may present antioxidant activity (Pellegrini et al. 2003).

The effect of the roasting process on the variety *P. huayllabambana* at temperatures from 100 °C to 160 °C for time periods from 10 to 30 minutes were also tested by Cisneros et al. (2014). These authors determined the effect of the seeds' heat treatments on the fatty acids, tocopherols, phytosterols, phenolic compounds, free fatty acids, peroxide, and *p*-anisidine values of the seed oil. They demonstrated that for *P. huayllabambana* seeds, roasting temperatures higher than 120 °C produce a negative effect on the contents of the studied components, and that the treatment of 100 °C for 10 minutes maintained the bioactive compounds in the seeds and the quality of the oil by increasing its oxidative stability.

7.7.2 Sacha Inchi Oils Supplemented with Antioxidant Extracts

The most common strategy to improve the oxidative stability of oils is the addition of antioxidant substances. Generally, they are synthetic compounds (butylated hydroxyanisole [BHA], butylated hydroxytoluene [BHT], tertiary butylhydroquinone [TBHQ], etc.), although they can also be natural plant extracts which inhibit or delay oil oxidation through different, well-studied mechanisms. Thus, they eliminate free radicals, chelate pro-oxidant metals, or quench singlet oxygen (Choe and Min 2009).

The synthetic antioxidant BHT and different extracts from tara (*Caesalpinia spinosa*) pods were added to sacha inchi (*P. volubilis*) oils and their effects were evaluated under accelerated conditions at 60 °C for 15 days (Herman et al. 2020). The oxidation of the sacha inchi oil was monitored by induction time, peroxide value, conjugated dienoic acid, *p*-anisidine value, total unsaturated fatty acids, and α-linolenic acid contents and also by the MIR spectroscopy coupled with chemometric tools. The results revealed that extracts from tara pods, mainly those partially hydrolyzed, were more efficient than BHT against oil oxidation for up to 7 days.

Castro et al. (2020) prepared several formulations containing mixtures of commercial antioxidants of a bio-based specialty polyether polyol (ascorbyl palmitate and propyl gallate) named Ecoprol 2020, and tocopherol, in order to find the optimal concentration to be added to sacha inchi oils. The authors found that an oil formulation with α-tocopherol (150 $mg \cdot kg^{-1}$ of oil) and Ecoprol 2020 (1000 $mg \cdot kg^{-1}$ of oil) displayed the lowest peroxide values (2.6 ± 0.1 $mEq\ O_2 \cdot kg^{-1}$ of oil) after 20 days of storage at 8 °C. Data are within those established in the regulation (NTP 151.400 2009), where the maximum for extra virgin sacha inchi oil is 10 $mEq\ O_2 \cdot Kg^{-1}$ of oil. Even more, it was able to reduce approximately 50% of fatty acid oxidation in relation to the control.

7.7.3 Microencapsulation of Sacha Inchi Oils

Microencapsulation is an effective technique which is widely used in the food industry to protect labile compounds from adverse conditions, such as

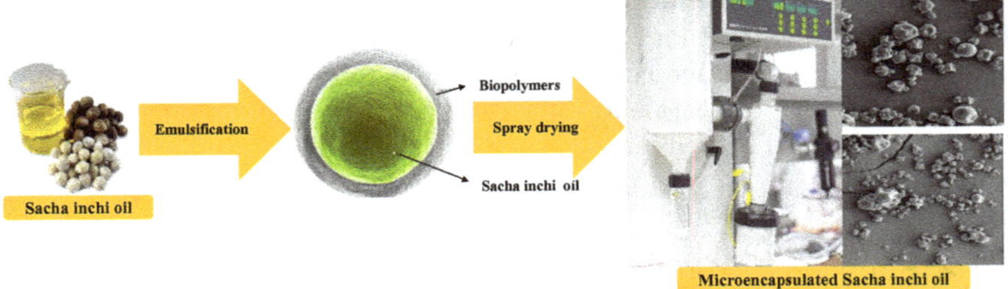

Figure 7.2 Microencapsulation of sacha inchi (*P. huayllabambana*) oil by spray-drying.

light, moisture, and oxygen, helping to maintain their biological and functional properties. The technique successfully emerged from the pharmaceutical industry during the last century and employs the novelties in the fields of materials and processes. The food industry took longer to incorporate this process, but today there are many food preparations that contain active principles or ingredients incorporated as microcapsules, such as baby milk preparations. In this context, sacha inchi oils were good candidates to be processed with the microencapsulation technique with the aim of being added to food while keeping their highly valuable compositions. The microencapsulation of sacha inchi oils is very recent, as research began in last decade but a great diversity of encapsulation materials is continually tested (Quispe-Condori et al. 2011; Pastuña-Pullutasig et al. 2016; Vicente et al. 2017; Sánchez-Reinoso and Gutiérrez 2017). Maltodextrin (MD) and Arabic gum (AG) are the most widely used coating materials, but new ones are also being tested alone or in binary and even ternary mixtures. Corn zein, whey protein isolate (WPI), ovalbumin, pectin, xanthan gum, Capsul and HI-CAP and their mixtures were also studied as coating materials in order to find the one with the best protection. The procedure for the oil encapsulation consists of obtaining a stable water:oil emulsified encapsulation material. Once this emulsion is obtained it is necessary to eliminate the water, frequently using spray-drying or freeze-drying (Figure 7.2).

The efficiency of the encapsulation process, the physicochemical characteristics of the microparticles, the particle size distribution and the morphology, along with oxidative stability, are the most frequent determinations to monitor the process. Chasquibol et al. (2019) determined how a lab-scale microencapsulation process with spray-drying and several wall materials affected the fatty acid composition as well as the minor unsaponifiable components, sterols, and tocopherols of sacha inchi oils, both the *P. huayllabambana* L. and *P. volubilis* L. species. The authors obtained the free and the encapsulated fractions of microencapsulated oils, and analyzed their fatty acid compositions separately in order to determine differences. With respect to the minor compounds, few losses were observed in the sterols with respect to the original oils for the wall materials assayed. Nevertheless, unexpected results were obtained when WPI was used, as a new peak appeared in the GC-FID of sterols. The study by GC-MS confirmed that it was cholesterol and quantitative results using α-cholestanol as internal standard indicated values higher than 1000 mg $\cdot$ kg^{-1}. WPI is an isolate that comes from an animal source and hence the presence of this animal sterol.

Regarding the tocopherols, quantities of 2660–4393 mg $\cdot$ kg^{-1} were present in the sacha inchi oils before microencapsulation. It was determined that for all the wall material studied, tocopherols suffered losses of 22%–30% during the encapsulation and the losses of γ-tocopherol were higher than δ-tocopherol (Chasquibol et al. 2019). Those results are in agreement with those reported for microencapsulated sunflower oils, where losses of 12.4%–29.1% were found (Holgado et al. 2019).

Recently, Alarcón et al. (2019) studied the effect of four wall materials (GA, HI-CAP, MD, and WPI) and two combinations (GA+MD and GA+MD+WPI) on the shelf life of sacha inchi *P. huayllabambana* and *P. volubilis* oils. For that purpose, the Rancimat equipment was used at several temperatures: 70 °C, 80 °C, 90 °C, and 100 °C. The extrapolations of the stabilities at 25 °C correspond, in theory, to the shelf life, which for microencapsulated oils was much longer than for oils without the application of this protecting process. The results also determined the best formulation of wall material to protect the sacha inchi oils and allow them to be used in preparing functional foods to be consumed at any time during their market period. Oxidative stability and induction periods confirm that the microencapsulation of sacha inchi oils extends the shelf life of the oils by up to 90% and 200% for *P. huayllabambana* and *P. volubilis*, respectively. Every wall material used in the approach protected the oils from the lipid oxidation process, with HI-CAP being the best one, followed by the ternary mixture of materials (GA+MD+WPI) for both ecotypes (Alarcón et al. 2019). This information is in agreement with that obtained by Landoni et al. (2020), who studied the physicochemical parameters and the oxidative stability of microencapsulated edible sacha inchi seed oil using spray-drying. The authors also obtained sacha inchi oil microcapsules from *P. volubilis* and *P. huayllabambana*, with different biopolymers as wall materials. The physicochemical characteristics such as encapsulation efficiency, particle size, morphology and oxidation onset temperature (OOT) were analyzed. The authors concluded that microcapsules made of 100% HI-CAP presented the best values for oxidative stability and the highest encapsulation efficiency (93.3%–96.5%) compared to the other formulations.

7.8 CREATIVE AND HEALTHY GASTRONOMIC PROPOSALS

Creative cuisine is fashionable, especially that which preserves the legacy of ancestral peoples since its main dishes are made with unique ingredients that go from the mountains to the coast. The most widely consumed dishes are prepared with a variety of herbs to provide unique flavors, and are served garnished with rice, fried potatoes, fresh salads, a variety of chili peppers, fried plantains, and sauces. Sacha inchi oils have become part of the new food preparations. The idea is to satisfy the needs of people who are looking for edible products for daily use which in turn favor their health. In that way, a mayonnaise containing sacha inchi oil as its main ingredient has been commercialized and its properties have been studied. This type of foods, thanks to its chemical composition, contributes to what people are seeking. The interest in these mayonnaises obtained from an oil rich in polyunsaturated fatty acids has increased and the producers are concerned because it may become rancid quickly (Rodríguez et al. 2015; Jacobsen et al. 2001; Di Mattia et al. 2015). Medina-Marroquín et al. (2018) studied the effect of the different parameters which may affect the sacha inchi mayonnaise during its storage and estimated its shelf life. The authors showed that peroxide values increased over time for all storage temperatures studied. Considering the limit of 7.96 $meqO_2 \cdot kg^{-1}$ oil established sensorially for mayonnaise, the estimated shelf life was 40, 30.4, and 12 days for storage temperatures of 12 °C, 22 °C, and 32 °C, respectively. The authors consider their data similar to the data obtained with other mayonnaise formulations made with soybean or sunflower oils.

Other formulations based on sacha inchi oils were also studied. Thus, Castro et al. (2020) developed a factorial method to prepare a fresh cheese enriched with quantities of 1%–4% of sacha inchi oil. In their formulations an adequate quantity equal to 2.5% of sacha inchi oils supplemented with antioxidants served to increase the cheese's shelf life from 7 to 16 days under refrigeration.

Currently, famous Peruvian chefs are developing a creative gastronomic proposal for healthy, nutritious, and functional foods with microencapsulated sacha inchi oils and products of biodiversity. Integrating science and gastronomy into their recipes, the CCori organization, led by its chef Palmiro Ocampo, prepares sweet and salty snacks with microencapsulated sacha inchi.

The rehydrated, powdered raw material serves as an energetic beverage as well. It was prepared at Universidad de Lima by extracting antioxidants from the peel of camu camu (*Myrciaria dubia* (HBK) McVaugh) and mango (*Mangifera indica*) fruits, and the extracts were microencapsulated together with sacha inchi (*P. huayllabambana*) oil and presented in a sachet. The powdered beverage contains 300 mg of ω-3 fatty acids per sachet. Its solubility is more than 90.7%, its bioavailability is above 65%, and its shelf life is around 6 months.

Figure 7.3 shows details of the new dishes and the powdered beverage which are prepared including microencapsulated sacha inchi oils as ingredients. To complete this project, fruit jellybeans were designed containing mango fruit pulp or red prickly pears and microcapsules of sacha inchi (*P. huayllabambana*) oil, as detailed in Figure 7.3. Each gram of fruit jellybeans contains 6 mg of ω-3, with no artificial colors or flavors, and is shown to have a shelf life of 3 months.

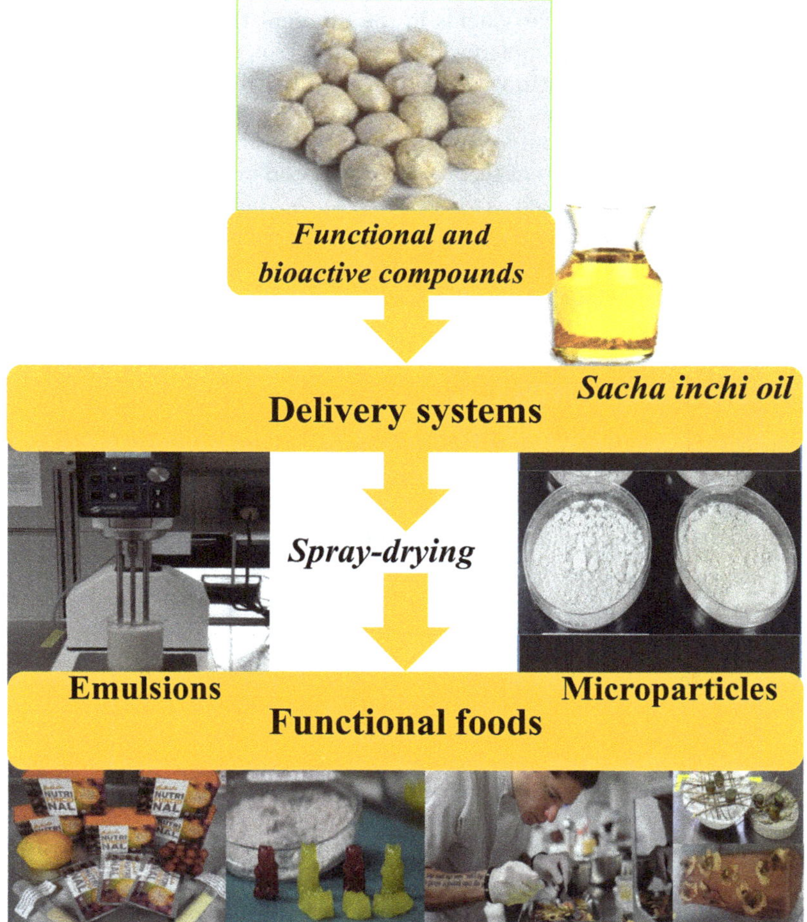

Figure 7.3 Functional foods with microencapsulated sacha inchi (*Plukenetia huayllabambana*) oil.

7.9 CONCLUSIONS

Sacha inchi (*P. volubilis* and *P. huayllabambana*) seeds yield oils with around 50% by the cold pressure process. The oil has a light-yellow color and a very pleasant fruity smell and has been consumed raw since ancient times, without any additional treatment. Research in animal models and also in humans not only demonstrates that they are harmless but that they provide important health benefits. From the point of view of their chemical composition, sacha inchi oils are rich in the essential fatty acid ω-3-linolenic (>44.7%) and ω-6-linoleic (>24%), which meets the expectations of people who are looking for a heart-healthy oil. The combination of their fatty acids in the different triacylglycerols, with the major molecular specie dilinolenoyl-linoleoylglycerol, LnLnL, makes them different from other polyunsaturated oils such as chia or linseed oil, where LnLnLn is the major one. Sacha inchi oils are also characterized by a high ratio between the sterols stigmasterol/campesterol, which is different for the majority of the edible vegetable oils. Different techniques have been proposed to protect sacha inchi oils from oxidation such as roasting the seeds, the addition of natural antioxidants, and microencapsulation, which guarantee a longer shelf life. Sacha inchi oils are currently involved in a gastronomic proposal for healthy, nutritious, and functional foods where the microencapsulation process is present.

ACKNOWLEDGMENTS

The authors thank the National Program of Agricultural Innovation (PNIA) of the Ministry of Agriculture and Irrigation, INNOVATE PERU of the Ministry of Production-Peru, Instituto de la Grasa-CSIC, Sevilla, Spain and the Universidad de Lima, Perú, for the financial support and for all facilities.

REFERENCES

Alarcón Rivera, Rafael, M. Carmen Pérez Camino, and Nancy Chasquibol Silva. 2019. "Evaluación de la Vida Útil de los Aceites de Sacha Inchi (*Plukenetia huayllabambana y Plukenetia volubilis*) Microencapsulados." *Revista de la Sociedad Química del Perú* 85 (3): 327–337. www.scielo.org.pe/scielo.php?script=sci_arttext&pid=S1810-634X2019000300005&lng=es&nrm=iso.

Alayón, Alicia Norma, and Isabella J. Echeverri. 2016. "Sacha Inchi (*Plukenetia volubilis* linneo): A wasted Ancestral Experience? Clinical Evidence Related to its Consumption." *Revista Chilena de Nutrición* 43 (2): 167–171. http://doi.org/10.4067/S0717-75182016000200009.

Berto, Bruna María, Rita Kasia Almeida Garcia, Gabriel D. Fernandes, Daniel Barrera-Arellano, and Gustavo Pereira. 2020. "Linseed Oil: Characterization and Study of Its Oxidative Degradation." *Grasas Aceites* 71 (1): e337. https://doi.org/10.3989/gya.1059182.

Bondioli, Paolo, Laura Della-Bella, and Petra Rettke. 2006. "Alpha Linolenic Acid Rich Oils. Composition of *Plukenetia volubilis* (Sacha Inchi) Oil from Peru." *Rivista Italiana Delle Sostanze Grasse* 83: 120–123. www.researchgate.net/publication/264514466_Alpha_linolenic_acid_rich_oils_Composition_of_Plukenetia_volubilis_Sacha_Inchi_oil_from_Peru.

Bussmann, Rainer W., Carolina Téllez, and Ashley Glenn. 2009. "*Plukenetia huayllabambana* sp. nov. (*Euphorbiaceae*) from the Upper Amazon of Peru." *Nordic Journal of Botany* 27: 313–315. https://doi.org/10.1111/j.1756-1051.2009.00460.x.

Castro, Juan Pablo, Carlos Francisco Vaca, Edson José Soto, and Tarsila Tuesta. 2020. "Sacha inchi oil (Plukenetia volubilis) stabilized with antioxidants for addition in fresh cheese". *African Journal of Food, Agriculture, Nutrition and Development* 16638–16651. https://doi.org/10.18697/ajfand.94.18595.

Chasquibol, Nancy A., Chellah del Aguila, J. Carlos Yaćono, Ángeles Guinda, Wenceslao Moreda, Raquel B. Gómez-Coca, and M. Carmen Pérez-Camino. 2014. "Characterization of Glyceridic and Unsaponifiable Compounds of Sacha Inchi (*Plukenetia huayllabambana* L.) Oils." *Journal of Agricultural and Food Chemistry* 62: 10162–10169. https://doi.org/10.1021/jf5028697.

Chasquibol, Nancy A., Raquel B., Gómez-Coca, J. Carlos Yácono, Ángeles Guinda, Wenceslao Moreda, Chellah del Aguila, and M. Carmen Pérez-Camino. 2016. "Markers of Quality and Genuineness of Commercial Extra Virgin Sacha Inchi Oils." *Grasas y Aceites* 67 (4): e169. http://doi.org/10.3989/gya.0457161.

Chasquibol, Nancy A., Gabriela Gallardo, Raquel B. Gómez-Coca, Diego Trujillo, Wenceslao Moreda, and M. Carmen Pérez-Camino 2019. "Glyceridic and Unsaponifiable Components of Microencapsulated Sacha Inchi (*Plukenetia huayllabambana* L. and *Plukenetia volubilis* L.) Edible Oils." *Foods* 8: 671. https://doi.org/10.3390/foods8120671.

Chirinos, Rosana, Gledy Zuloeta, Romina Pedreschi, Eric Mignolet, Yvan Larondelle, and David Campos. 2013. "Sacha Inchi (*Plukenetia volubilis*): A Seed Source of Polyunsaturated Fatty Acids, Tocopherols, Phytosterols, Phenolic Compounds and Antioxidant Capacity." *Food Chemistry* 141: 1732–1739. https://doi.org/10.1016/j.foodchem.2013.04.078.

Chirinos, Rosana, Romina Pedreschi, Gilberto Domínguez, and David Campos. 2015. "Comparison of the Physico-Chemical and Phytochemical Characteristics of the Oil of Two *Plukenetia* Varieties." *Food Chemistry* 173: 1203–1206. https://doi.org/10.1016/j.foodchem.2014.10.120.

Choe Eunok, and David B. Min. 2009. "Mechanisms of Antioxidants in the Oxidation of Foods" *Comprehensive Reviews in Food Science and Food Safety* 8: 345–358. https://ift.onlinelibrary.wiley.com/doi/pdfdirect/10.1111/j.1541-4337.2009.00085.x.

Cisneros, Fausto H., Daniel Paredes, Adrian Arana, and Luis Cisneros-Zevallos. 2014. "Chemical Composition, Oxidative Stability and Antioxidant Capacity of Oil Extracted from Roasted Seeds of Sacha-Inchi (*Plukenetia volubilis* L.)." *Journal of Agricultural and Food Chemistry* 62: 5191–5197. https://doi.org/10.1021/jf500936j.

Di Mattia, Carla, Federica Balestra, Giampiero Sacchetti, Lilia Neri, Dino Mastrocola, and Paola Pittia. 2015. "Physical and Structural Properties of Extra Virgin Olive Oil Based Mayonnaise." *LWT-Food Science and Technol*ogy 62 (1): 764–770. https://doi.org/10.1016/j.lwt.2014.09.065.

Fanali, Chiara, Laura Dugo, Francesco Cacciola, Marco Beccaria, Simone Grasso, Marina Dach, Paola Dugo, and Luigi Mondello. 2011. "Chemical Characterization of Sacha Inchi (Plukenetia volubilis L.) Oil." *Journal of Agricultural and Food Chemistry* 59: 13043–13049. http://doi.org/10.1021/jf203184y.

Garmendia, Fausto, Rosa Pando, and Gerardo Ronceros. 2011. "Efecto del Aceite de Sacha Inchi (*Plukenetia volubilis* L) sobre el Perfil Lipídico en Pacientes con Hiperlipoproteinemia." *Revista Peruana de Medicina Experimental y Salud Pública.* 28 (4): 628–632. www.scielo.org.pe/scielo.php?script=sci_arttext&pid=S1726-46342011000400009&lng=es&nrm=iso.

Gonzales, Gustavo, and Carla Gonzales. 2014. "A Randomized, Double-Blind Placebo-Controlled Study on Acceptability, Safety and Efficacy of Oral Administration of Sacha Inchi Oil (*Plukenetia volubilis* L.) in Adult Human Subjects." *Food Chemical Toxicology* 65: 168–176. https://pubmed.ncbi.nlm.nih.gov/24389453/.

Gorriti, Arilmi, Jorge Arroyo, Fredy Quispe, Braulio Cisneros, Martín Condorhuamán, Yuan Almora, and Víctor Chumpitaz. 2010. "Oral Toxicity at 60 Days of Sacha Inchi Oil (*Plukenetia volubilis* L.) and Linseed (*Linum usitatissimum* L.), and Determination of Lethal Dose in Rodents." *Revista Peruana de Medicina Experimental y Salud Pública* 27: 352–360. https://scielosp.org/article/ssm/content/raw/?resource_ssm_path=/media/assets/rpmesp/v27n3/a07v27n3.pdf.

Guillén, María D., Ainhoa Ruiz, Nerea Cabo, Rosana Chirinos, and Gloria Pascual. 2003. "Characterization of Sacha Inchi (*Plukenetia volubilis* L.) Oil by FTIR Spectroscopy and 1H NMR. Comparison with Linseed oil." *Journal of the American Oil Chemists' Society* 80: 755–762. https://doi.org/10.1007/s11746-003-0768-z.

Herchi, Wahid, Saleh Bahashwan, Hajer Trabelsi, Sadok Boukhchina, Habib Kallel, Sophie Rochut, and Claude Pepe. 2014. "Changes in Proximate Composition and Oil Characteristics During Flaxseed Development." *Grasas Aceites* 65 (2): e022. http://doi.org/10.3989/gya.097713.

Herman Christelle, Darly Pompeu, David Campos, Yvan Larondelle, Herve Rogez, and Vincent Baeten. 2020. "Monitoring of the oxidation of the oil from sacha inchi (*Plukenetia volubilis*) seeds supplemented with extracts from tara (*Caesalpinia spinosa*) pods, using conventional and MIR techniques". *Grasas y Aceites* 71(2): e359. https://doi. org/10.3989/gya.0228191.

Holgado, Francisca, Gloria Márquez-Ruiz, M. Victoria Ruiz-Méndez, and Joaquín Velasco. 2019. "Effects of the Drying Method on the Oxidative Stability of the Free and Encapsulated Fractions of Microencapsulated Sunflower Oil." *International Journal of Food Science and Technology* 54 (8): 2520–2528. https://doi.org/10.1111/ijfs.14162.

Huamán, Juan, Katterine Chávez, Erdwin Castañeda, Santiago Carranza, Tania Chávez, Yuri Beltrán, Carlos Caffo, Rómulo Cadillo, and Jeff Cadenillas. 2008. "Effect of *Plukenetia volubilis* L. (Sacha inchi) on Postprandial Triglycerides." *Anales de la Facultad de Medicina* 69 (4): 263–266. https://doi.org/10.15381/anales.v69i4.1128.

IOC, International Olive Council. 2019. "Trade Standard Applying to Olive Oils and Olive Pomace Oils." COI/T.15/NC No 3/Rev, June 13. www.internationaloliveoil.org/wp-content/uploads/2019/11/COI-T.15-NC.-No-3-Rev.-13-2019-Eng.pdf.

Ixtaina, Vanesa Y., Marcela L. Martínez, Viviana Spotorno, Carmen M. Mateo, Damian M. Maestri, Bernd W. K. Diehl, Susana M. Nolasco, and Mabel C. Tomas. 2011. "Characterization of Chia Seed Oils Obtained by Pressing and Solvent Extraction." *Journal of Food Composition and Analysis* 24: 166–174. http://doi.org/10.1016/j.jfca.2010.08.006.

Jacobsen, Charlotte, Maike Timm, and Anne S. Meyer. 2001. "Oxidation in Fish Oil Enriched Mayonnaise: Ascorbic Acid and Low pH Increase Oxidative

Deterioration." *Journal of Agricultural and Food Chemistry* 49 (8): 3947–3956. https://doi.org/10.1021/jf001253e.

Kastelein, John P., Kevin C. Maki, Andrey Susekov, Marat Ezhov, Borge G. Nordestgaard, Ben N. Machielse, Douglas Kling, and Michael H. Davidson. 2014. "Omega-3 Free Fatty Acids for the Treatment of Severe Hypertriglyceridemia: The EpanoVa for Lowering Very High Triglycerides (EVOLVE) Trial." *Journal of Clinical Lipidology* 8 (1): 94–106. https://doi.org/10.1016/j.jacl.2013.10.003.

Kodahl, Nete. 2020. "Sacha inchi (*Plukenetia volubilis* L.)—From Lost Crop of the Incas to Part of the Solution to Global Challenges?." *Planta* 251: 80. https://doi.org/10.1007/s00425-020-03377-3.

Kulczyński, Bartosz, Joanna Kobus-Cisowska, Maciej Taczanowski, Dominik Kmiecik, and Anna Gramza-Michałowska. 2019. "The Chemical Composition and Nutritional Value of Chia Seeds—Current State of Knowledge." *Nutrients.* 11 (6): 1242–1258. https://doi.org/10.3390/nu11061242.

Landoni, Lourdes, Rafael Alarcón, Laida Vilca, Nancy Chasquibol, M. Carmen Pérez-Camino, and G. Gabriela Gallardo. 2020. "Physicochemical Characterization and Oxidative Stability of Microencapsulated Edible Sacha Inchi Seed Oil by Spray Drying." *Grasas y Aceites* 71 (4): e387. https://doi.org/10.3989/gya.1028192.

Liu, Qiang, You-Kai Xu, Ping Zhang, Z. Na, T. Tang, and Y. X. Shi. 2014. "Chemical Composition and Oxidative Evolution of Sacha Inchi (*Plukenetia volubilis* L.) Oil from Xishuangbanna (China)." *Grasas Aceites* 65 (1): e012. http://doi.org/10.3989/gya.075713.

Maurer, Natalie, Beatriz Hatta-Sakoda, Gloria Pascual-Chagman, and Luis E. Rodriguez-Saona. 2012. "Characterisation and Authentication of a Novel Vegetable Source of Omega-3 Fatty Acids, Sacha Inchi (*Plukenetia volubilis* L.) Oil." *Food Chemistry* 134 (2): 1173–1180. https://doi.org/10.1016/j.foodchem.2012.02.143.

Medina-Marroquín, Luis, Anibal Vásquez-Chicata, and Omar Bellido-Valencia. 2018. Estimating the Shelf-Life of Mayonnaise Made from Sacha Inchi (*Plukenetia volubilis* L.) Oil and Duck (*Anas platyrhynchos* L.) Egg Yolk." *Journal of Food Science and Technology* 10 (4): 16–22. www.researchgate.net/publication/330353576.

Mosquera, Tatiana, Paco Noriega, Wilson Tapia, and Silvia H. Pérez. 2012. "Effectiveness Evaluation of Cosmetic Creams elaborated from Oils extracted from Amazon Plants: *Mauritia flexuosa* (Morete), *Plukenetia volubilis* (Sacha Inchi) and *Oenocarpus bataua* (Ungurahua)." *LA GRANJA, Revista de Ciencias de la Vida* 16 (2): 14–22. https://doi.org/10.17163/lgr.n16.2012.02.

Mozaffarian, Dariush, and Jason H. Y. Wu. 2011. "Omega-3 Fatty Acids and Cardiovascular Disease." *Journal of the American College of Cardiology* 58 (20): 2047–2067. https://doi.org/10.1016/j.jacc.2011.06.063.

Nasiff-Hadad, Alfredo, and Erardo Meriño-Ibarra. 2003. "Ácidos Grasos Omega-3: Pescados de Carne Azul y Concentrados de Aceites de Pescado. Lo Bueno y lo Malo." *Revista Cubana de Medicina* 42 (2): 49–55. http://scielo.sld.cu/scielo.php?script=sci_arttext&pid=S0034-75232003000200008&lng=es&nrm=iso. ISSN:0034-752.

NTP. 2009. *Norma Técnica Peruana 151.400.* Lima, Perú: Requisitos Aceite Sacha Inchi, INDECOPI, 2010.

NTP. 2018. *Norma Técnica Peruana 151.400, Amendment to NTP 151.400, 2014: Sacha Inchi Oil. Requirements.* Lima, Peru: R.D. N_047-2018-INACAL/DN.

Pastuña-Pullutasig, Alex, Orestes Lopéz-Hernández, Alexis Debut, Andrea Vaca A., Eduardo Rodríguez-Leyes, Roxana Vicente, Victor Gonzalez, María

González-Sanabia, and Fausto Tapia-Hernández. 2016. "Microencapsulación de Aceite de Sacha Inchi (*Plukenetia volubilis* L.) mediante Secado por Aspersion." *Revista Colombiana de Ciencias Químicas y Farmaceutica* 45 (3): 422–437. http://doi.org/10.15446/rcciquifa.v45n3.62029.

Pellegrini Nicoletta, Mauro Serafini, Barbara Colombi, Daniele Del Rio, Sara Salvatore, Marta Bianchi, and Furio Brighenti. 2003. "Total antioxidant capacity of plant foods, beverages and oils consumed in Italy assessed by three different in vitro assays". *Journal of Nutrition* 133(9): 2812–2819. doi: 10.1093/jn/133.9.2812.

Quispe-Condori, Sócrates, Marleny D. A. Saldaña, and Feral Temelli. 2011. "Microencapsulation of Flax Oil with Zein Using Spray and Freeze Drying." *LWT-Food Science and Technology* 44: 1880–1887. https://doi.org/10.1016/j.lwt.2011.01.005.

Ramos-Escudero, Fernando, Ana María Muñoz-Jauregui, Mónica Ramos-Escudero, Adriana Viñas-Ospino, María Teresa Morales, and Agustín G. Asuero. 2019. "Characterization of Commercial Sacha Inchi Oil According to Its Composition: Tocopherols, Fatty Acids, Sterols, Triterpene and Aliphatic Alcohols." *Journal of Food Science and Technol*ogy 56 (10): 4503–4515. https://doi.org/10.1007%2Fs13197-019-03938-9.

Rodríguez, Ángel, Mike Corazon-Guivin, Danter Cachique, Kember Mejía, Dennis Del Castillo, Jean-François Renno, and Carmen García-Dávila. 2010. "Diferenciación Morfológica y por ISSR (Inter Simple Sequence Repeats) de Especies del Género *Plukenetia* (*Euphorbiaceae*) de la Amazonía Peruana: Propuesta de una Nueva Especie." *Revista Peruana de Biología* 17 (3): 325–330. http://sisbib.unmsm.edu.pe/BVRevistas/biologia/biologiaNEW.htm.

Rodríguez, Gilbert, Eudes Villanueva, Patricia Glorio, and Mery Baquerizo. 2015. "Estabilidad Oxidativa y Estimación de la Vida Útil del Aceite de Sacha Inchi (*Plukenetia volubilis*)." *Scientia Agropecuaria* 6 (3): 155–163. http://doi.org/10.17268/sci.agropecu.2015.03.02.

Sánchez-Reinoso, Zain, and Luis Felipe Gutiérrez. 2017. "Effects of the Emulsion Composition on the Physical Properties and Oxidative Stability of Sacha Inchi (*Plukenetia volubilis* L.) Oil Microcapsules Produced by Spray Drying." *Food and Bioprocess Technology* 10: 1354–1366. https://doi.org/10.1007/s11947-017-1906-3.

Sathe, Shridhar K., Bruce R. Hamaker, Kar Wai Clara Sze-Tao, and Mahesh Venkatachalam. 2002. "Isolation, Purification and Biochemical Characterization of a Novel Water Soluble Protein from Inca Peanut (*Plukenetia volubilis*)." *Journal of Agricultural and Food Chemistry* 50 (17): 4906–4908. https://doi.org/10.1021/jf020126a.

Uncu, Oguz, Banu Ozen, and Figen Tokatli. 2019. "Mid-Infrared Spectroscopic Detection of Sunflower Oil Adulteration with Safflower Oil." *Grasas Aceites* 70 (1): e290. https://doi.org/10.3989/gya.0579181.

Vicente, Juarez, Taylana de Souza Cezarino, Luciano José Barreto Pereira, Elisa Pinto da Rocha, Guilherme Raymundo Sá, Ormindo Domingues Gamallo, Mario Geraldo de Carvalho, and Edwin Elard García-Rojas. 2017. "Microencapsulation of Sacha Inchi Oil Using Emulsion Based Delivery Systems." *Food Research International* 99: 612–622. https://doi.org/10.1016/j.foodres.2017.06.039.

Wang, Sunan, Fan Zhu, and Yukio Kakuda. 2018. "Sacha Inchi (*Plukenetia volubilis* L.): Nutritional Composition, Biological Activity, and Uses." *Food Chemistry* 265: 316–328. https://doi.org/10.1016/j.foodchem.2018.05.055.

Chapter 8

Action of Amaranth Peptides on the Cardiovascular System

María C. Añón, Alejandra V. Quiroga, Adriana A. Scilingo, Valeria A. Tironi, Ana C. Sabbione, Agustina E. Nardo, Santiago E. Suárez, and Susan F. García Fillería

CONTENTS

8.1 GENERAL SCOPE

The cardiovascular system (CVS), also called circulatory system, has the function of distributing blood in the human body. It comprises the heart, which acts as a driving pump, and a system of closed elastic vessels: arteries, veins, and capillaries. The CVS can be affected by several disorders that damage the heart and blood vessels, collectively called cardiovascular diseases (CVDs). They are associated with lipid deposits in arteries, formation of blood clots, and damage to the brain, heart, or kidney arteries. CVDs are one of the chronic noncommunicable diseases that have led deaths worldwide for four decades. In 2017, out of a total of 56 million deaths, 31% were attributed to CVDs, particularly due to heart attacks and strokes (WHO 2016, 2018). The origin of CVDs is not clear, but a number of risk factors have been identified. These include high blood pressure, endothelial dysfunction, prothrombotic potential, obesity/insulin resistance, oxidative stress, and the inflammatory state. Other key

DOI: 10.1201/9781003087618-8

factors involved are social habits such as smoking, alcohol intake, lack of physical activity, and unhealthy diets (Iwaniak et al. 2018).

The relationship between diet and disease has been known since ancient times. The concept of food as medicine has been mentioned in the Vedic texts of ancient India, in traditional Chinese medicine, and in other ancient civilizations such as Greece. More recently, in 1984, the Ministry of Education, Science and Culture of Japan started a national project to study the connection between food and human health, while the Ministry of Health and Welfare promoted a program for the regulation and approval of foods beneficial to the health and well-being of consumers, focused on the older population (Arai 1996). A few years later, in 1987, the term *nutraceutical* (nutrition + pharmaceutical) was defined, and in 1993 the concept of *functional food* appeared in the journal *Nature* (Arai 1996; Swinbanks and O'Brien 1993). Both terms quickly became popular and were adopted worldwide. The European Commission Concerted Action on Functional Food Science in Europe (FUFOSE), coordinated by the International Institute of Life Sciences, ILSI (European Commission 1995), defined functional food as that which has beneficial effects for the health on one or more functions of the body, beyond its nutritional properties, so that it improves the general state of health and/or reduces the risk of contracting diseases. These foods contain one or more ingredients with recognized physiological activity, known as bioactive compounds, such as fibers, pigments, oligosaccharides, fatty acids, flavonoids, polyphenols, vitamins, and peptides, among others.

This chapter focuses on bioactive peptides, which in recent years have been intensively studied and have sparked interest in the clinical sector. These peptides are encrypted in the sequence of different food proteins (milk, egg, meat, and different plants, among others) and must be released to exert their biological function either by exogenous proteases, fermentation processes, or during gastrointestinal digestion (Orona-Tamayo et al. 2019). Bioactive peptides display different physiological activities: antimicrobial, antioxidant, antihypertensive, antithrombotic, hypocholesterolemic, antiproliferative/antitumor, antidiabetic (dipeptidyl-peptidase IV and alpha-amylase inhibitors), and immunomodulatory (Orona-Tamayo et al. 2019). The conventional approach to study these activities includes *in vitro, in vivo*, and *ex vivo* assays and clinical trials. This approach allows the identification of bioactive peptides from different protein sources and their mechanisms of action, but presents several limitations which can be partly corrected by the use of *in silico* assays (Li-chan 2015).

Amaranth is an ancestral crop that has many advantages from an agronomic point of view, and its grains have a unique and attractive agronomic and nutritional composition. In particular, amaranth grain storage proteins have a balanced amino acid composition, higher than that of most proteins of plant origin, and good techno-functional properties (Velarde-Salcedo et al. 2018). Silva-Sánchez et al. (2008) showed through bioinformatic tools that amaranth is a potential source of bioactive peptides. In more recent years, it has been shown that amaranth proteins contain peptide sequences that have antihypertensive, antithrombotic, and antioxidant properties. These activities, as previously indicated, may favorably reduce the risk of developing CVDs. The supporting evidence for this argument/claim is described in this chapter.

8.2 ANTIHYPERTENSIVE ACTIVITY

Blood pressure (BP) is defined as the pressure exerted by circulating blood against the walls of the arteries. Normal BP values in adults are approximately 120/80 mmHg, whereas values greater than 130–139/80–89 mmHg are characteristic of hypertension. Hypertension is the main risk factor for developing CVDs. It is estimated that there are more than 1.1 billion people with hypertension in the world, mostly in low- and middle-income countries (WHO 2016, 2018). Increased hypertension has been associated with a family history of high BP, age, and coexisting diseases such as diabetes and kidney disease. In addition, the consumption of unhealthy diets with a high content of fats, trans fatty acids, and excessive salt and low intake of fruits and vegetables favors its development and the incidence of CVDs.

BP regulation is a very complex process, in which different mechanisms participate: the baroreceptor reflex, the antidiuretic hormone also known as vasopressin, and the RAAS (renin-angiotensin-aldosterone system; Zhuo et al. 2013). Short-term regulation is mediated by the former mechanism, while the other two are involved in long-term control. RAAS is a multifunctional system that controls fluid and electrolyte balance and BP through systemic actions. The RAAS components exist in circulation but are also found in many tissues (brain, kidneys, heart, blood vessels) and inside cells (Zhuo et al. 2013), these local systems can act in a completely independent manner. In the last two decades, significant progress has been made in the knowledge and complexity of the RAAS, and currently four axes can be identified:

- *Axis 1*: classic renin–angiotensin converting enzyme (ACE)–angiotensin II–AT1 and AT2 receptors. The only precursor of all the angiotensins described so far is angiotensinogen (Agt), secreted by the liver. This peptide is transformed into angiotensin I (AngI), and this into angiotensin II (AngII) by the action of renin and ACE, respectively. ACE is also capable of hydrolyzing bradykinin, a powerful vasodilator. AngII exerts its action through its binding to receptors AT1 and AT2. The activation of AT1 produces vasoconstriction and promotes

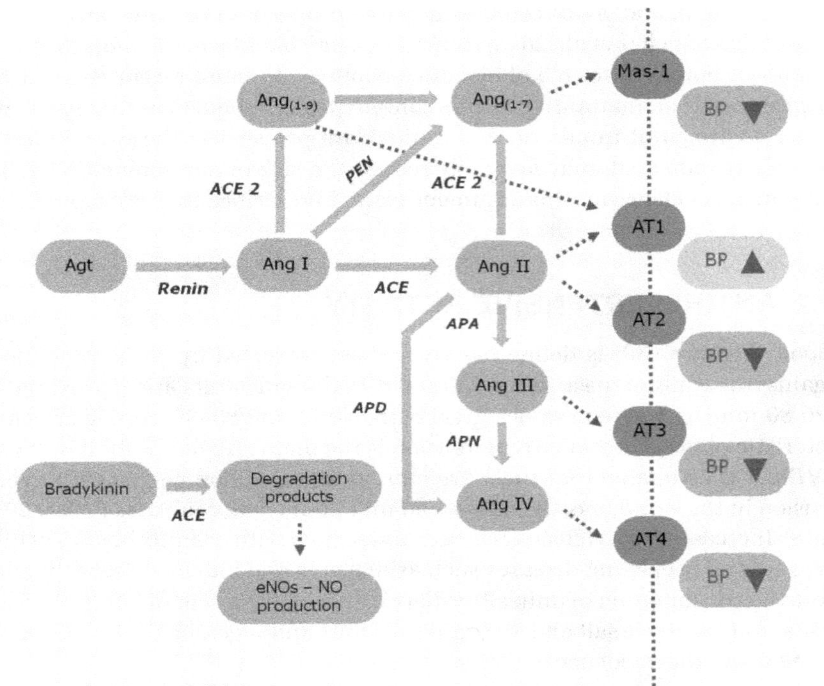

Figure 8.1 Scheme of the renin-angiotensin-aldosterone system (RAAS) responsible for regulating blood pressure (BP).

Abbreviations: Agt = angiotensinogen; Ang I, II, III, IV, (1–9), (1–7) = angiotensin I, II, III, IV, (1–9), (1–7); ACE = angiotensin converting enzyme; APA, APN, APD = aminopeptidases A, N, and P; PEN = neutral peptidases; Mas-1 = Mas-1 receptor; AT2 = angiotensin II type 2 receptor; AT3 = angiotensin III receptor; AT4 = angiotensin IV receptor.

oxidative stress, neuroinflammation, apoptosis, and cell proliferation; while the activation of AT2 induces vasodilation and angiogenesis as well as an antioxidant, antiproliferative, and anti-apoptotic effect.

- *Axis 2*: ACE2–angiotensin (1–7)–Mas-1 receptor. ACE2, an enzyme discovered in 2000, is capable of cleaving AngI into two new peptides, Ang (1–9) that binds to the AT2 receptor and Ang (1–7). ACE2 is also capable of transforming AngII into Ang (1–7). The latter angiotensin binds to the Mas-1 receptor, opposing the damaging actions of AngII, promoting vasodilation, endothelial protection, and inhibition of growth and fibrosis.

- *Axis 3*: angiotensin IV–AT4 receptor–insulin-regulated aminopeptidase (IRAP). Different aminopeptidases, N and P, can cleave AngII into AngIII and AngIV. AngIII is able to bind to AT1 and AT2 receptors, while AngIV binds to a recently discovered self-receptor, AT4. This receptor is expressed in different tissues and its activation causes an increase in tubular sodium reabsorption and natriuresis as well as renal cortical blood flow, among others.
- *Axis 4*: prorenin–(pro)renin receptor–MAP kinase. A receptor for renin and its precursor prorenin called PRR has been identified in different tissues. This receptor performs a role mediated by AngII and other independent of AngII. In the former, the binding of renin to the receptor increases its ability to hydrolyze Agt and the binding of prorenin converts it to its active form, thus managing to participate in the RAAS cascade. In the latter, PRR triggers its own signaling pathway and induces pro-oxidative effects similar to those induced by the AT1 receptor.

Given the relevance of hypertension worldwide, in the 1950s the pharmaceutical industry started to develop different drugs to reduce cardiac output and/or peripheral resistance of blood vessels, several of them specifically directed to the RAAS components. Considering that all these drugs have some side effects and the fact that there is a close relationship between hypertension and diet, the scientific community began to study the action of peptides from different food sources on BP regulation. In the particular case of amaranth proteins, the objective of primary studies of most of the research groups was the ACE inhibition, which, as indicated, promotes the formation of AngII, the largest RAAS effector and vasoconstrictor action. The use of *in silico* tools showed that amaranth is a rich source of antihypertensive peptides (ACE inhibitors), preferentially located in the globulin fraction (Silva-Sánchez et al. 2008; Vecchi and Añón 2009). In general, these sequences contain 2–10 amino acids and are rich in hydrophobic amino acids, proline, valine, leucine, and histidine. Vecchi and Añón (2009) identified two tetrapeptides, ALEP and VIKP, which had a binding energy to the ACE active site similar to that of captopril, a drug that inhibits this enzyme, commonly used in medical practice. Peptides synthesis via chemical methods confirmed the predicted activity obtaining IC_{50} values of 3.05 mg/mL and 89.2 ng/mL for ALEP and VIKP, respectively. It should be noted that this is the only work able to identify amaranth peptide sequences with experimentally corroborated antihypertensive capacity.

The release of encrypted peptides contained in a protein can be achieved, as previously indicated, by different approaches: *in vitro* and *in vivo*. *In vitro* simulation of the gastrointestinal process (GID) exhibits the bioaccessibility of the peptides under study, and consequently their resistance capacity to gastric and intestinal proteases. Conversely, *in vivo* assays provide information on the bioavailability of peptides, implying their capacity to cross the intestinal wall and the possibility to exert its physiological action on the target organ.

Different studies reported the use of several proteases to release amaranth peptides, such as alcalase (Tovar-Pérez et al. 2009; Tiengo et al. 2009; Fritz et al. 2011), trypsin (Silva-Sánchez et al. 2008; Fritz et al. 2011), pronase, papain, and chymotrypsin (Fritz et al. 2011). In this way, hydrolysates from the main protein fractions of the grain (albumin and 11S globulins; Tovar-Pérez et al. 2009) and glutenins (Silva-Sánchez et al. 2008), concentrates (Tiengo et al. 2009) and protein isolates (Fritz et al. 2011) of *Amaranthus hypochondriacus, A. mantegazzianus, A. caudatus*, and *A. cruentus* were obtained. The degrees of hydrolysis (HD) achieved were high (35%–56%) and the ability to inhibit ACE was diverse, with IC_{50} values in the range of 1.16–0.01 mg/mL. Ayala-Niño et al. (2019a) used the combined action of alcalase and flavourzyme and obtained hydrolysates with high bioactivity (% inhibition: 49.49 ± 1.47, 39.77 ± 2.15, 58.53 ± 2.58 for alcalase, flavourzyme, and alcalase-flavourzyme, respectively). HPLC fractionation of the most active hydrolysates allowed the separation of fractions with higher activity and in some cases the identification of ACE inhibitor peptide sequences (Ayala-Niño et al. 2019a). Since the potential bioactive sequences were not synthesized and tested, their real activity remains unknown. The quantity and quality of the released peptides depends on the protein(s) used as substrate, the enzyme(s), and the hydrolysis conditions. ACE inhibitory peptides are characterized by being short, which requires extensive hydrolysis. According to studies carried out by different authors, alcalase is one of the most efficient proteases to release short peptides. This evidence led Ramírez-Torres et al. (2017) to optimize the reaction conditions for this enzyme in terms of pH, temperature, enzyme/substrate, and time. The HD obtained under optimal conditions and the percentage of ACE inhibition reached were higher than most of those achieved in previous studies, obtaining hydrolysates with a significantly lower IC_{50}.

The IC_{50} values of peptides or peptide mixtures obtained so far is significantly higher than that of captopril. This indicates that the antihypertensive capacity of peptides is still very far from that of one of the most popular drugs used for inhibition of ACE.

Quiroga et al. (2017) studied for the first time the capacity of amaranth peptides to inhibit renin. The authors obtained alcalase hydrolysates from protein isolates of *A. hypochondriacus* with an average peptide length of 4–6 amino acids. These peptides showed a renin inhibition capacity of 60% and an IC_{50} of 0.6 mg/ml. The hydrolysate was subsequently fractionated; the most active fraction showed a renin inhibition capacity of 67.5%. Six peptides were identified in this fraction: QAFEDGFEWVSKF, AFEDGFEWVSKF, SFNLPILR, FNLPILR,

SFNLPIL, and VNVDDPSKA. Recently, by bioinformatics analysis, Nardo et al. (2020) showed that the peptides aforementioned are capable of binding to the renin active site, as well as to other sites of the molecule and that certain renin residues participate more frequently in the interaction. The AFEDGFEWVSKF and SFNLPILR peptides were those that best interacted with the active site of the enzyme. However, renin inhibition tests performed *in vitro* with the synthesized peptides showed that FNLPILR, followed by AFEDGFEWVSKF and SFNLPILR (IC_{50} 0.41, 1.47, and 2.50 mM, respectively) were the most potent peptides. These results anticipate that predictive models do not always represent what was observed in experimental assays.

In order to assess the bioaccessibility of amaranth antihypertensive peptides, different research groups have used *in vitro* simulation of the GID process. The first approach was made by Tiengo et al. (2009) who subjected amaranth concentrates and alcalase hydrolysates to the action of pepsin and pancreatin. The hydrolysates obtained after simulating the digestive process had lower ACE inhibition capacity than the hydrolysates obtained with alcalase before or after having been subjected to the simulated GID process (IC_{50} 0.439 ± 0.018 mg protein/mL, 0.176 ± 0.014 mg protein/mL, and 0.137 ± 0.002 mg protein/mL, respectively), suggesting that the peptides released by this enzyme are resistant to the action of pepsin and pancreatin and would be good candidates to be used in the formulation of functional foods. Although this conclusion is correct, the fact that the enzymes involved in the simulated GID process may release new peptides with ACE inhibitory activity equivalent to that of the peptides released by alcalase cannot be discarded. Quiroga et al. (2012) used a protocol equivalent to that used by Tiengo et al. (2009) and showed that the hydrolysates obtained from partially purified 7S globulin and 11S globulin isolated from *A. hypochondriacus* had a similar ACE inhibitory capacity (IC_{50} 0.17 ± 0.03 mg and 0.17 ± 0.01 mg protein/mL, respectively). These authors also showed the presence of ACE inhibitors in different subunits belonging to 7S globulin, mainly di- and tripeptides. These results would indicate that both, 7S and 11S globulin fractions, are source of antihypertensive peptides with equivalent potentiality. The conditions used by different researchers to simulate the GID process of different food proteins have been very dissimilar, which makes the analysis and comparison of results difficult. Within the framework of the INFOGEST project, a protocol was agreed and it has been adopted by most of the research groups simulating the physiological conditions of the oral, gastric, and intestinal stages (Minekus et al. 2014). Vilcacundo et al. (2019) used this protocol to obtain different hydrolysates from a protein concentrate of *A. caudatus*. These authors analyzed, among other bioactivities, the inhibitory capacity of ACE at the beginning of the simulated digestion process, after the conclusion of the gastric stage and in the middle and at the end of the intestinal stage. The IC_{50}

values obtained were all very similar, with the exception of the one corresponding to the gastric stage, which suggests the generation, by the action of pepsin, of bioactive peptides sensitive to the action of pancreatin (79.13 ± 1.08, 39.00 ± 2.99, 81.00 ± 10.53 and 88.01 ± 13.9 μg peptides/mL, respectively). In this work, it is striking the fact that the non-hydrolyzed sample has an IC_{50} value equivalent to the final digested. Some authors have reported antihypertensive activity in non-hydrolyzed isolates, but with much lower activity than the corresponding hydrolyzed samples (Fritz et al. 2011; % ACE inhibitor 10.8 ± 2.1 and 74.5 ± 2.0 for non-hydrolyzed and alcalase-hydrolyzed isolate, respectively), while others indicated that intact proteins had no activity (Tiengo et al. 2009). It is possible that, since a protein concentrate has been used, with 53.6% protein, other inhibitory non-protein nature components or free ACE inhibitor peptides are present in the initial raw material. These authors also identified in the final digest, by LC/MS/MS, 2–8 amino acid peptides containing leucine, proline, valine, or histidine residues of relevance to the potential inhibitory activity of ACE.

The studies carried out to date with other proteins of different origins show that they are partially hydrolyzed during the gastric stage, being much more sensitive to the action of intestinal proteases.

Although simulation of the GID process provides more information regarding the potential activity of peptides, nothing still ensures that they can cross the intestinal wall and exert their function in the internal environment. For this reason, it is necessary to carry out *in vitro* or *in vivo* tests with an animal model system and/or with human beings. To date, only three *in vivo* studies have been carried out in which spontaneously hypertensive Wistar rats (SHR) were administered with amaranth proteins. No studies involving human trials are recorded, which would be essential for making an hypotensive health claim on functional foods, containing amaranth proteins.

Lado et al. (2015) found that the intake of a normal diet for rats, in which approximately 10% of the total protein was replaced by amaranth protein isolate, had a hypotensive effect after 4 weeks, suggesting the release of bioactive peptides during *in vivo* gastrointestinal digestion.

Fritz et al. (2011) showed that intragastric administration of alcalase hydrolysates (45% HD, 1.5 g hydrolysate/kg of body weight) to SHR is effective in reducing the blood pressure of animals. This reduction was dose-dependent, reaching normal pressure values (of ~150/90 mmHg) after 1 hour of ingestion. Furthermore, they showed that the hypotensive effect persisted for 7 hours after ingestion. The reduction in blood pressure can, principally, be associated with both cardiac output and peripheral resistance of the vessels. Through *ex vivo* assays with papillary muscles isolated from the hearts of rats, Fritz et al. (2011) concluded that there was no significant inotropic effect, suggesting that through *in vivo* studies the heart of animals would not be affected by amaranth intake. Therefore, in principle, the antihypertensive effect would be associated with the ACE inhibitory activity detected *in vitro*.

The administration of proteins in different forms: protein isolate, pure peptides, and protein isolate + hydrolysate used in the formation of oil:water emulsions also caused a significant reduction in the BP of Wistar SHR rats (Suárez et al. 2020). The greatest reduction in systolic BP was achieved after animals were administered with the emulsion alone or with the addition of the VIKP peptide. This reduction was equivalent to that achieved by the animals administered with captopril which corroborates the results obtained in the *in vitro* assays. The rest of the samples analyzed were also able to reduce the BP of the animals but to a lesser extent.

The results discussed so far indicate that amaranth proteins contain peptides released by the use of exogenous enzymes or during the *in vivo* gastrointestinal digestion process, capable of inhibiting ACE and renin.

Fritz et al. (2011) and Suárez et al. (2020) also performed *ex vivo* assays using aortic rings from animals administered with the different protein samples assayed. These rings were exposed to vasoconstrictor substances such as potassium and norepinephrine. The addition of a constant amount of alcalase hydrolysate to the contracted rings by the action of different amounts of norepinephrine, caused a significant reduction in contractile force indicating vasorelaxing activity of this hydrolysate (Fritz et al. 2011). In contrast, Suárez et al. (2020) showed that aortic rings corresponding to animals administered with isolate, hydrolysate, VIKP, and emulsion added or not with VIKP, required a lower force to maintain the light of the ring when it was faced with a constant amount of potassium, suggesting a relaxing effect. This ion causes depolarization of the plasma membrane and the opening of L-type calcium channels, allowing the entry of this ion into the cellular system and the contraction of the myofibrillar system (Rinaldi 1990). Therefore, the results obtained suggest that amaranth peptides present in the different samples analyzed may act not only at the level of the circulating RAAS system but also at the local or tissue RAAS system level. This hypothesis was confirmed when norepinephrine was used; in this case, a vasodilator effect was only detected in the rings of the animals administered with VIKP and the emulsion added with this tetrapeptide. Norepinephrine mainly binds to alpha-adrenergic receptors allowing a rapid release of intracellular calcium and an influx of extracellular calcium, causing vascular contraction (Rinaldi et al. 1991). The results obtained suggest a specific effect of VIKP on these receptors and therefore a potential action on the tissue RAAS system.

Other evidence that suggests the action of amaranth peptides beyond the circulating RAAS system was provided by the work carried out by Barba de la Rosa et al. (2010). These authors showed that peptides present in tryptic hydrolysates of the glutenin fraction were capable of inhibiting ACE (IC_{50} 0.250 mg/mL) and also stimulating the endothelial production of nitric oxide (NO, 52% compared to the control). Bradykinin (BK) is known to stimulate vasodilation of vessels modulated by NO production. The vasodilator effect of the peptides was verified in *ex vivo* experiments, with isolated rat aortic rings. In the presence of glutelin hydrolysate, 0.1 mg/mL, 83% dilatation of the arterial vessel was recorded. This effect was reduced to 50% when the rings used did not have endothelium, suggesting a partial dependence on the endothelium. The vasodilator effect was inhibited by HOE-140, a bradykinin receptor agonist, suggesting an important role of this compound in the mechanism of action of these peptides.

The previously discussed results indicate that amaranth peptides with antihypertensive activity may have different action targets in the BP regulation system. To date, at least three have been identified: the two main enzymes of the circulating RAAS, the production of NO mediated by bradykinin, and the level of intracellular Ca^{+2} through channels operated by voltage or by adrenergic receptors.

Antihypertensive peptides have also been the subject of study in other pseudocereals such as quinoa, buckwheat, and chia. In chia grains, Segura-Campos et al. (2013a) showed that peptides from a protein-rich fraction extensively hydrolyzed by alcalase-flavourzyme presented a high ACE inhibitory activity that increased when the peptide size was reduced. It was possible to isolate a very active fraction (IC_{50} 3.97 μg/mL) that contained peptides of 7–12 amino acids. Segura-Campos et al. (2013b) incorporated the hydrolysates produced by enzymatic hydrolysis of chia protein-rich fractions, previously described, in the formulation of white bread (1 and 3 mg hydrolysate/g of flour) and carrot cream (2.5 and 5 mg of hydrolysate/g of carrot). Both products showed significant antihypertensive activity, promising for the production of new functional foods. In contrast, Nieman et al. (2012) reported that dietary supplementation with chia grains in overweight patients did not influence the BP values.

Regarding buckwheat, Li et al. (2002) obtained hydrolysates from a dispersion of flour subjected to the action of pepsin followed by chymotrypsin and trypsin, with ACE inhibitory activity (IC_{50} 0.140 mg/ml). Buckwheat proteins were also very resistant to pepsin treatment. Oral administration of these digests to SHR (100 mg/Kg of weight) reduced, after 2 hours, the systolic BP of animals (ΔBP 27.0 ± 7.6 mmHg), an effect that was maintained for 4 hours after ingestion. From the digests, a fraction constituted by active short peptides (IC_{50} 0.17 mg protein/mL), dipeptides, tripeptides, and a tetrapeptide was identified, which were assigned a preponderant contribution to antihypertensive activity. In particular, Ma et al. (2006) identified a tripeptide, GPP, with relevant ACE inhibitory activity (IC_{50} 6.25 μg protein/mL) from an alkaline aqueous extract of buckwheat flour.

Different authors have also studied the presence of ACE inhibitor peptides from quinoa storage proteins. Aluko and Monu (2003) obtained hydrolysates from quinoa concentrates by the action of exogenous enzymes that were later purified and enriched by ultrafiltration, and showed that inhibition capacity increases as the size of the peptides is reduced, a fact also demonstrated for other protein sources. Zheng et al. (2019) obtained hydrolysates of quinoa albumin fraction that were purified by gel filtration and LC-MS/MS, which allowed the identification of a dozen peptide sequences. They identified the heptapeptide RGQVIYVL which has a relevant ACE inhibitory capacity *in vitro* (IC_{50} 38.16 μM) and was able to reduce systolic BP of SHR rats when administered orally (100–150 mg/kg of weight).

The discussed results show that both amaranth proteins and those of other pseudo-cereals are a source of antihypertensive peptides and can be used in the formulation of functional foods.

8.3 ANTITHROMBOTIC ACTIVITY

Thrombosis is an underlying cause of many cardiovascular events. It is considered a multicausal pathology associated with hemodynamic disruption, intrinsic hypercoagulability, and endothelial damage (Monie and DeLoughery 2017). In this pathological process, hemostasis mechanisms are activated by non-physiological stimuli and, instead of doing it for restorative purposes after traumatic injury, it obstructs blood vessels. The phenomenon can occur anywhere in the circulatory system: heart, arteries, veins, and capillaries, causing partial or permanent circulatory blockages, with diverse consequences depending on the location.

Hemostasis, on the other hand, is the process that maintains the integrity of the circulatory system under non-pathological conditions after generating vascular damage. Vessel injuries along with extravasation of blood would rapidly initiate events in the vessel wall and blood in charge of sealing the gap, stopping the bleeding, and initiating injury repair. Vascular endothelium, circulating platelets, and coagulation factors co-work in a delicate balance that ensures the fluidity of the blood. The process initiates with vessel vasoconstriction followed by primary hemostasis which involves platelets recognizing the vascular injury site, releasing agonists, adhering to each other, and forming the platelet plug through different receptors (McNicol and Israels 2003). Afterward, secondary hemostasis, commonly called coagulation, is triggered by collagen intrinsic pathway and tissue factor (TF) extrinsic pathway. The process involves orderly activation and amplification of clotting factors, which normally are proteases synthesized in the liver as zymogens that circulate inactive in the blood.

Activation occurs on the platelets surface and they participate in many linked enzymatic reactions known as the coagulation cascade. Extrinsic and intrinsic pathways converge in the common pathway, where thrombin (coagulation factor IIa), a serine protease at the end of the coagulation system, proteolyzes fibrinogen, a soluble glycoprotein present in the blood (Li et al. 2011). Hydrolysis of fibrinogen into fibrin enables the formation of macromolecular aggregates since fibrin is an insoluble protein capable of interlacing with other similar molecules. Moreover, thrombin activates fibrin stabilizing factor (coagulation factor XIIIa) which cross-links and stabilizes fibrin reinforcing the friable platelet plug into a tough fibrin three-dimensional meshwork containing platelets, other proteins, water, salts, and even blood cells trapped (Cheng et al. 2019). In this regard, hemostasis is considered a complex feedback process since thrombin is a potent platelet-activating factor and activated platelets would strengthen the coagulation system to produce more thrombin.

The importance of using anticoagulant and antiplatelet treatments for thrombosis prevention in patients with CVDs has become increasingly recognized. Recently, the research on antithrombotic compounds from edible natural sources has increased swiftly by demand, showing remarkable progress. Amaranth seed proteins could be used as functional ingredients. To our knowledge, the antithrombotic effect of amaranth proteins is one of the less-studied biological activities. Some authors reported that amaranth peptides exhibited *in vitro* antithrombotic activity with assays that use plasma to evaluate intrinsic, extrinsic, and common pathways (Sabbione et al. 2015), platelet aggregation (Montoya-Rodríguez et al. 2015), and simplified assays using thrombin and fibrinogen where turbidity gets associated with the formation of a fibrin clot (Ayala-Niño et al. 2019b; Sabbione et al. 2015).

Among *in vitro* antithrombotic activity studies, Sabbione et al. (2015) described amaranth hydrolysates able to inhibit thrombin (EC 3.4.21.5), the enzyme that exerts the main role at the end of the coagulation cascade. However, no inhibition of coagulation was found in the protein isolates (Sabbione et al. 2016a, 2015). Nevertheless, Ayala-Niño et al. (2019a, 2019b; see Table 8.1) evidenced a small percentage of coagulation inhibition when testing amaranth proteins, with a significant increase of the inhibitory activity when proteins were hydrolyzed by bacterial fermentation (Ayala-Niño et al. 2019b). Moreover, Sabbione et al. (2015) studied the inhibition of coagulation of amaranth protein fractions and their hydrolysates. Glutelin exhibited the lowest IC_{50} value of 0.080 ± 0.003 mg/ml. Its hydrolysate, obtained after performing alcalase and trypsin proteolysis steps, presented a lower IC_{50} value when compared with the hydrolysate of amaranth isolate (IC_{50} 10.87 ± 1.00 mg/ml, Table 8.1). Coagulation studies reported so far evidence that hydrolysis is necessary to

TABLE 8.1 ANTITHROMBOTIC ACTIVITY OF AMARANTH PEPTIDES DETERMINED BY *IN VITRO* MICROPLATES METHODOLOGY

Sample	% of inhibition	Protein concentration (mg/ml)	IC_{50} (mg/ml)	References
Protein isolate	–		–	Sabbione et al. (2015, 2016b)
	11.90 ± 10.10	50		Ayala-Niño et al. (2019a)
	4.87 ± 2.43	40	–	Ayala-Niño et al. (2019b)
GID hydrolysate	100	0.76	0.23 ± 0.02	Sabbione et al. (2016b)
Active fraction from GID hydrolysate	100	0.16	0.07 ± 0.01	Sabbione et al. (2016b)
Isolate fermented with *Lactobacillus casei* Shirota	85.46 ± 3.01	40	–	Ayala-Niño et al. (2019b)
Isolate fermented with *Streptococcus thermophilus* 54102	75.60 ± 4.87	40	–	Ayala-Niño et al. (2019b)
Isolate fermented with *L. casei* Shirota + *S. thermophilus*	93.49 ± 3.75	40	–	Ayala-Niño et al. (2019b)
Endogenous protease hydrolysate	90.65 ± 1.02	13	5.90 ± 0.10	Sabbione et al. (2016a)
Alcalase and trypsin hydrolysate	75.93 ± 7.17	13.6	10.87 ± 1.00	Sabbione et al. (2015)
Alcalase and flavourzyme hydrolysate (AFH)	92.85 ± 3.36	50	–	Ayala-Niño et al. (2019a)
Fraction 22 of alcalase and flavourzyme hydrolysate	–	–	0.167*	Ayala-Niño et al. (2019a)
Fraction 28 of alcalase and flavourzyme hydrolysate	–	–	0.135*	Ayala-Niño et al. (2019a)
Fraction 45 of alcalase and flavourzyme hydrolysate	–	–	0.155*	Ayala-Niño et al. (2019a)

Values are shown as mean ± standard deviation.

*Authors only presented mean values without standard deviation.

Abbreviations: GID hydrolysate: protein isolate subjected to simulated gastrointestinal digestion; IC_{50}, concentration that inhibits 50% of coagulation.

increase antithrombotic activity in amaranth proteins (Sabbione et al. 2015, 2016a; Ayala-Niño et al. 2019a). In this sense, amaranth proteins isolate subjected to simulated GID showed antithrombotic activity (Sabbione et al. 2016b; see Table 8.1). The HD informed by the authors after performing GID was over 50%, implying the presence of small peptides that could be associated with a higher inhibition of the coagulation.

Table 8.1 presents coagulation inhibition values of amaranth peptides obtained by using an *in vitro* turbidimetric assay called the microplates method (Sabbione et al. 2015). The assay studies the coagulation cascade common pathway in a simplified manner using thrombin and fibrinogen reagents to evaluate fibrin clot formation. Table 8.1 shows that the isolate submitted to GID releases potential antithrombotic peptides with an active fraction (AF), obtained by size-exclusion chromatography, which resulted three times more potent than its original sample (Sabbione et al. 2016b). This AF exerts the greatest potential coagulation inhibitory capacity ever reported in the bibliography. Sabbione et al. (2016b) sequenced the AF and potentially bioactive peptides, in particular thrombin inhibitors, were found since it is one of the possible mechanisms of action reflected by the microplates method. Furthermore, the absorption capacity of the AF through intestinal epithelium was studied using Caco2-TC7 cell cultures, evidencing that a portion of these peptides was able to cross the Caco2-TC7 monolayer cell (Sabbione et al. 2016b). In this regard, bioactive peptides were selected as those with potential antithrombotic activity considering their interaction with thrombin. Most of these peptides were located on the surface of their native proteins, hence more exposed to proteolytic attacks. The results described are promising and indicate that amaranth proteins could be used as a potential functional ingredient in diets with antithrombotic peptides released after ingestion. These diets could offer a convenient and effective way of preventing and mitigating CVD incidence (Cheng et al. 2019).

The aforementioned results refer to *in vitro* assays that constitute a first approximation of amaranth antithrombotic studies. *In vivo* and *ex vivo* assays provide a closer approach to the evaluation of this biological activity. To our knowledge, only Sabbione et al. (2016c) described the antithrombotic effect of amaranth proteins using *in vivo* assays, studying blood coagulation of male Wistar rats fed with amaranth proteins. The authors used a control group (CG), which ingested a control diet (AIN-93F); an amaranth group (AG) that was fed with the amaranth diet; and a heparin group (HG), which consumed the control diet, and after the 2 weeks of the experiment the animals were intraperitoneally injected with sodium heparin. Once the feeding period was completed, primary hemostasis of animals was assessed by measuring bleeding time (BT). A small cut on the tip of the tail of the animals was undertaken, and bleeding was recorded by touching the area of the incision with a filter paper strip every 30 s. BT was defined as the time required to completely stop the bleeding. This parameter is an *in vivo* screening test that studies the interaction between platelets and the blood vessel wall, it increases when alterations in platelets adhesion or aggregation occur. The average BT reported was 360 ± 24.5 s for HG, 270 ± 21 s for AG, and 230 ± 10 s for CG. HG showed the longest bleeding time, significantly different compared to AG and CG. There was no significant difference between AG and control rats, but results showed a strong tendency, suggesting that amaranth proteins could be involved in the inhibition of platelet plug formation. Since measuring BT implies cutting an arteriole from the tail triggering smooth muscle layer vasoconstriction after injury, the presence of vasodilators could partially counteract the injury vasoconstriction, causing a prolongation of BT. Heparinized animals highest BT observed is related to thrombin's ability to interact with protease-activated receptors (PAR family) present in platelets surface, causing the activation of these receptors and leading to shape changes that promote platelet aggregation. In this sense, heparin inhibition of thrombin prevents interaction with PAR receptors and hence platelet aggregation, prolonging BT.

Hemostatometer assay allows the dynamic study of different clotting parameters while blood is flowing; the equipment follows the descriptive details given by various authors (Shimizu et al. 2009; Satake et al. 2007; Sabbione et al. 2016c). Blood extracted from the animal's abdominal aorta artery is placed into a 37 °C thermostated chamber. The blood is displaced by liquid paraffin pumped at a constant flow through polyethylene tubing. The basal pressure of the system is set at 60 mmHg and a pressure transducer records pressure changes in the system over time, generating hemostatograms. While blood is flowing, the tube is pierced with a needle 90 s after blood withdrawal, resulting in "bleeding" from the tube. Phenomena tending to occlude the injury of the tube stop the hemorrhage due to platelet hemostatic plug formation in the holes, causing recovery of the initial pressure in the system. Subsequently, blood flow in the tube stops and the pressure increases, indicating total clotting. The pressure recovery area of hemostatograms assesses platelet plug formation while the time required to reach total coagulation, defined as clotting time, involves the coagulation process.

Sabbione et al. (2016c) performed hemostatometer studies and described an average recovery area of AG animals (13,618 ± 1454 mm Hg.s) higher compared to CG (6422 ± 1208 mm Hg.s), whereas significant differences were not observed between AG and HG (12,613 ± 836 mm Hg.s). Results evidenced that the inclusion of amaranth protein isolate in the diet produced an increase in this parameter, similar to that generated by the use of heparin anticoagulant. The recovery area values obtained by Sabbione et al. (2016c) resulted comparable to others informed in previous works, however, none of the authors studied amaranth proteins nor similar food matrices. For example, Shimizu et al. (2009) showed recovery area values from 5000 to 7000 mm Hg.s for control rat blood, and values between 10,000 and 12,000 mm Hg.s when blood was blended with a protein hydrolysate from pig muscle.

The clotting time parameter is associated with secondary hemostasis, where thrombin enzyme generates fibrin monomers to reach total coagulation of the blood. Sabbione et al. (2016c) described significant differences in average clotting times obtained for the three groups in study: the highest clotting time corresponded to HG (828.5 ± 46.8 s), an expectable behavior because heparin is a thrombin inhibitor anticoagulant. AG clotting time was 612.1 ± 32.4 s, while CG presented the lowest value, 330.9 ± 49.5 s. In concordance with these results, Satake et al. (2007) found differences between the clotting times of animal's control group and those administered with *Centella asiatica* extracts (500 s and 600–750 s, respectively). However, other authors did not detect clotting time variations when studying potentially antithrombotic samples compared to a control group (Shimizu et al. 2009).

Clotting tests are clinical functional assays that evaluate the rate of clot formation once the coagulation cascade gets activated. These tests are commonly used to identify defects of the extrinsic, intrinsic, and final common pathways of the coagulation cascade. Rats have been a widely used animal model over the last years to study different clotting parameters showing great variability in the results depending on the kits and rats used. The prothrombin time (PT), activated partial thromboplastin time (APTT), and thrombin time (TT) measurements are some of the clotting tests, performed using clinically commercial kits with plasma separated from an aliquot of animals' blood. The TT evaluates the conversion of fibrinogen to fibrin in the common pathway of the coagulation cascade, evidencing the presence of thrombin inhibitors; the PT detects alterations of the extrinsic pathway factors as it records the clotting time of a citrated plasma in presence of thromboplastin (tissue factor and phospholipids from platelet membranes); the APTT studies changes in the intrinsic pathway components and consists in the measurement of the clotting time of a citrated plasma in presence of thromboplastin phospholipid components (partial thromboplastin), Ca^{2+} and a plasma activator used to eliminate the variability associated with glass activators. Sabbione et al. (2016c) described a significant increase of TT of plasma from AG rats (113.1 ± 5.6 s) considering that CG presented a lower TT (96.4 ± 4.3 s). However, this increment was not comparable to that detected in the HG (>600 s). TT of AG showed a 17% increase compared to animals from CG, higher extension than that informed by You et al. (2004), who studied the effect of an enzyme from snake venom. APTT values described by Sabbione et al. (2016a) showed significant differences between the animal groups. HG showed the highest APTT (89.8 ± 8 s), followed by AG (49.4 ± 6 s) with a 14% increment over CG (43.4 ± 2.9 s). However, PT from the plasma of AG rats showed no significant difference compared with CG, while PT from HG resulted significantly different. *Ex vivo* clotting test data described by Sabbione et al. (2016c) would indicate that amaranth peptides released during rats' digestion process, produce alterations in the common pathway of coagulation (fibrin formation phase), and could also induce modifications over the intrinsic pathway. The results are in concordance with hemostatometer clotting time results due to an increase in clot formation time of AG animals compared with CG. Platelet aggregation effects cannot be discarded according to the recovery area and bleeding time findings. Moreover, inhibitory clotting effects from amaranth peptides were observed using an *in vitro* assay that assesses common pathway (Sabbione et al. 2015, 2016b; Ayala-Niño et al. 2019a, 2019b).

8.4 ANTIOXIDANT ACTIVITY AND PREVENTION OF THE LOW-DENSITY LIPOPROTEINS (LDL) OXIDATION

Atherosclerosis, one of the main causes of CVDs, is a chronic, progressive, and multifactorial disorder characterized by accumulation of lipids and fibrous elements in middle-sized and large arteries (Frostegård 2013). Oxidative stress, endothelial dysfunction, and inflammation represent the key triad in its development and progression, involving macrophages, endothelial cells, and smooth muscle cells. LDL oxidation in arterial walls is crucial in early stages of atherogenic plaque development. The LDL particles, which have apo-B100 as their main constitutive apoprotein, transport cholesterol from the liver to peripheral cells where its incorporation is mediated by apo-B100 receptors in a regulated and saturable manner. Macrophages have a non-regulated pathway of cholesterol incorporation, but it is mediated by unspecific receptors capable of recognizing LDL modified mainly by oxidation (LDLox). This pathway can lead to an abnormal accumulation of cholesterol by macrophages that degenerate to foamy characteristic cells of young atherogenic plaque (Lusis 2010). Other atherogenic characteristics are monocytes chemotaxis, the capacity to enhance the monocytes adhesion and differentiation to macrophages, cytotoxicity against endothelial and smooth muscle cells, and immunogenicity (Young and Parthasarathy 1994). During LDL oxidation, both lipids and proteins can undergo oxidative changes. *In vivo* LDL oxidation occurs mostly in sub-endothelial space and involves transition metals (iron and copper) and several enzyme systems (Young and McEneny 2001). Polyunsaturated fatty acids of LDL are the major targets of oxidation, being the hydroperoxides and conjugated dienes (CD) the first oxidation products. Further oxidation results in short-chain aldehyde or carboxy derivatives, 4-hydroxynonenal (HNE), and malondialdehyde (MDA) among them. The oxidation lipid products can react with the apo-B100 fraction generating fragmentation and negative charge. When there is a minor damage, only the lipid fraction gets affected resulting in a minimally modified LDL, which preserves the affinity for the specific LDL receptor, has a small negative charge, and induces inflammatory changes. With the increasing inflammatory state, a continuous oxidation of LDL is produced affecting lipids, and proteins and leading to the highly modified LDL these particles are not recognized by LDL receptors but they are recognized by the unspecific receptors of macrophages (Levitan et al. 2010). Dietary antioxidants are potentially anti-atherogenic agents capable of reducing the risk of CVD (Shahidi and Zhong 2015). The anti-atherogenic capacity can be assessed through the ability to prevent the LDL oxidation, which can be evaluated by diverse methodologies including *ex vivo* LDL resistance to an induced oxidative stress model, and *in vivo* biomarkers such as the determination of circulating LDLox and the presence of autoantibodies against LDLox (Winklhofer-Roob et al. 2017). In *ex vivo* assays, LDL isolated from plasma are oxidatively modified by catalytic amounts of metal ions (iron or copper) or, in the presence of H_2O_2, metal ions catalyze the formation of OH· radical, which is one of the most potent biological oxidants. One of the most commonly used procedures for measuring the oxidation resistance in these models is the determination of CD formation. Kinetics of LDL oxidation presents an initial lag phase (LT), followed by a propagation phase characterized by a propagation rate (PR), with a maximum slope, that starts presumably after the endogenous antioxidants have been consumed (Jialal and Devaraj 1996). In other *ex vivo* study models, LDL oxidation is initiated by arterial wall cell cultures, mainly macrophages, and is mediated by mechanisms involving free radicals, thiols, and peroxynitrites (Salvayre et al. 2016). Montoya-Rodríguez et al. (2014) studied the anti-atherosclerotic potential of pepsin-pancreatin hydrolysates from unprocessed and extruded amaranth in THP-1 lipopolysaccharide-induced human macrophages and found that extruded amaranth hydrolysate showed potential anti-atherosclerotic effect by reducing the expression of proteins associated with the lectin-like oxidized LDL receptor-1 (LOX-1) signaling pathway. In a similar study, three amaranth pure peptides (HGSEPFGPR, RPRYPWRYT, and RDGPFPWPWYSH) also reduced the expression of proteins associated with LOX-1 signaling pathway, but this action did not correlate with their capability to act as antioxidants (Montoya-Rodríguez and González de Mejía 2015). *A. mantegazzianus* protein isolates were subjected to simulated GID processes. The gastrointestinal digest and some constituent peptide fractions were analyzed through different methodologies for the evaluation of the antioxidant activity. These studies included *in vitro* methods of scavenging or inhibition of ROS, and *ex vivo* methods to analyze the prevention of LDL oxidation. The effect of amaranth peptides on LDL oxidation was studied using an *ex vivo* model system where the LDL oxidation was induced by a combination of Cu^{2+}/H_2O_2. Several aspects were evaluated: the CD time-evolution, the formation of lipid oxidation secondary products (2-thiobarbituric acid reactive substances assay, TBARS), the formation of lipid oxidation products-protein adducts (fluorescence time-evolution), and changes in LDL charge (agarose gel electrophoresis). The digest obtained from amaranth protein isolate was able to retard the beginning of lipid oxidation (LT) and decrease the PR of CD formation in a protein concentration-dependent way. Protein modifications were also delayed although there was some degree of alteration indicated by a certain increase in fluorescence; however, these protein modifications were not enough to change the electrophoretic mobility (García Fillería and Tironi 2017). Orsini Delgado et al. (2015) demonstrated that gastrointestinal digest has the ability to inhibit the formation of OH· radicals (by metal chelators peptides, HORAC assay) as well as to neutralize ROO· radicals (chain breaking by donation of a hydrogen atom, ORAC assay), which could be potential action mechanisms in the retardation of lipid and protein LDL oxidation. Among all molecules present in gastrointestinal digest, the most active fractions in the prevention and/or retardation of LDL oxidation were those with molecular weights (MW) between 0.59–1.4 kDa (5–13 amino acids). Seven fractions separated (gel filtration FPLC) within this MW range were able to completely inhibit CD formation and to diminish TBARS content. Some of these fractions (0.75–1.1 kDa fractions containing both hydrophilic and hydrophobic molecules according to RP-HPLC analysis) were also capable of completely inhibiting the fluorescence increment and preventing changes in the electrophoretic mobility. The remaining fractions (1.1–1.4 kDa with high content of hydrophobic molecules but low proportion of hydrophilic molecules; and 0.59–0.75 kDa with high content of hydrophilic compounds but very low proportion of hydrophobic molecules) caused a partial effect on these parameters. The activity of these fractions did not present a direct correlation with their peptide concentrations, indicating that the higher antioxidant effect was associated with the type of peptides and not with the total concentration. Some of the most active fractions against LDL oxidation presented also the highest activity by ORAC and HORAC assays, while other ones showed low activity by both methodologies. In contrast, fractions of low MW (<0.4 kDa) presented a very high ORAC activity but low HORAC activities, and did not present a good capacity to prevent LDL oxidation. A combination of peptides with different characteristics would expand the possible action mechanisms and the ability to act in different environments; while more hydrophilic molecules could act on water soluble free radicals, other more hydrophobic peptides could interact in the water-lipid interface and scavenge lipid or protein radicals (García Fillería and Tironi 2017).

Prevention of LDL oxidation by ten synthetic peptides (P1: TEVWDSNEQ, P2: IYIEQGNGITGM, P3: GDRFQDQHQ, P4: LAGKPQQEHSGEHQ, P5: YLAGKPQQEH, P6: LQAEQDDR, P7: HVIKPPSRA, P8: AWEEREQGSR, P9: AVNVDDPSK, P10: KFNRPETT) derived from GID of amaranth 11S globulin was also studied. These peptides identified by LC-MS/MS in fractions with high ORAC activity (Orsini Delgado et al. 2016), presented diverse behaviors: P1 showed a minor effect on the LT value; P8, P5, and P2 produced a decrease in the PR value; and P6 and P9 presented a minor effect on both parameters. P7, P3, and P10 were the most active peptides, and their activity was dependent on the peptide concentration. These three peptides exhibited a total inhibition of LDL oxidation at concentrations of 1 mg/mL, and LT increment as well as PR decrease at lower concentrations. They also reduced the TBARS and prevented changes in LDL mobility. Moreover, P3 and P7 completely inhibit the fluorescent products formation, while P10 produced a delay (García Fillería and Tironi 2017). Regarding ORAC activity, P8>P5~P2~P1 showed the highest activity, followed by P4>P7~P3 which presented a much lower activity. P6, P9, and P10 did not present activity in the ORAC assay (Orsini Delgado et al. 2016). For the HORAC assay, the order of activity was P8>P2>P5=P1>P4=P7=P6>P9, while P3 and P10 showed no activity (García Fillería 2019). In this sense, those peptides with the highest ORAC and HORAC activities (P8, P5, P2) were not the most active in prevention of LDL oxidation, and produced only a reduction in the PR of lipid oxidation. These peptides are negatively charged or close to neutrality at physiological pH, with a major proportion of acidic amino acids in P8, and hydrophobic amino acids in P5 and P2, being the last one the most hydrophobic. All of these peptides have at least one voluminous aromatic residue in their sequences. However, peptides with the highest activity against Cu^{2+}/H_2O_2-induced oxidation of LDL (P7, P3 and P10) did not show significant ROO· scavenging (ORAC) or inhibition of OH· formation (HORAC). Besides, these three peptides have diverse physicochemical properties: P7 is a positively charged peptide with a high proportion of aliphatic and basic amino acids and considerable hydrophobic character; P3 is a negatively charged peptide with a high proportion of acidic amino acids and hydrophilic character; and P10 is a positively charged peptide, with no significant predominance of any type of amino acid, with hydrophilic character. The activity of these peptides was explained by different properties: the cationic character (P7 and P10); the presence of the amino acid histidine (P7 and P3); the presence of lysine (P7 and P10; García Fillería 2019). Positively charged fractions with a high content of basic amino acids can act as chain terminators and prevent the spread of LDL oxidation by hydrogen donation to neutralize the lipid radicals (Pan et al. 2019). Moreover, the three peptides contained some hydrophobic amino acids which can enhance their antioxidant properties since they increase the accessibility to hydrophobic targets, such as LDL. In this regard, a possible explanation for the discrepancies between the behaviors in aqueous solutions and in the presence of LDL could be that the peptides probably interact with the LDL surface inducing conformational changes of the peptides and modifying the exposition of certain residues (García Fillería 2019).

In order to exert any effect on LDL oxidation, peptides must be able to enter the body. This capacity was evaluated through intestinal absorption simulation studies using a Caco2-TC7 cell model. An active fraction with molecules in the range of 1.1–1.25 kDa was partially modified by the action of brush border peptidases, and some of the resulting molecules were able to cross the monolayer. Moreover, another fraction (MW about 0.6 kDa) was modified in a lower proportion by extracellular peptidases and was absorbed in a significant manner (García Fillería and Tironi 2020). P5, P6, P8, and P9 were able to cross intestinal cells without previous modifications but they had a moderate or low activity against LDL oxidation. The peptides with the highest prevention of Cu^{2+}/H_2O_2-induced LDL oxidation (P7, P3, and P10) presented diverse behaviors. P7 was completely degraded by brush border peptidases. P10 was partially degraded into more hydrophilic molecules that were able to cross the cell monolayer but not the original molecule. P3 was partially degraded, and both the products generated and the remaining peptide could cross the monolayer, constituting a good candidate for the prevention of LDL oxidation (García Fillería 2019).

8.5 CONCLUSION

Bioactive peptides derived from amaranth storage proteins could play an important role in preventing the development of CVDs. Based on the *in silico, in vitro, ex vivo* and *in vivo* results described above, it can be inferred that amaranth proteins are potential antihypertensive, antithrombotic, and LDL oxidation prevention agents.

It has been shown that antihypertensive peptides could lower BP through at least three mechanisms: inhibition of ACE and renin, the main enzymes of the circulating RAAS, bradykinin-mediated NO production and the level of intracellular Ca^{2+} through voltage-operated channels or by adrenergic receptors. On the other hand, antithrombotic amaranth peptides have the ability to inhibit *in vitro* blood clot formation, exerting their action over thrombin enzyme or fibrinogen polymerization.

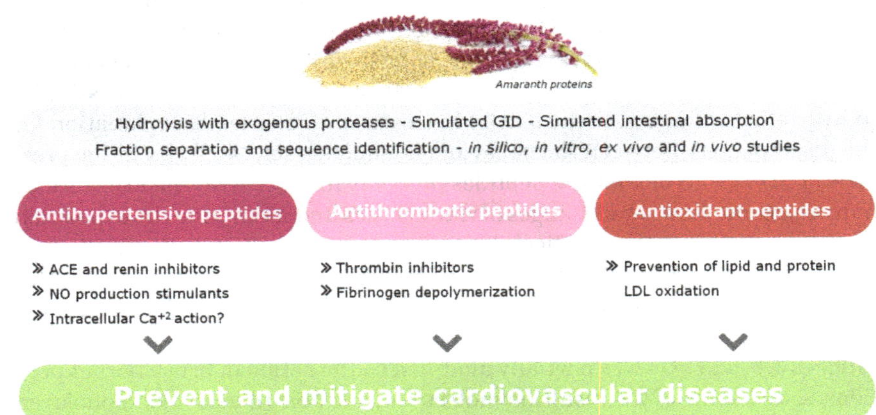

Figure 8.2 Effects of amaranth peptides on cardiovascular diseases.

Finally, amaranth protein digests have the ability to prevent or retard lipid and protein oxidation of LDL particles. Moreover, when amaranth proteins were incorporated into the diet of different animals, a significant reduction in BP (in hypertensive animals) and in clot formation (in normal animals) was observed, implying that these bioactive peptides were released during digestion, absorbed in the intestine, passed through the circulatory system, and exerted their biological effect.

Given the multiple activities exhibited by amaranth peptides, it would be important to analyze their potential multi-activity using an integrated methodological strategy: *in silico* analysis with *in vitro* and *in vivo* experiments. Furthermore, it would be important to obtain multifunctional hydrolysates in which the biological activities of peptides collaborate in reducing the risk of developing CVDs to incorporate them into the formulation of functional foods. The final validation of the physiological activity of amaranth proteins and peptides, as well as of the foods that may contain them, requires something not done so far: clinical trials.

ACKNOWLEDGMENTS

We are grateful to all the national and international institutions that have supported the studies carried out by our research group as well as those fellows and researchers who have participated in them.

REFERENCES

Aluko, Rotimi E., and Emefa Monu. 2003. "Functional and Bioactive Properties of Quinoa Seed Protein Hydrolysates." *Food Chemistry and Toxicology* 68 (4): 1254–1258. https://doi.org/10.1111/j.1365-2621.2003.tb09635.x.

Arai, Soichi. 1996. "Studies on Functional Foods in Japan-State of the Art." *Bioscience, Biotechnology and Biochemistry* 60 (1): 9–15. https://doi.org/10.1271/bbb.60.9.

Ayala-Niño, Alexis, Gabriela M. Rodríguez-Serrano, Luis G. González-Olivares, Elizabeth Contreras-López, Patricia Regal-López, and Alberto Cepeda-Saez. 2019a. "Sequence Identification of Bioactive Peptides from Amaranth Seed Proteins (Amaranthus Hypochondriacus Spp.)." *Molecules* 24 (17): 3033. https://doi.org/10.3390/molecules 24173033.

Ayala-Niño, Alexis, Gabriela M. Rodríguez-Serrano, Ruben Jiménez-Alvarado, Mirandeli Bautista-Avila, José A. Sánchez-Franco, Luis G. González-Olivares, and Alberto Cepeda-Saez. 2019b. "Bioactivity of Peptides Released During Lactic Fermentation of Amaranth Proteins with Potential Cardiovascular Protective Effect: An in Vitro Study." *Journal of Medicinal Food* 22 (10): 976–981. https://doi.org/10.1089/jmf.2019.0039.

Barba de la Rosa, Ana P., Adriana Barba Montoya, Pedro Martínez-Cuevas, Blanca Hernández-Ledesma, Ma Fabiola León-Galván, Antonio De León-Rodríguez, and Carmen González. 2010. "Tryptic Amaranth Glutelin Digests Induce Endothelial Nitric Oxide Production through Inhibition of ACE: Antihypertensive Role of Amaranth Peptides." *Nitric Oxide—Biology and Chemistry* 23 (2): 106–111. https://doi.org/10.1016/j.niox.2010.04.006.

Cheng, Shuzheng, Maolin Tu, Hanxiong Liu, Guanghua Zhao, and Ming Du. 2019. "Food-Derived Antithrombotic Peptides: Preparation, Identification, and Interactions with Thrombin." Critical *Reviews in Food Science and Nutrition* 59 (Suppl 1): S81–S95. https://doi.org/10.1080/10408398.2018.1524363.

European Commission. 1995. "Functional Food Science in Europe: FUFOSE." https://cordis.europa.eu/project/id/FAIR950572/es. Accessed September 25, 2020.

Fritz, Mariana, Bruno Vecchi, Gustavo Rinaldi, and María C. Añón. 2011. "Amaranth Seed Protein Hydrolysates Have in Vivo and in Vitro Antihypertensive Activity." *Food Chemistry* 126 (3): 878–884. https://doi.org/10.1016/j.foodchem.2010.11.065.

Frostegård, Johan. 2013. "Immunity, Atherosclerosis and Cardiovascular Disease." *BMC Medicine* 11 (1): 117. https://doi.org/10.1186/1741-7015-11-117.

García Fillería, Susan F. 2019. "Evaluación de aislado proteico de amaranto como fuente de péptidos antioxidantes: Estudios in vitro e in vivo." Ph.D. diss., Facultad de Ciencias Exactas, Universidad Nacional de La Plata, La Plata.

García Fillería, Susan F., and Valeria A. Tironi. 2017. "Prevention of in Vitro Oxidation of Low Density Lipoproteins (LDL) by Amaranth Peptides Released by Gastrointestinal Digestion." *Journal of Functional Foods* 34: 197–206. https://doi.org/10.1016/j.jff.2017.04.032.

García Fillería, Susan F., and Valeria A. Tironi. 2020. "Intracellular Antioxidant Activity and Intestinal Absorption of Amaranth Peptides Released by Gastrointestinal Digestion in Caco-2 TC7 Cells." *Food Bioscience* (in press).

Iwaniak, Anna, Małgorzata Darewicz, and Piotr Minkiewicz. 2018. "Peptides Derived from Foods as Supportive Diet Components in the Prevention of Metabolic Syndrome." *Comprehensive Reviews in Food Science and Food Safety* 17 (1): 63–81. https://doi.org/10.1111/1541-4337.12321.

Jialal, Ishwarlal, and Sridevi Devaraj. 1996. "Low-Density Lipoprotein Oxidation, Antioxidants, and Atherosclerosis: A Clinical Biochemistry Perspective." *Clinical Chemistry 42* (4): 498–506. https://doi.org/10.1093/clinchem/42.4.498.

Lado, María Belén, Julieta Burini, Gustavo Rinaldi, María C. Añón, and Valeria A. Tironi. 2015. "Effects of the Dietary Addition of Amaranth (*Amaranthus mantegazzianus*) Protein Isolate on Antioxidant Status, Lipid Profiles and Blood Pressure of Rats." *Plant Foods for Human Nutrition* 70 (4): 371–379. https://doi.org/10.1007/s11130-015-0516-3.

Levitan, Irena, Suncica Volkov, and Papasani V. Subbaiah. 2010. "Oxidized LDL: Diversity, Patterns of Recognition, and Pathophysiology." *Antioxidants & Redox Signaling* 13 (1): 39–75. https://doi.org/10.1089/ars.2009.2733.

Li, Chun Hui, Toshiro Matsui, Kiyoshi Matsumoto, Rikio Yamasaki, and Terukazu Kawasaki. 2002. "Latent Production of Angiotensin I-Converting Enzyme Inhibitors from Buckwheat Protein." *Journal of Peptide Science* 8 (6): 267–274. https://doi.org/10.1002/psc.387.

Li, Jian, Allen Schantz, Maureen Schwegler, and Gopi Shankar. 2011. "Detection of Low-Affinity Anti-Drug Antibodies and Improved Drug Tolerance in Immunogenicity Testing by Octet® Biolayer Interferometry." *Journal of Pharmaceutical and Biomedical Analysis* 54 (2): 286–294. https://doi.org/10.1016/j.jpba.2010.08.022.

Li-chan, Eunice C. Y. 2015. "Bioactive Peptides and Protein Hydrolysates: Research Trends and Challenges for Application as Nutraceuticals and Functional Food Ingredients." *Current Opinion in Food Science* 1: 28–37. https://doi.org/10.1016/j.cofs.2014.09.005.

Lusis, Aldons J. 2010. "Atherosclerosis." *Nature* 407 (6801): 233–241. https://doi.org/10.1038/35025203.

Ma, Min Suk, Young Bae In, Gyu Lee Hyeon, and Cha Bum Yang. 2006. "Purification and Identification of Angiotensin I-Converting Enzyme Inhibitory Peptide from Buckwheat (Fagopyrum Esculentum Moench)." *Food Chemistry* 96 (1): 36–42. https://doi.org/10.1016/j.foodchem.2005.01.052.

McNicol, Archibald, and Sara J. Israels. 2003. "Platelets and Anti-Platelet Therapy." *Journal of Pharmacological Sciences* 93 (4): 381–396. https://doi.org/10.1254/jphs.93.381.

Minekus, Mans, Marie Alminger, Paula Alvito, Simon Ballance, Torsten Bohn, Claire Bourlieu, Frédéric Carrière et al. 2014. "A Standardised Static in Vitro Digestion Method Suitable for Food—an International Consensus." *Food Function* 5 (5): 1113–1124. https://doi.org/10.1039/c3fo60702j.

Monie, Dileep D., and Emma P. DeLoughery. 2017. "Pathogenesis of Thrombosis: Cellular and Pharmacogenetic Contributions." *Cardiovascular Diagnosis and Therapy* 7 (S3): S291–S298. https://doi.org/10.21037/cdt.2017.09.11.

Montoya-Rodríguez, Álvaro, Mario A. Gómez-Favela, Cuauhtémoc Reyes-Moreno, Jorge Milán-Carrillo, and Elvira González de Mejía. 2015. "Identification of Bioactive Peptide Sequences from Amaranth (*Amaranthus hypochondriacus*) Seed Proteins and Their Potential Role in the Prevention of Chronic Diseases." *Comprehensive Reviews in Food Science and Food Safety* 14 (2): 139–158. https://doi.org/10.1111/1541-4337.12125.

Montoya-Rodríguez, Álvaro, and Elvira González de Mejía. 2015. "Pure Peptides from Amaranth (*Amaranthus hypochondriacus*) Proteins Inhibit LOX-1 Receptor and Cellular Markers Associated with Atherosclerosis Development in Vitro." *FRIN* 77: 204–214. https://doi.org/10.1016/j.foodres.2015.06.032.

Montoya-Rodríguez, Álvaro, Jorge Milán-Carrillo, Vermont P. Dia, Cuauhtémoc Reyes-Moreno, and Elvira González de Mejía. 2014. "Pepsin-Pancreatin Protein Hydrolysates from Extruded Amaranth Inhibit Markers of Atherosclerosis in LPS-Induced THP-1 Macrophages-Like Human Cells by Reducing Expression of Proteins in LOX-1 Signaling Pathway." *Proteome Science* 12 (30): 1–13. https://doi.org/10.1186/1477-5956-12-30.

Nardo, Agustina E., María C. Añón, and Alejandra V. Quiroga. 2020. "Identification of Renin Inhibitors Peptides from Amaranth Proteins by Docking Protocols." *Journal of Functional Foods* 64: 103683, January. https://doi.org/10.1016/j.jff.2019.103683.

Nieman, David C., Nicholas Gillitt, Fuxia Jin, Dru A. Henson, Krista Kennerly, R. Andrew Shanely, Brandon Ore, Mingming Su, and Sarah Schwartz. 2012. "Chia Seed Supplementation and Disease Risk Factors in Overweight Women: A Metabolomics Investigation." *Journal of Alternative and Complementary Medicine* 18 (7): 700–708. https://doi.org/10.1089/acm.2011.0443.

Orona-Tamayo, Domancar, María E. Valverde, and Octavio Paredes-López. 2019. "Bioactive Peptides from Selected Latin American Food Crops—A Nutraceutical and Molecular Approach." *Critical Reviews in Food Science and Nutrition* 59 (12): 1949–1975. https://doi.org/10.1080/10408398.2018.1434480.

Orsini Delgado, María C., Mónica Galleano, María C. Añón, and Valeria A. Tironi. 2015. "Amaranth Peptides from Simulated Gastrointestinal Digestion: Antioxidant Activity against Reactive Species." *Plant Foods for Human Nutrition* 70 (1): 27–34. https://doi.org/10.1007/s11130-014-0457-2.

Orsini Delgado, María C., Agustina E. Nardo, Marija Pavlovic, Hélène Rogniaux, María C. Añón, and Valeria A. Tironi. 2016. "Identification and Characterization of Antioxidant Peptides Obtained by Gastrointestinal Digestion of Amaranth Proteins." *Food Chemistry* 197 (B): 1160–1167. https://doi.org/10.1016/j.foodchem.2015.11.092.

Pan, Mengshi, Yanjiao Huo, Chengtao Wang, Yanchun Zhang, Zhiyong Dai, and Bo Li. 2019. "Positively Charged Peptides from Casein Hydrolysate Show Strong Inhibitory Effects on LDL Oxidation and Cellular Lipid Accumulation in Raw264.7 Cells." *International Dairy Journal* 91: 119–128. https://doi.org/10.1016/j.idairyj.2018.09.011.

Quiroga, Alejandra V., Paula Aphalo, Agustina E. Nardo, and María C. Añón. 2017. "In Vitro Modulation of Renin—Angiotensin System Enzymes by Amaranth (*Amaranthus hypochondriacus*) Protein-Derived Peptides: Alternative Mechanisms Different from ACE Inhibition." *Journal of Agricultural and Food Chemistry* 65 (34): 7415–7423. https://doi.org/10.1021/acs.jafc.7b02240.

Quiroga, Alejandra V., Paula Aphalo, Jorge L. Ventureira, E. Nora Martínez, and María C. Añón. 2012. "Physicochemical, Functional and Angiotensin Converting Enzyme Inhibitory Properties of Amaranth (*Amaranthus hypochondriacus*) 7S Globulin." *Journal of the Science of Food and Agriculture* 92 (2): 397–403. https://doi.org/10.1002/jsfa.4590.

Ramírez-Torres, Giovanni, Noé Ontiveros, Verónica Lopez-Teros, Jesús Ibarra-Diarte, Cuauhtémoc Reyes-Moreno, Edith Cuevas-Rodríguez, and Francisco Cabrera-Chávez. 2017. "Amaranth Protein Hydrolysates Efficiently Reduce Systolic Blood Pressure in Spontaneously Hypertensive Rats." *Molecules* 22 (11): 1–8. https://doi.org/10.3390/molecules22111905.

Rinaldi, Gustavo. J. 1990. "Norepinephrine-Sensitive Calcium Pools in Tail Arteries of 22 Normotensive and Spontaneously Hypertensive Rats: Effects of Ryanodine and Caffeine." *Acta Physiologica et Pharmacologica Latinoamericana: Organo de La Asociacion Latinoamericana de Ciencias Fisiologicas y de La Asociacion Latinoamericana de Farmacologia* 40 (3): 339–355. www.ncbi.nlm.nih.gov/pubmed/2094167.

Rinaldi, Gustavo J., Grand Carlos, and Horacio E. Cingolani. 1991. "Characteristics of the Fast and Slow Components of the Response to Noradrenaline in Rat Aorta." *The Canadian Journal of Cardiology* 7 (7): 316–322. http://europepmc.org/abstract/MED/1933640.

Sabbione, Ana C., Sabrina M. Ibañez, E. Nora Martínez, María C. Añón, and Adriana A. Scilingo. 2016a. "Antithrombotic and Antioxidant Activity of Amaranth Hydrolysate Obtained by Activation of an Endogenous Protease." *Plant Foods for Human Nutrition* 71 (2): 174–182. https://doi.org/10.1007/s11130-016-0540-y.

Sabbione, Ana C., Agustina E. Nardo, María C. Añón, and Adriana A. Scilingo. 2016b. "Amaranth Peptides with Antithrombotic Activity Released by Simulated Gastrointestinal Digestion." *Journal of Functional Foods* 20 (January): 204–214. https://doi.org/10.1016/j.jff.2015.10.015.

Sabbione, Ana C., Gustavo Rinaldi, María C. Añón, and Adriana A. Scilingo. 2016c. "Antithrombotic Effects of *Amaranthus hypochondriacus* Proteins in Rats." *Plant Foods for Human Nutrition* 71 (1): 19–27. https://doi.org/10.1007/s11130-015-0517-2.

Sabbione, Ana C., Adriana Scilingo, and María C. Añón. 2015. "Potential Antithrombotic Activity Detected in Amaranth Proteins and Its Hydrolysates." *LWT—Food Science and Technology* 60 (1): 171–177. https://doi.org/10.1016/j.lwt.2014.07.015.

Salvayre, Robert, Anne Negre-Salvayre, and Caroline Camaré. 2016. "Oxidative Theory of Atherosclerosis and Antioxidants." *Biochimie* 125: 281–296. https://doi.org/10.1016/j.biochi.2015.12.014.

Satake, Toshiko, Kohei Kamiya, Yin An, Tomomi Oishi, and Junichiro Yamamoto. 2007. "The Anti-Thrombotic Active Constituents from *Centella Asiatica*." *Biological & Pharmaceutical Bulletin* 30 (5): 935–940. https://doi.org/10.1248/bpb.30.935.

Segura Campos, Maira R., Fanny Peralta González, Luis Chel Guerrero, and David Betancur Ancona. 2013a. "Angiotensin I-Converting Enzyme Inhibitory Peptides of Chia (*Salvia hispanica*) Produced by Enzymatic Hydrolysis." *International Journal of Food Science* 2013: 1–8. https://doi.org/10.1155/2013/158482.

Segura-Campos, Maira R., Ine M. Salazar-Vega, Luis A. Chel-Guerrero, and David A. Betancur-Ancona. 2013b. "Biological Potential of Chia (*Salvia hispanica* L.) Protein Hydrolysates and Their Incorporation into Functional Foods." *LWT—Food Science and Technology* 50 (2): 723–731. https://doi.org/10.1016/j.lwt.2012.07.017.

Shahidi, Fereidoon, and Ying Zhong. 2015. "Measurement of Antioxidant Activity." *Journal of Functional Foods* 18: 757–781. https://doi.org/10.1016/j.jff.2015.01.047.

Shimizu, Muneshige, Naoko Sawashita, Fumiki Morimatsu, Jun Ichikawa, Yasuki Taguchi, Yoshinobu Ijiri, and Junichiro Yamamoto. 2009. "Antithrombotic Papain-Hydrolyzed Peptides Isolated from Pork Meat." *Thrombosis Research* 123 (5): 753–757. https://doi.org/10.1016/j.thromres.2008.07.005.

Silva-Sánchez, Cecilia, Ana P. Barba de la Rosa, M. Fabiola León-Galván, Ben O. de Lumen, Antonio de León-Rodríguez, and Elvira González de Mejía. 2008. "Bioactive Peptides in Amaranth (*Amaranthus hypochondriacus*) Seed." *Journal of Agricultural and Food Chemistry* 56 (4): 1233–1240. https://doi.org/10.1021/jf072911z.

Suárez, Santiago, Paula Aphalo, Gustavo Rinaldi, Maria C. Añón, and Alejandra Quiroga. 2020. "Effect of Amaranth Proteins on the RAS System: In Vitro, in Vivo and Ex Vivo Assays." *Food Chemistry* 308: 125601, March. https://doi.org/10.1016/j.foodchem.2019.125601.

Swinbanks, David, and John O'Brien. 1993. "Japan Explores the Boundary Between Food and Medicine." *Nature Publishing Group*, July 15. https://doi.org/10.1038/364180a0.

Tiengo, Andréa, Mario Faria, and Flavia Maria Netto. 2009. "Characterization and ACE-Inhibitory Activity of Amaranth Proteins." *Journal of Food Science* 74 (5): H121–H126. https://doi.org/10.1111/j.1750-3841.2009.01145.x.

Tovar-Pérez, Erik G., Isabel Guerrero-Legarreta, Amelia Farrés-González, and Jorge Soriano-Santos. 2009. "Angiotensin I-Converting Enzyme-Inhibitory Peptide Fractions from Albumin 1 and Globulin as Obtained of Amaranth Grain." *Food Chemistry* 116 (2): 437–444. https://doi.org/10.1016/j.foodchem.2009.02.062.

Vecchi, Bruno, and María C. Añón. 2009. "ACE Inhibitory Tetrapeptides from *Amaranthus hypochondriacus* 11S Globulin." *Phytochemistry* 70 (7): 864–870. https://doi.org/10.1016/j.phytochem.2009.04.006.

Velarde-Salcedo, Aída J., Evelyn Regalado-Rentería, Rodrigo Velarde-Salcedo, Bertha I. Juárez-Flores, Alberto Barrera-Pacheco, Elvira González De Mejía, and Ana P. Barba De La Rosa. 2018. "Consumption of Amaranth Induces the Accumulation of the Antioxidant Protein Paraoxonase/Arylesterase 1 and Modulates Dipeptidyl Peptidase IV Activity in Plasma of Streptozotocin-Induced Hyperglycemic Rats." *Journal of Nutrigenetics and Nutrigenomics* 10: 181–193. https://doi.org/10.1159/000486482.

Vilcacundo, Rubén, Cristina Martínez-Villaluenga, Beatriz Miralles, and Blanca Hernández-Ledesma. 2019. "Release of Multifunctional Peptides from Kiwicha (*Amaranthus caudatus*) Protein Under in Vitro Gastrointestinal Digestion." *Journal of the Science of Food and Agriculture* 99 (3): 1225–1232. https://doi.org/10.1002/jsfa.9294.

WHO. 2016. *World Health Statistics 2016: Monitoring Health for the SDGs, Sustainable Development Goals*. Geneva, Switzerland: WHO Press.

WHO. 2018. *World Health Statistics 2018: Monitoring Health for the SDGs, Sustainable Development Goals*. Geneva, Switzerland: WHO Press.

Winklhofer-Roob, Brigitte M., Gernot Faustmann, and Johannes M. Roob. 2017. "Low-Density Lipoprotein Oxidation Biomarkers in Human Health and Disease and Effects of Bioactive Compounds." *Free Radical Biology and Medicine* 111: 38–86, October. https://doi.org/10.1016/j.freeradbiomed.2017.04.345.

You, Weon-Kyoo, Won-Seok Choi, You-Seok Koh, Hang-Cheol Shin, Yangsoo Jang, and Kwang-Hoe Chung. 2004. "Functional Characterization of Recombinant Batroxobin, a Snake Venom Thrombin-like Enzyme, Expressed from Pichia Pastoris." *FEBS Letters* 571 (1–3): 67–73. https://doi.org/10.1016/j.febslet.2004.06.060.

Young, Ian S., and Jane McEneny. 2001. "Lipoprotein Oxidation and Atherosclerosis." *Biochemical Society Transactions* 29 (2): 358–362. https://doi.org/10.1042/bst0290358.

Young, Stephen G., and Sampath Parthasarathy. 1994. "Why Are Low-Density Lipoproteins Atherogenic?" *The Western Journal of Medicine* 160 (2): 153–164. www.ncbi.nlm.nih.gov/pubmed/8160466.

Zheng, Yajun, Xian Wang, Yongliang Zhuang, Yan Li, Hailong Tian, Panqi Shi, and Guifeng Li. 2019. "Isolation of Novel ACE-Inhibitory and Antioxidant Peptides from Quinoa Bran Albumin Assisted with an in Silico Approach: Characterization, in Vivo Antihypertension, and Molecular Docking." *Molecules* 24 (24): 4562. https://doi.org/10.3390/molecules24244562.

Zhuo, Jia L., Fernanda M. Ferrao, Yun Zheng, and Xiao C. Li. 2013. "New Frontiers in the Intrarenal Renin-Angiotensin System: A Critical Review of Classical and New Paradigms." *Frontiers in Endocrinology* 4: 166, November. https://doi.org/10.3389/fendo.2013.00166.

Chapter 9

Effect of Amaranth Bioactive Peptides on the Gastrointestinal System

Valeria A. Tironi, María C. Añón, Adriana A. Scilingo, Alejandra V. Quiroga, and Ana C. Sabbione

CONTENTS

9.1 GENERAL SCOPE

The gastrointestinal system involves integrated physiological processes to provide effective digestion and absorption of nutrients, excretion of toxic or unusable compounds, and defense response against pathogens. Digestion and absorption processes involve mechanical transport; the exertion of environmental conditions (pH, salts, etc.); the activity of digestive enzymes; interactions between water, lipids, membranes, and partially hydrolyzed molecules; neuroendocrine control system activity; secretory processes; and transport through the intestinal wall. The defense mechanisms of the intestine include the physical barrier of the mucosa, which regulates paracellular and transcellular transport and the entry of luminal antigens, and the immune system through innate and adaptive immunity, mucosa-associated lymphatic tissues (MALT), and tolerance to commensal microbiota (Guerra et al. 2012).

DOI: 10.1201/9781003087618-9

In the human gut there are trillions of microorganisms (microbiota) of considerable biodiversity (approximately 1000 bacterial species) with a density of 10^{13}–10^{14}. In the colon there are around 160 to 500 different bacterial species, implying a wide range of specific microbiological characteristics. The microbiota develop various immune and protective functions, stimulate cell renewal, and participate in the host's metabolism by regulating gene expression and providing enzymes for the synthesis or modification of many substances, including vitamins, non-digestible carbohydrates and glycoconjugates, undigested proteins, and bile acids. The fermentation of saccharides produces short-chain fatty acids (SCFAs) that play an essential role in maintaining a healthy mucosa and the production of anti-inflammatory interleukins (Tomasello et al. 2016).

In health, homeostasis occurs between the mucosa (complete barrier), the microbiota (eubiosis), and the immune system. However, certain environmental and genetic factors modify these conditions, leading to pathological states (Vindigni et al. 2016). When the intestinal barrier is disrupted, luminal toxins and antigens will penetrate sub-epithelial tissues through the barrier, causing a mucosal oxidative stress and a systemic inflammatory response. In addition, overproduction of reactive oxygen and nitrogen species (ROS and RNS) and pro-inflammatory cytokines disrupts the intestinal barrier. In consequence, oxidative stress is involved in the development of intestinal diseases such as inflammatory bowel disease (IBD), irritable bowel syndrome, and colon cancer (Zhuang et al. 2019). IBD is a recurrent inflammatory condition that is increasing worldwide and presents a multifactorial pathogenesis: genetic, environmental factors (diet, stress, sleep patterns, hygiene, antibiotics, tobacco, etc.) and microbiota incidence. IBD includes two clinical entities: Crohn's disease (CD), which affects the entire thickness of the wall and can involve ileum and colon; and ulcerative colitis (UC), characterized by mucosal inflammation mainly limited to the rectum (Zhang et al. 2015). During the progression of chronic inflammation of the intestine, colitis develops. Furthermore, IBD has been linked to colorectal cancer development. ROS and RNS produced by inflammatory cells can affect the regulation of genes that encode factors which prevent carcinogenesis (e.g., p53, DNA mismatch repair proteins, DNA base excision-repair proteins), transcription factors (e.g., nuclear factor κB), or signaling proteins (e.g., cyclooxygenases; Ullman and Itzkowitz 2011).

Diet plays a very important role in intestinal function and immune activity, and in the microbial community as a source of metabolizable substances (Dolan and Chang 2017). It has been extensively demonstrated that diet has a great influence on the composition of the microbiota, estimated as 57% in comparison with the 12% caused by genetic factors. It is believed that the Western-style diet, high in refined carbohydrates, simple sugars, saturated fatty acids, and omega-6 fatty acids, and poor in fruits and vegetables, may be a trigger for IBD. Excessive consumption of animal protein is associated with an increased risk of developing CD, while the consumption of large amounts of fatty acids is associated with UC (Tomasello et al. 2016). The role of the diet (especially low fruit and vegetable consumption) and other habits (lack of physical activity, alcohol, tobacco) in cancer promotion has also been widely recognized. In contrast, there is growing evidence indicating that some components of different foods, such as peptides, proteins, and secondary metabolites, can prevent the development of this disease (Ferlay et al. 2019).

Bioavailability and bioaccessibility are important aspects in the study of potentially bioactive compounds (Alegría et al. 2015). Even though *in vivo* assays are considered the best methodology to simulate gastrointestinal process, *in vitro* methods present some advantages since they are faster, cheaper, and mainly without ethical restrictions. Moreover, *in vitro*–simulated gastrointestinal digestion (SGID) assays are suitable for mechanistic studies and hypothesis construction (Minekus et al. 2014). *In vitro* methods simulate physiological conditions and the sequence of events produced in the gastrointestinal tract. Static methodologies are the simplest ones and include two or three digestion steps (oral, gastric, intestinal), all occurring in the same reactor. These methods do not simulate physical processes, nor do they simulate absorption, hormonal and nervous controls, immunological system, and microbiota actions (Alegría et al. 2015). Many static SGID protocols have been widely used. In order to homogenize results, the INFOGEST action has proposed a consensus protocol based on physiological conditions that include oral, gastric, and intestinal phases containing complete electrolytic solutions, specific enzymes, and bile salts in the intestinal phase. This protocol considers the digestion of all food components and can be applied to diverse matrices (Minekus et al. 2014).

Food-derived peptides generated by gastrointestinal digestion (GID) could exert local biological actions at the intestinal lumen and/or cellular level and/or inside the organism. Once in contact with enterocytes, peptides can experiment modifications caused by the attack of brush border peptidases and/or can pass through these cells (Picariello et al. 2015). These processes can be simulated using intestinal cells cultures monolayers (Caco-2 and Caco-2 TC7 cells among them). There is much experimental evidence in the literature indicating that peptides with more than three amino acids could be absorbed intact through the intestinal epithelium, depending on several factors such as molecular mass, hydrophobicity, net charge, or tendency to aggregate (Xu et al. 2019).

Diverse research studies have demonstrated that amaranth proteins contain peptide sequences with potential antioxidant and antitumor activities. Experimental evidence in relation to the physiopathology of intestinal cells is presented in this chapter. In addition, the effect of amaranth proteins and peptides on the cholesterol absorption and, consequently, their hypocholesterolemic activity, is described.

9.2 ANTIOXIDANT ACTIVITY AND REDOX BALANCE IN THE INTESTINE

The intestinal mucosa is constantly exposed to oxidants, mutagens, and carcinogens from the diet, and to endogenously generated ROS and RNS because mitochondrial oxidation processes, enzymatic action, and defense mechanisms against microorganisms are sources of their continuous production in cells. In order to preserve their cellular integrity and tissue homeostasis, intestinal cells have various defense mechanisms including antioxidant enzymes (e.g., superoxide dismutase [SOD], catalase [CAT], glutathione peroxidase [GPx]) and low-molecular-weight compounds capable of scavenging ROS, such as reduced glutathione (GSH; Piechota-Polanczyk and Fichna 2014). An array of inducible defense genes that result in the neutralization of oxidative stress events is present. The signal transduction pathways responsible for sensing oxidative stress and activating the appropriate defense genes involve the transcription factor Nrf2, which is activated by ROS. Nrf2 recognizes the antioxidant response element (ARE) that participates in the activation of direct antioxidant enzymes and phase II detoxifying enzymes (conjugants). These enzymes are classified as "indirect" antioxidants based on their role in maintaining the redox balance and thiol homeostasis (Finley et al. 2011). The enzymatic defense can be induced or consumed by the presence of a stressor factor, increasing or decreasing their activity level, respectively. Antioxidant enzymes activity and the content of GSH are used as biomarkers in the study of several diseases related to oxidative stress since they are the first to indicate the cellular redox state (Constantini 2019).

Despite the existence of many defense systems, oxidative stress and redox imbalance can occur, for example, in the presence of lipid peroxide concentrations that are commonly found in foods cooked in fats or oils, which affect intestinal metabolic homeostasis (Aw 1999). Since oxidative or nitrosative stress can produce modifications in lipids, proteins and DNA, and lead to inflammatory processes, apoptosis, or transformation into tumor cells, they have been implicated in inflammatory bowel diseases and colorectal cancer. The basal level of ROS in small intestine enterocytes is lower than in colon cells, implying a greater contribution to colon diseases. Furthermore, the colonocytes have a higher concentration of CAT, SOD, and GPx compared to the small intestine (Piechota-Polanczyk and Fichna 2014). Increased oxidative stress and reduced antioxidant defenses have been observed in biopsies of people with IBD, which is generated by oxidative respiratory "blasts" of immune cells in the chronic inflammation phase (Zhang et al. 2015).

The incorporation of antioxidant dietary supplements is necessary to prevent the oxidative stress state. Although ascorbic acid, tocopherols, and polyphenols are the most studied antioxidant components, protein hydrolysates and peptides could also exert antioxidant effects through different mechanisms

(Esfandi et al. 2019). Antioxidant activity at the cellular level could be exercised by direct mechanisms such as ROS scavenging by hydrogen or electrons donation and/or ROS formation inhibition by metal chelation. In addition, some substances function as inducers and/or cell signals that lead to changes in gene expression, which result in the activation of enzymes that individually or synergistically eliminate ROS and/or toxins to protect cell integrity (Finley et al. 2011). In order to exert the intracellular antioxidant activity, peptides should be able to interact with the cells or penetrate them, which is a feature that highly depends on their physicochemical and structural characteristics (Wang et al. 2016).

The antioxidant activity of peptides derived from amaranth proteins has been studied through *in silico, in vitro*, and *in vivo* approaches. The first studies carried out showed that amaranth seed (*Amaranthus mantegazzianus*) contains peptides/polypeptides with moderate *in vitro* activity (ABTS$\cdot^+$ scavenging, prevention of the linoleic acid oxidation) which was found in the protein isolate as well as in albumin, globulin, and glutelin fractions, but not in globulin P fraction. The activities of all these samples strongly increased after extensive hydrolysis with alcalase (protein hydrolysis degrees, HD, between 20% and 30%), including globulin P fraction (Tironi and Añón 2010). Other authors informed that alcalase-hydrolysates from *A. hypochondriacus* albumins and globulins presented scavenging activity of DPPH· and ABTS$\cdot^+$ radicals, high chelation of Fe^{2+}, low chelation of Cu^{2+}, and iron-reducing power (FRAP assay; Soriano-Santos and Escalona-Buendía 2015). Moreover, Ayala-Niño et al. (2019) described ABTS$\cdot^+$ scavenging activity in hydrolysates of *A. hypochondriacus* obtained by a treatment with alcalase and flavourzyme. The authors identified potentially antioxidant peptide sequences with molecular weights in the range of 500 to 1400 Da.

Antioxidant activity of amaranth proteins (*A. mantegazzianus*) showed an important increase after performing a simplified protocol of SGID process optimized for amaranth protein isolate digestion (Orsini Delgado et al. 2011). This gastrointestinal digest was able to act through different ROS and RNS. Scavenging of peroxyl radicals (ORAC assay, IC_{50} = 0.024 mg protein/mL; electron spin resonance, IC_{50} = 0.93 mg protein/mL), inhibition of hydroxyl radicals' formation by metal chelation (HORAC assay, IC_{50} = 0.97 mg protein/mL), scavenging of hydroxyl radical (electron spin resonance, IC_{50} = 0.93 mg protein/mL), and scavenging of peroxynitrites (IC_{50} = 0.73 mg protein/mL) was evidenced. Moreover, it was demonstrated that a previous hydrolysis with alcalase did not improve the antioxidant activity after the SGID process (Orsini Delgado et al. 2015). In addition, after the emergence of a consensus SGID protocol (INFOGEST; Minekus et al. 2014), this protocol was applied to the amaranth protein isolate and a comparison with the in-house method previously mentioned was made. The modification of digestion conditions produced only small differences in the molecular composition, but did not affect the proteolysis degree (~60%) and the antioxidant activity by ORAC and HORAC assays, reaffirming the ability of these peptides to act against ROS in acellular *in vitro* assays. All the digest fractions separated by gel filtration fast protein liquid chromatography (FPLC) according to their molecular weight showed ORAC activity, with the most active fractions corresponding to molecular masses between 0.11 and 3.5 kDa. Furthermore, fractions with molecular weights between 0.43 and 3.5 kDa presented HORAC activity (Rodríguez et al. 2020).

Orsini Delgado et al. (2016) identified ten potentially antioxidant peptide sequences after performing SGID of the protein isolate (P1: TEVWDSNEQ, P2: IYIEQGNGITGM, P3: GDRFQDQHQ, P4: LAGKPQQEHSGEHQ, P5: YLAGKPQQEH, P6: LQAEQDDR, P7: HVIKPPSRA, P8: AWEEREQGSR, P9: AVNVDDPSK, P10: KFNRPETT) derived from the amaranth 11S globulin. According to Li and Li (2013), the location of bulky hydrophobic amino acids in the C-terminus region, polar or charged amino acids at the C-terminal, as well as residues with low electronic properties at positions 1 and 2 of the N-terminal, contributed to increase the ORAC activity. When this activity was evaluated by Orsini Delgado et al. (2016), the most active peptides were P8, P5, P2, and P1 (IC_{50} = 0.0067, 0.016, 0.017, and 0.020 mg peptide/mL, respectively). These peptides contain one aromatic amino acid (W or Y) in their sequence and at least one acidic amino acid (E or D). Regarding P8, the presence of a charged amino acid (K) in C1, and P and V in the second N-terminal position could also contribute to ORAC activity. Regarding the chelating activity of metals, the importance of the presence of E, D, H, and K amino acid residues was demonstrated in fractions of rice bran proteins (Phongthai et al. 2018). Furthermore, increased chelating activity due to the presence of a Q amino acid residue at the N-terminus has been reported. Since Q contains a carbamoyl group ($-CONH_2$), the carbonyl group (CO) is able to function as a ligand facilitating the formation of a stable complex with the metal ion (Egusa Saiga and Nishimura 2013). The most active peptides in HORAC assay (P8, P2, P5, and P1; OH· inhibition % = 79, 66, 57, and 56% for concentrations of 0.020 mg peptide/mL, respectively) contained E; P1 also contains D and Q (in the N-terminus), and P5 also contains K and H (García Fillería 2019). According to molecular dynamic analysis, none of the ten peptides has a conformation with ordered structure or folds suggesting a favorable condition for amino acid residues to interact with other substances (free radicals, metals, etc.). The peptides with the highest activity observed in both, ORAC and HORAC methods (P8, P5, P2, and P1) have a negative charge or are close to neutrality, and belong to the acid subunit of 11S globulin. Two of them (P8 and P1) have a large proportion of acidic amino acids, while P5 and P2 have hydrophobic amino acids in a higher proportion. P2 is the only one of the four peptides with hydrophobic character and with a higher proportion of hydrophobic surface area exposed to the solvent. P1, P2, and P5 presented hydrophilicity with a higher proportion of exposed polar surface area (García Fillería 2019). Vilcacundo et al. (2019) obtained digests with antioxidant capacity evidenced in ORAC assay after sequential incubation of kiwicha (*A. caudatus*) protein with pepsin and pancreatin. These digests presented also other biological activities, including cytotoxicity against colon cancer cells. Authors found that low-molecular-weight peptides (<5 kDa) were the main responsible for the radical scavenging activity, while larger-size peptides (>5 kDa) were determinant in the cytotoxic effects against colon cancer cells. Three of the thirteen peptides identified (NRPET, HVIKPPS, and ASANEPEDEN) were within the sequence of peptides (FNRPETT, HVIKPPSRA, and ITASANEPEDN) previously identified by Orsini Delgado et al. (2016).

The effect of peptides generated by SGID in the prevention of oxidative stress of intestinal cells was studied in H_2O_2-induced-Caco-2 TC7 cells. In this sense, cells were treated with 500 μmol/L H_2O_2 for 1 hour, inducing an increment in the content of ROS and GSH without increasing the lipid oxidation products or causing damage to the mitochondrial or plasma membranes. In these conditions, a previous treatment with the amaranth protein isolate gastrointestinal digest or its FPLC fractions reduced the intracellular ROS content; some of the most active fractions (distributed in a wide range of molecular weight, between 0.27 and 8.4 kDa) that were able to keep the cells at their baseline ROS content were detected. In addition to the ROS decrease, diverse behaviors (no changes, decreases or increases) over the activity of the antioxidant enzymes SOD and GPx, and the content of GSH were observed (García Fillería and Tironi 2020). Moreover, the effect of the ten synthetic peptides previously identified (P1–P10, Orsini Delgado et al. 2016) was analyzed in the induced-cell cultures. All these peptides, except P8 and P9, were able to reduce the intracellular ROS content. P1, P2, P3, P4, P7, and P10 were the most active ones, capable of keeping the cells at their baseline ROS content. In concordance with the complete digest and its FPLC fractions, the reduction of ROS was accompanied by diverse behaviors of the antioxidant enzymes SOD and GPx, and the content of GSH (García Fillería 2019).

Results obtained in Caco-2 TC7 cells suggested that different amaranth peptides could exert antioxidant activity through diverse mechanisms, such as (1) entering the cell and acting as direct ROS scavengers inside the cell; (2) entering the cell and producing effect on signaling pathways that lead to induction of enzymes or antioxidant compounds; (3) interacting with the plasma membrane and inducing signaling pathways that activate enzymes or antioxidant compounds (García Fillería and Tironi 2020; García Fillería 2019).

The *in vitro* acellular and cellular activities of amaranth peptides did not present correlation. P6, P9, and P10 showed very little or no activity in the ORAC and HORAC assays; P9 was also unable to reduce the level of intracellular ROS, but P6 and P10 produced a diminution of intracellular ROS. Furthermore, P8 was the most active in ORAC and HORAC assays but exhibited no intracellular activity (Orsini Delgado et al. 2015; García Fillería 2019). There are several facts that could explain the lack of correlation between cellular and acellular activities. Peptides, as well as other food components, could act by mechanisms different from the direct inhibition of ROS in the cells such as indirect effects on the Nrf2 system previously described (Finley et al. 2011). Amaranth peptides could exert this mechanism of action that could be reflected in changes in SOD and GPx activities, and in the GSH level. Seven of the ten peptides analyzed were able to induce a significant increase in the GSH content of cells. However, five of these peptides also produced a reduction of the ROS content to baseline levels while the other two peptides were those unable to decrease ROS (P8 and P9). Another possible explanation for the differences observed between acellular and cellular activities is related to the fact that once in contact with intestinal cells, peptides could experiment modifications produced by the brush border peptidases that could change the activity of the molecules. García Fillería and Tironi (2020) showed that certain fractions with high intracellular antioxidant activity (approximately 8 kDa, 1 kDa, and <0.6 kDa) were partially modified by extracellular peptidases. The remaining original molecules would perform a direct scavenging and/or inhibition of ROS in the intestinal lumen since they presented also high ORAC and HORAC activity (Rodríguez et al. 2020); whereas original and/or modified peptides, could act at the cellular level since they produced a diminution of ROS and some other effects (García Fillería and Tironi 2020). Moreover, diverse behaviors related to the modification/passage by Caco-2 TC7 cells of the ten amaranth peptides were evidenced. There were some peptides that partially resisted degradation by extracellular peptidases (P2, P3, P4, P5, P6, P8, P9, and P10); when present in the intestinal lumen, they could contribute to maintaining the redox balance of intestinal mucosa cells. The peptides with high ORAC capacity (P2 and P1 were the most active) could exert a direct effect of ROS neutralization in the intestinal lumen. All the peptides, except P8 and P9, demonstrated intracellular antioxidant effect that could be exerted by the original and/or the modified peptides. They also evidenced passage through the cells to a certain extent (except P2, neither the original molecule nor any of its hydrolysis products), suggesting two possible mechanisms of action, one after entering into the cell and the other through the interaction with the plasma membrane. Apart from these studies, there is little evidence in the literature regarding the effect of amaranth peptides on intestinal cells. García-Nebot et al. (2014) demonstrated the ability of lunasin (a 43–amino acid peptide identified in several plants including amaranth) to reduce ROS and preserve cell viability in intestinal Caco-2 cells treated with hydrogen peroxide and tert-buthylhydroxide. The biological activities of this peptide will be described in detail in the next section of this chapter.

In vivo studies in which *A. mantegazzianus* protein isolate was incorporated in the diet of Wistar rats were also performed to evaluate its effect on the antioxidative status. In the first study, the diets were added with 1% w/w of cholesterol (to induce to a certain extent oxidative stress in the tissues) and with 2.5% w/w of protein isolate (about 10% of the total protein was from amaranth). After 28 days of this diet intake, rats presented a significant reduction of the total cholesterol in plasma and of cholesterol and triglycerides in liver. In addition, increment of the ferric reducing capacity of the plasma (FRAP) values, diminution of the lipid oxidation secondary products (TBA value) in plasma and liver, and diminution of SOD activity were produced by the protein isolate intake. Furthermore, an increase in fecal cholesterol excretion was evidenced, which could be one of the causes of improvement in the lipid profile and in the antioxidative status (Lado et al. 2015). In another study, Wistar rat diets were added with 2% w/w cholesterol/10% w/w porcine fat, and amaranth flour or protein isolate. These ingredients were administered in two ways: "acute" consumption (50% of protein replacement, 1 week after 4 weeks intake of the high fat diet), and "chronic" consumption (25% of protein replacement, 4 weeks). Different biomarkers of oxidative status were determined in plasma, liver, and cecum/colon after slaughter. The ingredient type (namely the presence of different components), as well as the dose and duration of administration had an effect on diverse parameters related to the antioxidant status of the intestine. Amaranth flour showed a higher effect when compared to the protein isolate, which could be attributed to the presence of non-digestible and non-absorbable (fiber) or partially absorbable components (polyphenols, squalene) in the flour that can act on the distal portions of the intestine. Regarding "chronic" administration (4 weeks), the presence of amaranth flour produced an increase in the GSH content while the intestinal cells ROS content remained unchanged, suggesting the presence of antioxidant compounds that would contribute to the maintenance of the redox balance. None of these effects were observed when rats were fed with the diet containing the protein isolate. When the rats were fed with higher content of amaranth ingredients for 1 week, the intake of amaranth flour produced an increase in SOD and GPx activities, suggesting an induction of these antioxidant enzymes and/or a lower consumption of them in the maintenance of cellular redox balance. The "acute" administration of protein isolate increased the GSH content in the intestinal cells, suggesting that increasing the peptides dose produced this acute cell response (Garcia Fillería, Rodríguez, and Tironi, unpublished). Although both diets provided comparable amounts of amaranth proteins (between 2.5 and 2.9% w/w), some differences in the proteolysis degree and peptide composition generated by SGID has been reported (Rodríguez et al. 2020). In summary, different effects were observed at the intestinal level depending on the amaranth ingredient, and on the administration manner, which could be associated with the change in the concentration of bioactive compounds to which the cells are exposed as well as with changes in cellular responses depending on the exposure time. A more

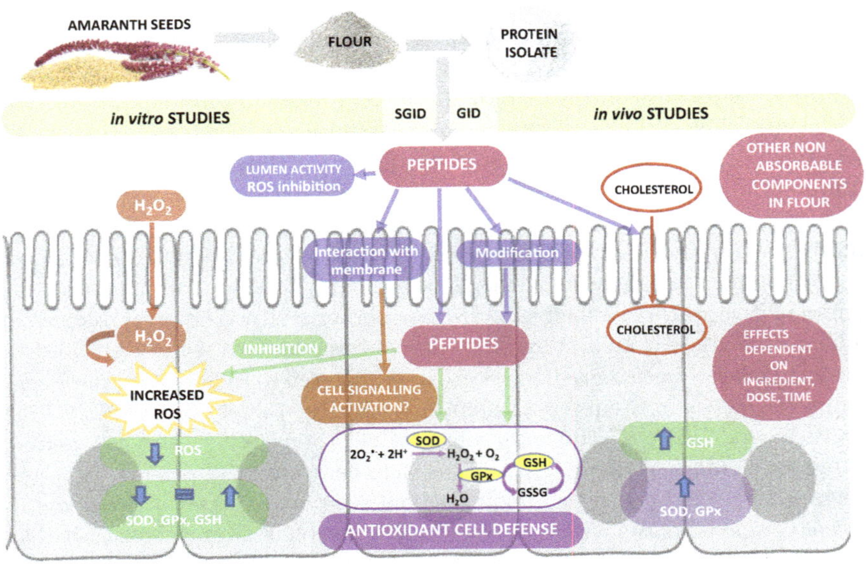

Figure 9.1 Schematic representation of *in vitro* and *in vivo* antioxidant activity of amaranth flour and protein isolate.

Abbreviations: SGID: simulated gastrointestinal digestion; GID: gastrointestinal digestion; SOD: superoxide dismutase; GPx: glutathione peroxidase.

detailed explanation of the effect of these diets on cholesterol excretion, which could be partially related to the effects observed on oxidative stress markers, especially in case of flour, appears in section 9.4 of this chapter.

An overview of the possible effects of amaranth peptides on the redox balance of intestinal cells is presented in Figure 9.1, according to the *in vitro* and *in vivo* evidence obtained to date.

9.3 ANTIPROLIFERATIVE AND ANTITUMOR EFFECTS

Non-communicable diseases (NCDs) are the leading cause of mortality in the world. Among them, cancer is one of the most important causes of death. A large percentage of NCDs is preventable and they share the same risk factors. The top five health risks are related to behavior and diet: high body mass index, low fruit and vegetable consumption, physical inactivity, tobacco use, and excessive alcohol intake. Successively, there is evidence that some components of different foods can protect against the development of this disease (peptides, proteins, secondary metabolites; Ferlay et al. 2019).

Several authors have reported antiproliferative and antitumor effects associated with pseudocereal proteins and peptides; among them, lunasin and lectins are the most recognized. As previously mentioned, lunasin is a 43-amino acid peptide (molecular weight 5.5 kDa) that was found for the first time in soybeans (Galvez and De Lumen 1999) and subsequently in wheat, barley, and other seed storage proteins like amaranth, and more recently in quinoa (Jeong et al. 2007a; Silva-Sánchez et al. 2008). Different studies carried out with this peptide have evaluated its action in different types of cancer, including colon cancer, using *in vitro* and *in vivo* models. Lunasin disrupts the cell cycle and in some cases induces apoptosis (Hsieh et al. 2018; Vuyyuri et al. 2018). Moreover, some authors have demonstrated that lunasin inhibits metastasis using *in vivo* models (Dia and González de Mejia 2013). Within lunasin sequence, four regions have been identified: three of them associated with the different mechanisms of action that have been described for this peptide. The fragment from amino acid 23 to 32 is associated with the ability to bind histones H3 and H4 due to their structural homology to a conserved region of chromatin binding protein. The third fragment formed only by three amino acids (arginine, glycine, aspartic) called RGD module, is a cell adhesion motif. This fragment competes with integrins to bind with the extracellular matrix and consequently suppresses the integrin-mediated signaling pathway that modulates proliferation, migration and apoptosis. The C-terminal end consists of eight aspartic acids (from 36 to 43) and is associated with the antimitotic capacity of this peptide (Hsieh et al. 2018; Ďúranová et al. 2019).

In addition to its recognized anticancer activity, hypocholesterolemic, immunomodulatory, and antioxidant activity has been associated with this peptide (Ďúranová et al. 2019; Hsieh et al. 2018). In order to be able to exert its effects, lunasin has to resist GID and to reach target tissues and organs in an intact and bioactive form. Studies performed with purified lunasin show that is rapidly digested. However, when studies were performed with crude protein extracts obtained from soybeans and *Solanum nigrum*, lunasin was able to resist SGID (Jeong et al. 2007b; Cruz-Huerta et al. 2015). These differences in lunasin behavior have been associated to the presence of protease inhibitors contained in crude extracts that protects it from digestion. In soybean, the addition of Bowman-Birk and Kunitz-trypsin inhibitors to purified lunasin and its protective effect against digestion has been studied, suggesting that the ingestion of whole grains or flour is better than the consumption of the purified peptide to conserve the chemopreventive effect of lunasin (Cruz-Huerta et al. 2015).

A lunasin-like peptide was found in the glutelin fraction of amaranth seed proteins by Silva-Sánchez et al. (2008). These authors reported a total lunasin concentration in amaranth from 5.9 to 8.7 μg/g of protein, content similar to barley. However, the molecular weight of lunasin-like peptide was higher (18.5 kDa) than soybean lunasin (5.5 kDa). A tryptic-digested glutelin showed apoptotic activity against HeLa cells (cervical cancer) probably associated to this peptide. Maldonado-Cervantes et al. (2010) reported that amaranth lunasin-like peptide is capable of reaching the nucleus, although the mechanism has not been elucidated. Amaranth lunasin-like peptide inhibited acetylation of histones H3 and H4, the same as soybean lunasin. It has been proposed that inhibition of acetylation of histones is probably an epigenetic mechanism for the cancer-preventive properties of lunasin (Jeong et al. 2007a, 2007b; Hsieh et al. 2018).

Recently, Ren et al. (2017) detected lunasin in quinoa. These authors evaluated the lunasin content in 15 quinoa samples and found differences among them (1.01×10^{-3} to 4.89×10^{-3} g/kg dry seed). Authors suggest that different environmental factors such as climate and soil conditions affect lunasin content. In addition, they evaluated *in vitro* antioxidant and anti-inflammatory activities. Purified quinoa lunasin exhibited a weak DPPH radical scavenging activity but a strong $ABTS^+$ radical scavenging activity (IC_{50} = 1.45 g/L). Furthermore, this peptide exhibited a dose-dependent activity in the ORAC assay, with a Trolox equivalent value of 40.06 μM/g quinoa lunasin. Anti-inflammatory activity of this peptide was studied by the production of NO and pro-inflammatory cytokines (TNF-α and IL-627), evidencing that quinoa lunasin inhibited the upregulation of these inflammatory markers in a dose-dependent manner, reaching 44.77%, 39.81%, and 33.50% of inhibition, respectively, at a concentration of 0.40 g/L. Although its anticancer activity has not been investigated yet, these results are promising and would suggest potential antitumor properties since these activities are related to cancer development.

As previously mentioned, another group of proteins with recognized antitumor activity are lectins. Plant lectins are proteins that may or may not be glycosylated, and are characterized by having at least one non-catalytic domain that can reversibly interact with specific carbohydrate structures. These proteins are involved in the recognition and binding of glycans from foreign organisms and play a role in the defense of the plant (Lannoo and Van Damme 2014). Based on this distinctive characteristic of lectins, their use as sensitive and selective markers of tumor cells has been investigated (Estrada-Martínez et al. 2017; Bhutia et al. 2019). These proteins are able to recognize in early stages, the development of cancer cells since they express a glycoside pattern different from that of normal cells on their cell surface (Bhutia et al. 2019). On the other hand, some studies carried out with plant lectins have demonstrated their antiproliferative effect through autophagy and apoptosis mechanisms (Bhutia et al. 2019). Furthermore, some plant lectins have even been applied in human clinical trials with success on different types of cancer, including colon cancer (Von Schoen-Anger et al. 2014; Steele et al. 2015).

Several authors have isolated, purified, and characterized amaranth lectins (Singh et al. 1994; Rinderle et al. 1990). These authors showed that lectins extracted from amaranth are homodimers with a molecular mass that varies in a range from 54 to 70 kDa, with a predominance of acidic amino acid residues (aspartic and glutamic). Their subunits have a molecular mass range from 35 kDa to 37 kDa. They have a specific binding site for N-acetyl-D-galactosamine. Transue et al. (1997) resolved the crystal structure of *A. caudatus* lectin bound to the Thomsen-Friedenreich antigen (T antigen; Galβ1, 3GalNAcα-O).

In regard to their potential antitumor effect, amaranth lectin was studied by several authors (Yu et al. 2001; Quiroga et al. 2015; Mengoni et al. 2016; Kaur et al. 2006). Particularly, Yu et al. (2001) and Mengoni et al. (2016) studied their effect on proliferation of human malignant gastrointestinal epithelial cells (HT-29 and Caco-2 TC7, respectively). These authors found opposite effects: while Yu et al. (2001) found an increase in cell proliferation, Mengoni et al. (2016) found an antiproliferative effect of a commercial lectin from *A. caudatus* reaching approximately 40% inhibition of cell growth at 0.25 mg/mL of protein concentration. These authors also studied the effect of isolated proteins from *A. mantegazzianus* and *A. hypochondriacus*, and a partially purified lectin (PPL) extracted from *A. hypochondriacus*. Both isolates showed significant antiproliferative activity with IC_{50} values of 0.32 ± 0.03 mg/mL and 0.77 ± 0.09 mg/mL, respectively. However, the activity of PPL was lower, and the inhibition occurred in a more gradual way, reaching a 20% inhibition at 3.18 mg/mL. In addition to lunasin and lectin found in amaranth proteins, Montoya-Rodríguez et al. (2015) have identified in amaranth 11S globulin sequence the tripeptide VVV, reported in the BIOPEP-UWM database as an anticarcinogenic peptide. An *in silico* analysis carried out on the 11S globulin sequence using different cutting enzymes such as pepsin, trypsin, chymotrypsin, and alcalase shows the release of several peptides containing this tripeptide (**VVV**PQN, GQL**VVV**PQNFAIVK, CVRGRGRIQIVNDQGQSVFDEELSRGQL**VVV**PQNF, DEELSRGQL**VVV**PQNFAIVKQAFE), suggesting a potential anticancer effect of amaranth proteins (Montoya-Rodríguez et al. 2015). However, these peptides were not isolated and tested in order to confirm their effect.

More recently, Sabbione et al. (2019) reported that *A. mantegazzianus* peptides released after SGID would exert a potential antiproliferative activity over the epithelial tumor cell line HT-29. These authors studied *in vitro* the effect of amaranth protein isolated before (API) and after SGID (API-D). While API showed a moderate effect on the inhibition of cell proliferation with an IC_{50} = 1.35 ± 0.12 mg soluble protein/mL, API-D presented the highest inhibition with an IC_{50} = 0.30 ± 0.07 mg soluble protein/mL. Both samples induced apoptosis and necrosis due to an increase in caspase-3 activity. This effect was more important in API-D treatment, suggesting that SGID releases potential bioactive peptides improving their antiproliferative effect.

Regarding the anticancer effect of quinoa proteins, scanty information has been reported in literature. Recently, Vilcacundo et al. (2018) have studied the effect of peptides released after SGID over colon cancer cell viability using three human colorectal cancer cell lines (Caco-2, HT-29, and HCT-116) with promising results. Before gastric digestion, the IC_{50} value reached was 0.843 ± 0.001 mg protein/mL, and after a complete digestion (gastric and duodenal digestion) IC_{50} value was lower, at 0.191 ± 0.003 mg protein/mL. These authors identified 17 peptides from 11S-globulin and other quinoa proteins. However, further work is needed to investigate the possible mechanism of action of these peptides as well as their bioavailability and possibility of reaching the target organ.

Ayyash et al. (2019) informed a moderate antiproliferative activity from fermented quinoa (FQ) against Caco-2 cells. These authors suggested that this activity could be related to the release of peptides during fermentation and to the degraded polysaccharides generated during bacterial growth. Nevertheless, complementary studies are necessary to associate the observed effects to the protein fraction.

A summary of the potential antiproliferative and antitumor effects of amaranth peptides is presented in Figure 9.2. All these data confirm that the consumption of Andean grains, especially amaranth, favors intestinal health, reduces the incidence of colon cancer, and improves the oxidant state.

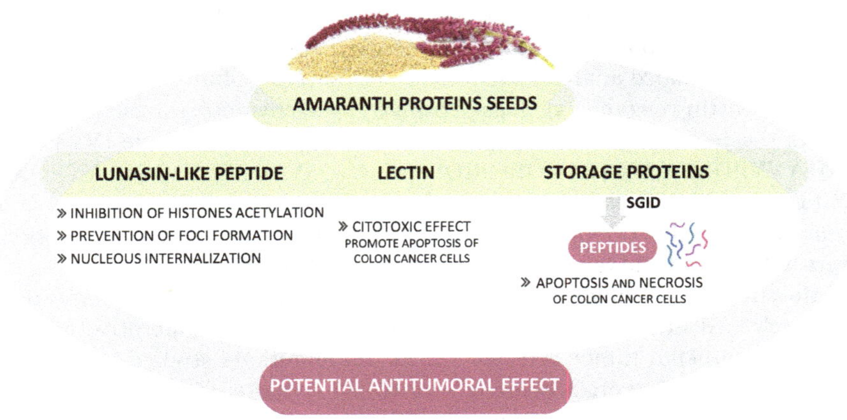

Figure 9.2 Effects of amaranth peptides on antitumor/antiproliferative activity.

Abbreviation: SGID, Simulated gastrointestinal digestion.

9.4 HYPOCHOLESTEROLEMIC ACTIVITY

Cholesterol is an unsaponificable lipid derived from cyclopentanoperhydrophenanthrene. It is contained in all human body cells and is essential for organism functioning, since it is part of cell membranes and is required in vitamin D metabolism, calcium and bile acids absorption, and hormone formation. Humans are capable of synthesizing liver cholesterol through a highly regulated complex process that begins in the mevalonate pathway, also known as the HMG-CoA reductase pathway. Diet constitutes the other source of cholesterol obtaining, which contributes 30% of the daily cholesterol, while 70% corresponds to *de novo* synthesis.

Intestine and liver are the organs responsible for cholesterol homeostasis. Intestine plays a main role in dietary cholesterol transport and absorption, as well as in the recirculation of bile acids and cholesterol metabolites through the enterohepatic circulatory pathway. Liver controls cholesterol levels in the blood and its elimination through the secretion and/or uptake of different lipoproteins.

The available evidence suggests that the intake of some food components may affect lipid metabolism, in particular cholesterol. Among these components are proteins/peptides and fiber. Both compounds could act at the intestinal level, while proteins/peptides could also exert its effect at different levels such as in adipose and liver tissue (Yao et al. 2018).

The absorption of dietary cholesterol through enterocytes requires its incorporation into micelles. The physiological function of micelles is to transport triglycerides, cholesterol, and other water-insoluble lipids to be absorbed by the intestinal epithelium. Micelles are *in vitro* or spontaneously *in vivo* produced when the concentrations of bile salts and phospholipids are higher than the critical micellar concentration. This creates a window of opportunity for peptides and fiber contained in the diet to have an effect on cholesterol metabolism. There is evidence that these peptides would inhibit micellar solubility of cholesterol and interact with bile acids through ionic and hydrophobic interactions (Boachie et al. 2018; Howard and Udenigwe 2013; Udenigwe and Rouvinen-Watt 2015; Yao et al. 2018). The ability to exert these activities would depend on the structure and composition of peptides, particularly on the presence of hydrophobic amino acids; for example, IIAEK peptide from a β-lactoglobulin hydrolysate decreases micellar solubility of cholesterol as it has a hydrophobic N-terminal end (IIA) and a hydrophilic C-terminal (EK), providing an amphipathic structure. This enables the peptide to interact with the phospholipids involved in micelles formation (Nagaoka et al. 2001). The peptide VAWWMY derived from soy glycinin can capture bile acids due to tryptophan and tyrosine hydrophobic amino acids contained, affecting micellar solubility of cholesterol (Choi et al. 2002).

Amaranth proteins showed potential cholesterol-lowering activity when their effect on cholesterol inclusion in a micelles model prepared according to Nagaoka et al. (2001) was studied. Micelles were obtained by sonicating mixtures of cholesterol, phosphatidylcholine, sodium taurocholate, and oleic acid, in the presence and absence of amaranth samples. Micelles prepared with amaranth seed proteins subjected to SGID (HD = 44%) were slightly larger than micelles without amaranth (maximum distribution size, $t_{max}d$, 82 and 53 nm, respectively) or than those formed in presence of unhydrolyzed amaranth proteins ($t_{max}d$ 71 nm). Meanwhile, micelles prepared with amaranth flour, before and after SGID (HD = 48.4%), were distributed in several populations of different sizes, 40, 80, 150–180, 700 nm and 40–60, 300–500 and 700–1400 nm, respectively. Possibly, the largest populations are produced by the aggregation of smaller micelles bond through the fiber, causing a lower uptake by the enterocytes. Amaranth increased the exclusion of cholesterol from the model micelles as the concentration of proteins increased (Sisti et al. 2019). Amaranth proteins inhibited to a greater extent the inclusion of cholesterol in micelles (IC_{50} = 0.2 ± 0.1 mg of protein/mL) when compared to peptides obtained by SGID (IC_{50} = 2.1 ± 0.03 mg of protein/mL), suggesting that peptides size is decisive for exerting bioactivity. An amaranth alcalase hydrolysate, HD = 20.0%, caused the displacement of 49.7 ± 3.6% of micelles cholesterol at a protein concentration of 1 mg/mL. This displacement resulted similar to that obtained with the amaranth isolate (50.7%) and higher than the produced with the digest (6.7%) at the same concentration, suggesting that the ability of amaranth proteins to displace cholesterol from model micelles strongly depends on the degree of hydrolysis achieved, and hence on its size (Sisti 2020). However, higher concentrations of the digest, 3 mg/mL, caused the displacement of 80% of the cholesterol from the model micelles, while the protein isolate reached a maximum inhibition nearby 60%. In addition to size and concentration effect of peptides, those that were more hydrophobic were more active, since they compete more effectively against cholesterol for its incorporation in the micelle; this effect was described for amaranth and other protein sources (Sisti 2020; Howard and Udenigwe 2013). Pastor Cavada (personal communication) showed that the most hydrophobic fraction present in an amaranth protein isolate and in an alcalase hydrolysate exhibited the highest cholesterol displacement of micelles capacity. According to the results obtained by the author, the hydrophobic fractions displaced 47.9 ± 0.6 and 55.6 ± 0.7% of the cholesterol in the micelles, respectively, while more hydrophilic fractions only displaced 32.4 ± 0.4 and 37.3 ± 1.6%, respectively. At low concentrations, greater hydrophobicity and a larger size of peptides correlate with a greater ability to inhibit cholesterol incorporation in the model micelles, whereas a higher hydrolysis degree has the opposite effect. The most effective amaranth polypeptides obtained after digestion would be those with a size enough to establish stable hydrophobic environments, able to interact with micelle components.

Similar behaviors have been described for other hydrolyzed vegetable proteins, such as sunflower and cowpea (Megías et al. 2009; Marques et al. 2015).

Amaranth seed flour, whose fiber content is close to that of proteins (fiber 9%–11%, proteins 13%–16%) exerts a dose-dependent effect on the cholesterol displacement of model micelles, with a higher potency than that of the digested protein fraction (IC_{50} 0.10 ± 0.3 and 0.71 ± 0.07 mg of protein/mL, respectively; Sisti et al. [2019]). Considering the same amount of proteins present in the amaranth isolate or flour (1 mg protein/mL), the cholesterol not incorporated in the micelles was higher when studying amaranth flour. The dietary fiber contained in amaranth flour caused a decrease of cholesterol absorption and could also be a source of short chain fatty acids produced after intestinal microbiota fermentation. Among these short-chain fatty acids is propionate, able to inhibit HMG-CoA reductase liver enzyme (Ötles and Ozgoz 2014), and to bind to various diet components, including cholesterol and bile acids, hence reducing their absorption.

Tiengo et al. (2011) have demonstrated *in vitro* the union between some amaranth components and different bile acids, attributing this capacity to proteins and dietary fiber soluble fraction contained in the flour and in a protein concentrate, which would have a direct effect on the formation of micelles. Although the interaction with taurocholic acid slightly increased, alcalase hydrolysis of the protein concentrate had no relevant effect on the ability to bind bile acids, showing values of approximately 20% of binding (positive control cholestyramine). It is difficult to elucidate whether proteins or peptides molecular size has a significant influence on this bioactivity since the hydrolysis degree was not reported. Other vegetable proteins and their hydrolysates exhibited higher activity compared to amaranth, such as lupin beans and their hydrolysate (between 40% and 68%; Yoshie-Stark and Wäsche 2004) or lentil beans and their hydrolysates, which had an ability to bind bile acids comparable to cholestyramine (Barbana et al. 2011). Barbana et al. (2011) suggested that bile acid excretion after binding to sequestrants could play a key role in coronary heart disease and colon cancer protection, showing that a high concentration of bile acids in the colon lumen, especially deoxycholic acid, enhances carcinogen-induced cell proliferation.

The lipid-lowering activity of amaranth and its components has been studied in hypercholesterolemic animals or in animals (such as rats, hamsters, and rabbits) fed with diets rich in lipids (Sánchez-Urdaneta et al. 2020; Sisti et al. 2019; Mendonça et al. 2009). These studies showed that amaranth modifies levels of sterols excretion in feces, cholesterol, and bile acids, total cholesterol and lipoproteins in blood, and hepatic lipid profiles. Various effects have been attributed to squalene (Shin et al. 2004), unsaturated fatty acids (Tiengo et al. 2011; de Castro et al. 2013), phytosterols, peptides released after digestion, and dietary fiber of amaranth seeds (Sisti et al. 2019; Grajeta 1999). Squalene, present in a high proportion in amaranth oil (6%–8% w/w), is capable of inhibiting the mevalonate pathway, blocking *de novo* cholesterol synthesis, acting on the HMG-CoA reductase enzyme (Shin et al. 2004). Phytosterols compete with cholesterol in the micelles formation in the intestine, promoting their exclusion (Chmelík et al. 2019). Amaranth proteins and fiber mechanisms of action are still under discussion. Sisti et al. (2019) studied the effect of amaranth seed components in Wistar rats fed with a high-fat and cholesterol-rich diet, finding a dose/time dependent effect, which varies for fiber and protein content. Animals fed for a short period (7 days) with diets containing a high proportion of amaranth protein from amaranth flour (casein/amaranth protein 50/50) showed a large increase in cholesterol and bile acids excretion in feces (77% and 74% over the control group), with an important decrease in liver cholesterol content (98%). These effects were not observed when proteins from amaranth protein isolates were used to replace 50% of the dietary casein. In longer feeding periods (30 days), even if lower amounts of amaranth protein were administered (casein/amaranth protein 75/25), a similar effect was observed when using amaranth flour as the protein source (increased fecal excretion of cholesterol and bile acids, 108% and 110%, respectively). Liver cholesterol showed greater decrease when rats were fed with amaranth protein isolate (96%) than when the flour was used as the protein source (53%). These results suggest that amaranth fiber develops an important effect at intestinal level, while amaranth proteins would also be involved in cholesterol metabolism modifications at liver level. A summary of the results obtained in *in vitro* and *in vivo* tests is presented in Figure 9.3.

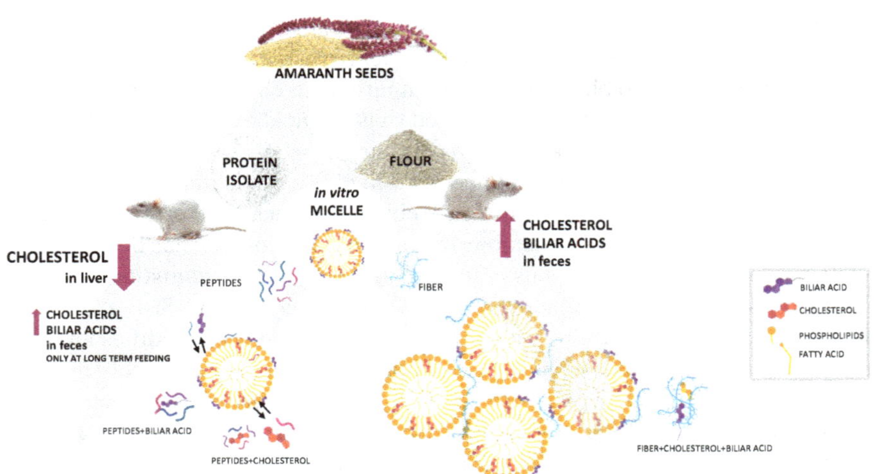

Figure 9.3 Schematic representation of *in vivo* and *in vitro* main effects on lipids produced by amaranth protein isolate and flour. The micelle components are shown in the right below square.

Similar results have been described by Lado et al. (2015) and Escudero et al. (2006), who fed Wistar rats with protein isolates or protein concentrates for 28 or 32 days, respectively. Escudero et al. (2006) used only amaranth as the protein source and found a reduction in LDL content and triacylglycerides in the plasma of the rats. Chaturvedi et al. (1993) reported a decrease in total lipids and cholesterol in the liver (38% and 44%, respectively) and in total cholesterol in plasma (52%) when they fed rats with amaranth flour as the only protein source for 80 days. In addition to the longer feeding period, using flour also increased the contribution of amaranth lipid fraction, which, as already mentioned, was able to modify lipid levels in animals (de Castro et al. 2013). Mendonça et al. (2009) described a decrease in plasma cholesterol and triacylglyceride levels, and an increase in cholesterol excretion via feces, when studying the effect of total or partial replacement of casein by amaranth protein isolate in hamsters fed for 28 days after consuming a hypercholesterolemic diet for 21 days. In contrast to previously mentioned results, Mendonça et al. (2009) did not use a high-fat diet and did not evidence an increase in bile acids in animal feces. Grajeta (1999) studied the effect of a diet containing whole amaranth seeds to feed Buffalo rats for 28 days. The diet consisted in 4% amaranth dietary fiber, 15% fat, amaranth protein replacing 20% of the casein in the control diet, and 0.5% cholesterol. The results showed a reduction in plasma and liver cholesterol (10.7% and 20%, respectively) and a decrease in triacylglycerides in the liver (10%) compared to rats fed with a control diet. This author attributed the hypolipidemic effect to the dietary fiber in amaranth seeds and considered that proteins, tocotrienols, unsaturated fatty acids, and vitamin E present in the lipid fraction could influence the hypocholesterolemic effect.

Amaranth fiber effect has been recently described by Sánchez-Urdaneta et al. (2020). These authors fed hypercholesterolemic and hyperglycemic rats with breads made with wheat flour and amaranth flour. They found that animals that ingested amaranth for 15 weeks had lower serum levels of glucose, total cholesterol, VLDL cholesterol, and triacylglycerides, and higher HDL cholesterol values compared to animals that did not ingest amaranth. The authors attributed the effects to amaranth fiber, although the amount of dietary fiber and protein from amaranth was similar. Few studies report human trials and results are still not conclusive to confirm or discard a positive effect of amaranth consumption in the control of lipid levels. While some authors describe positive effects (Maier et al. 2000), others do not detect them (Chávez-Jáuregui et al. 2010). Moreover, none of these studies focuses on intestine lipid levels control caused by amaranth consumption.

9.5 CONCLUSION

The consumption of amaranth proteins could provide a potential health benefit at intestinal level according to *in vitro* and *in vivo* (animal model) approaches since derived peptides showed antioxidant and antitumor activities. Peptides generated by SGID—namely, the complete digest, their fractions separated according to their molecular masses, and some identified sequences—are able to inhibit ROS in acellular systems, decrease the intracellular ROS content, and modify the activity of cellular antioxidant enzymes (SOD and GPx) and GSH content. Some peptides can be partially modified upon contact with intestinal cells, and some intact and/or modified peptides can cross the intestinal barrier. Antioxidant activity could be exerted at the intestinal lumen, after interaction with the cell membrane, and/or inside the cell. Additional studies should be carried out to establish the mechanisms of antioxidant action of the different molecules.

Regarding antitumor activity, peptides generated by SGID of amaranth proteins show an increase in apoptosis and necrosis of colon cancer cells. Moreover, lectin fraction promotes apoptosis of these types of cells, and a peptide present in several seeds including amaranth, called lunasin, evidences antitumor activity with several action mechanisms demonstrated (inhibition of histones acetylation, prevention of foci formation, nucleus internalization). Antioxidant activity has also been evidenced in these peptides.

In addition, amaranth peptides released during the GID process and amaranth fiber are capable of acting on the metabolism of cholesterol. These compounds would sequester cholesterol and/or bile acids, preventing their inclusion in the micelles and increasing cholesterol excretion via feces. The peptides could be also involved in the reduction of liver cholesterol.

In conclusion, amaranth seeds are a source of antioxidant, antitumor, and hypocholesterolemic compounds, but their action mechanisms have not yet been elucidated. Further studies are necessary to evaluate the inclusion of amaranth ingredients in potentially functional foods, which would ideally present multi-activities intending to maintain intestinal health. Finally, it will be imperative to evaluate the bioactivity of these functional foods in the human organism through clinical trials.

ACKNOWLEDGMENTS

We are grateful to all the national and international institutions that have supported the studies carried out by our research group as well as those fellows and researchers who have participated in them.

REFERENCES

Alegría, Amparo, Guadalupe Garcia-Llatas, and Antonio Cilla. 2015. "Static Digestion Models: General Introduction." In Kitty Verhoeckx, Paul Cotter, Iván López-Expósito, Charlotte Kleiveland, Tor Lea, Alan Mackie, Teresa Requena, Dominika

Swiatecka, and Harry Wichers (Eds.), *The Impact of Food Bioactives on Health*. Cham: Springer International Publishing, 3–12. https://doi.org/10.1007/978-3-319-16104-4.

Aw, Tak Yee. 1999. "Molecular and Cellular Responses to Oxidative Stress and Changes in Oxidation-Reduction Imbalance in the Intestine." *The American Journal of Clinical Nutrition* 70 (4): 557–565. https://doi.org/10.1093/ajcn/70.4.557.

Ayala-Niño, Alexis, Gabriela Mariana Rodríguez-Serrano, Luis Guillermo González-Olivares, Elizabeth Contreras-López, Patricia Regal-López, and Alberto Cepeda-Saez. 2019. "Sequence Identification of Bioactive Peptides from Amaranth Seed Proteins (*Amaranthus hypochondriacus* Spp.)." *Molecules* 24 (17): 3033. https://doi.org/10.3390/molecules24173033.

Ayyash, Mutamed, Stuart K. Johnson, Shao Quan Liu, Nouf Mesmari, Shaikhah Dahmani, Ayesha S. Al Dhaheri, and Jaleel Kizhakkayil. 2019. "*In Vitro* Investigation of Bioactivities of Solid-State Fermented Lupin, Quinoa and Wheat Using *Lactobacillus* Spp." *Food Chemistry* 275: 50–58, March. https://doi.org/10.1016/j.foodchem.2018.09.031.

Barbana, Chockry, Anne C. Boucher, and Joyce I. Boye. 2011. "*In Vitro* Binding of Bile Salts by Lentil Flours, Lentil Protein Concentrates and Lentil Protein Hydrolysates." *Food Research International* 44 (1): 174–180. https://doi.org/10.1016/j.foodres.2010.10.045.

Bhutia, Sujit K., Prashanta K. Panda, Niharika Sinha, Prakash P. Praharaj, Chandra S. Bhol, Debasna P. Panigrahi, Kewal K. Mahapatra, et al. 2019. "Plant Lectins in Cancer Therapeutics: Targeting Apoptosis and Autophagy-Dependent Cell Death." *Pharmacological Research* 144: 8–18, January. https://doi.org/10.1016/j.phrs.2019.04.001.

Boachie, Ruth, Shixiang Yao, and Chibuike C. Udenigwe. 2018. "Molecular Mechanisms of Cholesterol-Lowering Peptides Derived from Food Proteins." *Current Opinion in Food Science* 20: 58–63. https://doi.org/10.1016/j.cofs.2018.03.006.

Chaturvedi, Anurag, G. Sarojini, and Nanubala L. Devi. 1993. "Hypocholesterolemic Effect of Amaranth Seeds (*Amaranthus esculantus*)." *Plant Foods for Human Nutrition* 44 (1): 63–70. https://doi.org/10.1007/BF01088483.

Chávez-Jáuregui, Rosa Nilda, Raul Dias Santos, Alessandra Macedo, Ana Paula, Marte Chacra, Tania Leme Martinez, José Alfredo, and Gomes Arêas. 2010. "Effects of Defatted Amaranth (*Amaranthus caudatus*) Snacks on Lipid Metabolism of Patients with Moderate Hypercholesterolemia." *Ciência e Tecnologia de Alimentos* 30 (4): 1007–1010. https://doi.org/10.1590/S0101-20612010000400026.

Chmelík, Zdeněk, Michaela Šnejdrlová, and Michal Vrablík. 2019. "Amaranth as a Potential Dietary Adjunct of Lifestyle Modification to Improve Cardiovascular Risk Profile." *Nutrition Research* 72: 36–45, December. https://doi.org/10.1016/j.nutres.2019.09.006.

Choi, Seon Kang, Motoyasu Adachi, and Shigeru Utsumi. 2002. "Identification of the Bile Acid-Binding Region in the Soy Glycinin A1aB1b Subunit." *Bioscience, Biotechnology and Biochemistry* 66 (11): 2395–2401. https://doi.org/10.1271/bbb.66.2395.

Constantini, David. 2019. "Understanding Diversity in Oxidative Status and Oxidative Stress: The Opportunities and Challenges Ahead." *The Journal of Experimental Biology* 222 (13): jeb194688. https://doi.org/10.1242/jeb.194688.

Cruz-Huerta, Elvia, Samuel Fernández-Tomé, M. Carmen Arques, Lourdes Amigo, Isidra Recio, Alfonso Clemente, and Blanca Hernández-Ledesma. 2015. "The Protective Role of the Bowman-Birk Protease Inhibitor in Soybean Lunasin Digestion: The Effect of Released Peptides on Colon Cancer Growth." *Food and Function* 6 (8): 2626–2635. https://doi.org/10.1039/c5fo00454c.

de Castro, Luíla Ívini Andrade, Rosana Aparecida Manólio Soares, Paulo H. N. Saldiva, Roseli A. Ferrari, Ana M. R. O. Miguel, Claudia A. S. Almeida, and José Alfredo Gomes Arêas. 2013. "Amaranth Oil Increased Fecal Excretion of Bile Acid but Had No Effect in Reducing Plasma Cholesterol in Hamsters." *Lipids* 48 (6): 609–618. https://doi.org/10.1007/s11745-013-3772-8.

Dia, Vermont P., and Elvira González de Mejia. 2013. "Potential of Lunasin Orally-Administered in Comparison to Intraperitoneal Injection to Inhibit Colon Cancer Metastasis in Vivo." *Journal of Cancer Therapy* 4 (6). https://doi.org/10.4236/jct.2013.46a2005.

Dolan, Kyle T., and Eugene B. Chang. 2017. "Diet, Gut Microbes, and the Pathogenesis of Inflammatory Bowel Diseases." *Molecular Nutrition & Food Research* 61 (1): 1600129. https://doi.org/10.1002/mnfr.201600129.

Ďúranová, Hana, Veronika Fialková, Jana Bilčíková, Norbert Lukáč, and Zuzana Kňažická. 2019. "Lunasin and Its Versatile Health-Promoting Actions." *Journal of Microbiology, Biotechnology and Food Sciences* 8 (4): 1106–1110. https://doi.org/10.15414/jmbfs.2019.8.4.1106-1110.

Egusa Saiga, Ai, and Toshihide Nishimura. 2013. "Antioxidative Properties of Peptides Obtained from Porcine Myofibrillar Proteins by a Protease Treatment in an Fe (II)-Induced Aqueous Lipid Peroxidation System." *Bioscience, Biotechnology and Biochemistry* 77 (11): 2201–2204. https://doi.org/10.1271/bbb.130369.

Escudero, Nora L., Fanny Zirulnik, Nadia N. Gomez, S. I. Mucciarelli, and María S. Giménez. 2006. "Influence of a Protein Concentrate from *Amaranthus cruentus* Seeds on Lipid Metabolism." *Experimental Biology and Medicine* 231 (1): 50–59. https://doi.org/10.1177/153537020623100106.

Esfandi, Ramak, Mallory E. Walters, and Apollinaire Tsopmo. 2019. "Antioxidant Properties and Potential Mechanisms of Hydrolyzed Proteins and Peptides from Cereals." *Heliyon* 5 (4). https://doi.org/10.1016/j.heliyon.2019.e01538.

Estrada-Martínez, Laura E., Ulisses Moreno-Celis, Ricardo Cervantes-Jiménez, Roberto Augusto Ferriz-Martínez, Alejandro Blanco-Labra, and Teresa García-Gasca. 2017. "Plant Lectins as Medical Tools Against Digestive System Cancers." *International Journal of Molecular Sciences* 18 (7). https://doi.org/10.3390/ijms18071403.

Ferlay, Jacques, Murielle Colombet, Isabelle Soerjomataram, Colin Mathers, Donald M. Parkin, Marlon Piñeros, Ariana Znaor, and Freddie Bray. 2019. "Estimating the Global Cancer Incidence and Mortality in 2018: GLOBOCAN Sources and Methods." *International Journal of Cancer* 144 (8): 1941–1953. https://doi.org/10.1002/ijc.31937.

Finley, John W., Ah Ng Kong, Korry J. Hintze, Elizabeth H. Jeffery, Li Li Ji, and Xin Gen Lei. 2011. "Antioxidants in Foods: State of the Science Important to the Food Industry." *Journal of Agricultural and Food Chemistry* 59 (13): 6837–6846. https://doi.org/10.1021/jf2013875.

Galvez, Alfredo F., and Benito O. De Lumen. 1999. “A Soybean CDNA Encoding a Chromatin-Binding Peptide Inhibits Mitosis of Mammalian Cells.” *Nature Biotechnology* 17 (5). https://doi.org/10.1038/8676.

García Fillería, Susan F. 2019. “Evaluación de Aislado Proteico de Amaranto como Fuente de Péptidos Antioxidantes: Estudios in vitro e in vivo.” Ph.D. diss., Facultad de Ciencias Exactas, Universidad Nacional de La Plata, La Plata.

García Fillería, Susan F., and Valeria A. Tironi. 2020. “Intracellular Antioxidant Activity and Intestinal Absorption of Amaranth Peptides Released by Gastrointestinal Digestion in Caco-2 TC7 Cells.” *Food Bioscience* (in press).

García-Nebot, María J., Isidra Recio, and Blanca Hernández-Ledesma. 2014. “Antioxidant Activity and Protective Effects of Peptide Lunasin against Oxidative Stress in Intestinal Caco-2 Cells.” *Food and Chemical Toxicology* 65: 155–161. https://doi.org/10.1016/j.fct.2013.12.021.

Grajeta, Halina. 1999. “Effect of Amaranth and Oat Bran on Blood Serum and Liver Lipids in Rats Depending on the Kind of Dietary Fats.” *Nahrung—Food* 43 (2): 114–117. https://doi.org/10.1002/(SICI)1521-3803(19990301)43:2 < 114::AID-FOOD114 > 3.0.CO;2-%23.

Guerra, Aurélie, Lucie Etienne-Mesmin, Valérie Livrelli, Sylvain Denis, Stéphanie Blanquet-Diot, and Monique Alric. 2012. “Relevance and Challenges in Modeling Human Gastric and Small Intestinal Digestion.” *Trends in Biotechnology* 30 (11): 591–600. https://doi.org/10.1016/j.tibtech.2012.08.001.

Howard, Ashton, and Chibuike C. Udenigwe. 2013. “Mechanisms and Prospects of Food Protein Hydrolysates and Peptide-Induced Hypolipidaemia.” *Food and Function* 4 (1): 40–51. https://doi.org/10.1039/c2fo30216k.

Hsieh, Chia Chien, Cristina Martínez-Villaluenga, Ben O. de Lumen, and Blanca Hernández-Ledesma. 2018. “Updating the Research on the Chemopreventive and Therapeutic Role of the Peptide Lunasin.” *Journal of the Science of Food and Agriculture* 98 (6): 2070–2079. https://doi.org/10.1002/jsfa.8719.

Jeong, Hyung-Jin, Jin-Boo Jeong, Dae-Seop Kim, Jae-Ho Park, Jung-Bok Lee, Dae-Hyuk Kweon, Gyu-Young Chung, Eul-Won Seo, and Ben O. de Lumen. 2007a. “The Cancer Preventive Peptide Lunasin from Wheat Inhibits Core Histone Acetylation.” *Cancer Letters* 255 (1). https://doi.org/10.1016/j.canlet.2007.03.022.

Jeong, Jin-Boo, Hyung-Jin Jeong, Jae-Ho Park, Sun-Hee Lee, Jeong-Rak Lee, Hee-Kyeong Lee, Gyu-Young Chung, Jeong-Doo Choi, and Ben O. de Lumen. 2007b. “Cancer-Preventive Peptide Lunasin from Solanum Nigrum L. Inhibits Acetylation of Core Histones H3 and H4 and Phosphorylation of Retinoblastoma Protein (Rb).” *Journal of Agricultural and Food Chemistry* 55 (26): 10707–10713. https://doi.org/10.1021/jf072363p.

Kaur, N., V. Dhuna, S. S. Kamboj, J. N. Agrewala, and J. Singh. 2006. “A Novel Antiproliferative and Antifungal Lectin from *Amaranthus viridis* Linn Seeds.” *Protein and Peptide Letters* 13 (9): 897–905. https://doi.org/10.2174/092986606778256153.

Lado, María B., Julieta Burini, Gustavo Rinaldi, María C. Añón, and Valeria A. Tironi. 2015. “Effects of the Dietary Addition of Amaranth (*Amaranthus mantegazzianus*) Protein Isolate on Antioxidant Status, Lipid Profiles and Blood Pressure of Rats.” *Plant Foods for Human Nutrition* 70 (4): 371–379. https://doi.org/10.1007/s11130-015-0516-3.

Lannoo, Nausica Ã., and Els J. M. Van Damme. 2014. "Lectin Domains at the Frontiers of Plant Defense." *Frontiers in Plant Science* 5, August. https://doi.org/10.3389/fpls.2014.00397.

Li, Yao Wang, and Bo Li. 2013. "Characterization of Structure-Antioxidant Activity Relationship of Peptides in Free Radical Systems Using QSAR Models: Key Sequence Positions and Their Amino Acid Properties." *Journal of Theoretical Biology* 318: 29–43. https://doi.org/10.1016/j.jtbi.2012.10.029.

Maier, Susan M., Nancy D. Turner, and Joanne R. Lupton. 2000. "Serum Lipids in Hypercholesterolemic Men and Women Consuming Oat Bran and Amaranth Products." *Cereal Chemistry* 77 (3): 297–302. https://doi.org/10.1094/CCHEM.2000.77.3.297.

Maldonado-Cervantes, Enrique, Hyung Jin Jeong, Fabiola León-Galván, Alberto Barrera-Pacheco, Antonio De León-Rodríguez, Elvira González De Mejia, Ben O. De Lumen, and Ana P. Barba De La Rosa. 2010. "Amaranth Lunasin-like Peptide Internalizes into the Cell Nucleus and Inhibits Chemical Carcinogen-Induced Transformation of NIH-3T3 Cells." *Peptides* 31 (9): 1635–1642. https://doi.org/10.1016/j.peptides.2010.06.014.

Marques, Marcelo Rodrigues, Rosana Aparecida Manólio Soares Freitas, Amanda Caroline Corrêa Carlos, Érica Sayuri Siguemoto, Gustavo Guadagnucci Fontanari, and José Alfredo Gomes Arêas. 2015. "Peptides from Cowpea Present Antioxidant Activity, Inhibit Cholesterol Synthesis and Its Solubilisation into Micelles." *Food Chemistry* 168: 288–293. https://doi.org/10.1016/j.foodchem.2014.07.049.

Megías, Cristina, Justo Pedroche, María del Mar Yust, Manuel Alaiz, Julio Girón-Calle, Francisco Millán, and Javier Vioque. 2009. "Sunflower Protein Hydrolysates Reduce Cholesterol Micellar Solubility." *Plant Foods for Human Nutrition* 64 (2): 86–93. https://doi.org/10.1007/s11130-009-0108-1.

Mendonça, Simone, Paulo H. Saldiva, Robison J. Cruz, and José A. G. Arêas. 2009. "Amaranth Protein Presents Cholesterol-Lowering Effect." *Food Chemistry* 116 (3): 738–742. https://doi.org/10.1016/j.foodchem.2009.03.021.

Mengoni, Antonieta, Alejandra V. Quiroga, and María Cristina Añón. 2016. "Purificación y Caracterización de Una Lectina de *Amaranthus hypochondriacus*, Un Compuesto Antiproliferativo." *Innotec* 11: 27–35. http://ojs.latu.org.uy/index.php/INNOTEC/article/view/329.

Minekus, M., M. Alminger, P. Alvito, S. Ballance, T. Bohn, C. Bourlieu, F. Carrière, et al. 2014. "A Standardised Static in Vitro Digestion Method Suitable for Food-an International Consensus." *Food and Function* 5 (6): 1113–1124. https://doi.org/10.1039/c3fo60702j.

Montoya-Rodríguez, Alvaro, Mario A. Gómez-Favela, Cuauhtémoc Reyes-Moreno, Jorge Milán-Carrillo, and Elvira González de Mejía. 2015. "Identification of Bioactive Peptide Sequences from Amaranth (*Amaranthus hypochondriacus*) Seed Proteins and Their Potential Role in the Prevention of Chronic Diseases." *Comprehensive Reviews in Food Science and Food Safety* 14 (2): 139–158. https://doi.org/10.1111/1541-4337.12125.

Nagaoka, Satoshi, Yu Futamura, Keiji Miwa, Takako Awano, Kouhei Yamauchi, Yoshihiro Kanamaru, Kojima Tadashi, and Tamotsu Kuwata. 2001. "Identification of Novel Hypocholesterolemic Peptides Derived from Bovine Milkβ-Lactoglobulin."

Biochemical and Biophysical Research Communications 281 (1): 11–17. https://doi.org/10.1006/bbrc.2001.4298.

Orsini Delgado, María C., Mónica Galleano, María C. Añón, and Valeria A. Tironi. 2015. "Amaranth Peptides from Simulated Gastrointestinal Digestion: Antioxidant Activity Against Reactive Species." *Plant Foods for Human Nutrition* 70 (1): 27–34. https://doi.org/10.1007/s11130-014-0457-2.

Orsini Delgado, María C., Agustina E. Nardo, Marija Pavlovic, Hélène Rogniaux, María C. Añón, and Valeria A. Tironi. 2016. "Identification and Characterization of Antioxidant Peptides Obtained by Gastrointestinal Digestion of Amaranth Proteins." *Food Chemistry* 197: 1160–1167, April. https://doi.org/10.1016/j.foodchem.2015.11.092.

Orsini Delgado, María C., Valeria A. Tironi, and María C. Añón. 2011. "Antioxidant Activity of Amaranth Protein or Their Hydrolysates Under Simulated Gastrointestinal Digestion." *LWT—Food Science and Technology* 44 (8): 1752–1760. https://doi.org/10.1016/j.lwt.2011.04.002.

Ötles, Semih, and Selin Ozgoz. 2014. "Health Effects of Dietary Fiber." *Acta Scientiarum Polonorum, Technologia Alimentaria* 13 (2): 191–202. https://doi.org/10.17306/J.AFS.2014.2.8.

Phongthai, Suphat, Stefano D'Amico, Regine Schoenlechner, Wantida Homthawornchoo, and Saroat Rawdkuen. 2018. "Fractionation and Antioxidant Properties of Rice Bran Protein Hydrolysates Stimulated by in Vitro Gastrointestinal Digestion." *Food Chemistry* 240: 156–164, February. https://doi.org/10.1016/j.foodchem.2017.07.080.

Picariello, Gianluca, Beatriz Miralles, Gianfranco Mamone, Laura Sánchez-Rivera, Isidra Recio, Francesco Addeo, and Pasquale Ferranti. 2015. "Role of Intestinal Brush Border Peptidases in the Simulated Digestion of Milk Proteins." *Molecular Nutrition & Food Research* 59 (5): 948–956. https://doi.org/10.1002/mnfr.201400856.

Piechota-Polanczyk, Aleksandra, and Jakub Fichna. 2014. "Review Article: The Role of Oxidative Stress in Pathogenesis and Treatment of Inflammatory Bowel Diseases." *Naunyn-Schmiedeberg's Archives of Pharmacology* 387 (7): 605–620. https://doi.org/10.1007/s00210-014-0985-1.

Quiroga, Alejandra V., Daniel A. Barrio, and María C. Añón. 2015. "Amaranth Lectin Presents Potential Antitumor Properties." *LWT—Food Science and Technology* 60 (1). https://doi.org/10.1016/j.lwt.2014.07.035.

Ren, Guixing, Yingying Zhu, Zhenxing Shi, and Jianhui Li. 2017. "Detection of Lunasin in Quinoa (*Chenopodium Quinoa* Willd.) and the in Vitro Evaluation of Its Antioxidant and Anti-Inflammatory Activities." *Journal of the Science of Food and Agriculture*. https://doi.org/10.1002/jsfa.8278.

Rinderle, Stephen J., Irwin J. Goldstein, and Edward E. Remsen. 1990. "Physicochemical Properties of Amaranthin, the Lectin from *Amaranthus Caudatus* Seeds." *Biochemistry* 29 (46). https://doi.org/10.1021/bi00498a019.

Rodríguez, Mariela, Susan F. García Fillería, and Valeria A. Tironi. 2020. "Simulated Gastrointestinal Digestion of Amaranth Flour and Protein Isolate: Comparison of Methodologies and Release of Antioxidant Peptides." *Food Research International*, September. https://doi.org/10.1016/j.foodres.2020.109735.

Sabbione, Ana C., Fredrick Onyango Ogutu, Adriana Scilingo, Miao Zhang, María C. Añón, and Tai-Hua Mu. 2019. "Antiproliferative Effect of Amaranth Proteins and Peptides on HT-29 Human Colon Tumor Cell Line." *Plant Foods for Human Nutrition* 74 (1): 107–114. https://doi.org/10.1007/s11130-018-0708-8.

Sánchez-Urdaneta, Adriana Beatriz, Keyla Carolina Montero-Quintero, Pedro González-Redondo, Edgar Molina, Belkys Bracho-Bravo, and Rafael Moreno-Rojas. 2020. "Hypolipidemic and Hypoglycaemic Effect of Wholemeal Bread with Amaranth (*Amaranthus dubius* Mart. Ex Thell.) on Sprague Dawley Rats." *Foods* 9 (6): 707. https://doi.org/10.3390/foods9060707.

Shin, D. H., H. J. Heo, Y. J. Lee, and H. K. Kim. 2004. "Amaranth Squalene Reduces Serum and Liver Lipid Levels in Rats Fed a Cholesterol Diet." *British Journal of Biomedical Science* 61 (1): 11–14. https://doi.org/10.1080/09674845.2004.11732639.

Silva-Sánchez, Cecilia, Ana P. Barba de la Rosa, M. F. León-Galván, B. O. de Lumen, A. de León-Rodríguez, and Elvira González de Mejía. 2008. "Bioactive Peptides in Amaranth (*Amaranthus hypochondriacus*) Seed." *Journal of Agricultural and Food Chemistry* 56 (4): 1233–1240. https://doi.org/10.1021/jf072911z.

Singh, Jatinder, Kulwant Kaur Kamboj, Sukhdev Singh Kamboj, Sanjeev Shangary, and Rajindar Singh Sandhu. 1994. "*Amaranthus hypochondriacus* and A. Tricolor Lectins: Isolation and Characterization." *The Italian Journal of Biochemistry* 43 (5): 207–218.www.ncbi.nlm.nih.gov/pubmed/7698887.

Sisti, Martín S. 2020. "Proteínas y Fibra de Amaranto: Actividad Sobre El Metabolismo de Colesterol." Ph.D. diss., Facultad de Ciencias Exactas, Universidad Nacional de La Plata, La Plata.

Sisti, Martín S., Adriana Scilingo, and María C. Añón. 2019. "Effect of the Incorporation of Amaranth (*Amaranthus mantegazzianus*) into Fat- and Cholesterol-Rich Diets for Wistar Rats." *Journal of Food Science* 84 (11): 3075–3082. https://doi.org/10.1111/1750-3841.14810.

Soriano-Santos, Jorge, and Héctor Escalona-Buendía. 2015. "Angiotensin I-Converting Enzyme Inhibitory and Antioxidant Activities and Surfactant Properties of Protein Hydrolysates as Obtained of *Amaranthus hypochondriacus* L. Grain." *Journal of Food Science and Technology* 52 (4): 2073–2082. https://doi.org/10.1007/s13197-013-1223-4.

Steele, Megan L., Jan Axtner, Antje Happe, Matthias Kröz, Harald Matthes, and Friedemann Schad. 2015. "Use and Safety of Intratumoral Application of European Mistletoe (*Viscum album L*) Preparations in Oncology." *Integrative Cancer Therapies* 14 (2): 140–148. https://doi.org/10.1177/1534735414563977.

Tiengo, Andréa, Eliana M. Pettirossi Motta, and Flavia Maria Netto. 2011. "Chemical Composition and Bile Acid Binding Activity of Products Obtained from Amaranth (*Amaranthus cruentus*) Seeds." *Plant Foods for Human Nutrition* 66 (4): 370–375. https://doi.org/10.1007/s11130-011-0253-1.

Tironi, Valeria A., and María C. Añón. 2010. "Amaranth Proteins as a Source of Antioxidant Peptides: Effect of Proteolysis." *Food Research International* 43 (1): 315–322. https://doi.org/10.1016/j.foodres.2009.10.001.

Tomasello, Giovanni, Margherita Mazzola, Angelo Leone, Emanuele Sinagra, Giovanni Zummo, Felicia Farina, Provvidenza Damiani, et al. 2016. "Nutrition, Oxidative Stress and Intestinal Dysbiosis: Influence of Diet on Gut Microbiota in Inflammatory Bowel Diseases." *Biomedical Papers* 160 (4): 461–466. https://doi.org/10.5507/bp.2016.052.

Transue, Thomas R., Alexander K. Smith, Hanqing Mo, Irwin J. Goldstein, and Mark A. Saper. 1997. "Structure of Benzyl T-Antigen Disaccharide Bound to *Amaranthus caudatus* Agglutinin." *Nature Structural Biology* 4 (10): 779–783. https://doi.org/10.1038/nsb1097-779.

Udenigwe, Chibuike C., and Kirsti Rouvinen-Watt. 2015. "The Role of Food Peptides in Lipid Metabolism During Dyslipidemia and Associated Health Conditions." *International Journal of Molecular Sciences* 16 (5): 9303–913. https://doi.org/10.3390/ijms16059303.

Ullman, Thomas A., and Steven H. Itzkowitz. 2011. "Intestinal Inflammation and Cancer." *Gastroenterology* 140 (6): 1807–1816. https://doi.org/10.1053/j.gastro.2011.01.057.

Vilcacundo, Rubén, Cristina Martínez-Villaluenga, Beatriz Miralles, and Blanca Hernández-Ledesma. 2019. "Release of Multifunctional Peptides from Kiwicha (*Amaranthus caudatus*) Protein Under in Vitro Gastrointestinal Digestion." *Journal of the Science of Food and Agriculture* 99 (3): 1225–1232. https://doi.org/10.1002/jsfa.9294.

Vilcacundo, Rubén, Beatriz Miralles, Wilman Carrillo, and Blanca Hernández-Ledesma. 2018. "In Vitro Chemopreventive Properties of Peptides Released from Quinoa (*Chenopodium quinoa* Willd.) Protein Under Simulated Gastrointestinal Digestion." *Food Research International* 105: 403–411, March. https://doi.org/10.1016/j.foodres.2017.11.036.

Vindigni, Stephen M., Timothy L. Zisman, David L. Suskind, and Christopher J. Damman. 2016. "The Intestinal Microbiome, Barrier Function, and Immune System in Inflammatory Bowel Disease: A Tripartite Pathophysiological Circuit with Implications for New Therapeutic Directions." *Therapeutic Advances in Gastroenterology* 9 (4): 606–625. https://doi.org/10.1177/1756283X16644242.

Von Schoen-Angerer, Tido, Andreas Goyert, Jan Vagedes, Helmut Kiene, Harald Merckens, and Gunver S. Kienle. 2014. "Disappearance of an Advanced Adenomatous Colon Polyp After Intratumoural Injection with Viscum Album (European Mistletoe) Extract: A Case Report." *Journal of Gastrointestinal and Liver Diseases* 23 (4). https://doi.org/10.15403/jgld.2014.1121.234.acpy.

Vuyyuri, Saleha B., Chris Shidal, and Keith R. Davis. 2018. "Development of the Plant-Derived Peptide Lunasin as an Anticancer Agent." *Current Opinion in Pharmacology* 41: 27–33. https://doi.org/10.1016/j.coph.2018.04.006.

Wang, Liying, Long Ding, Zhipeng Yu, Ting Zhang, Shuang Ma, and Jingbo Liu. 2016. "Intracellular ROS Scavenging and Antioxidant Enzyme Regulating Capacities of Corn Gluten Meal-Derived Antioxidant Peptides in HepG2 Cells." *Food Research International* 90: 33–41, December. https://doi.org/10.1016/j.foodres.2016.10.023.

Xu, Qingbiao, Hui Hong, Jianping Wu, and Xianghua Yan. 2019. "Bioavailability of Bioactive Peptides Derived from Food Proteins across the Intestinal Epithelial Membrane: A Review." *Trends in Food Science and Technology* 86: 399–411. https://doi.org/10.1016/j.tifs.2019.02.050.

Yao, Shixiang, Dominic Agyei, and Chibuike C. Udenigwe. 2018. "Structural Basis of Bioactivity of Food Peptides in Promoting Metabolic Health." *Advances in Food and Nutrition Research* 84: 145–181. https://doi.org/10.1016/bs.afnr.2017.12.002.

Yoshie-Stark, Yumiko, and Andreas Wäsche. 2004. “In Vitro Binding of Bile Acids by Lupin Protein Isolates and Their Hydrolysates.” *Food Chemistry* 88 (2): 179–184. https://doi.org/10.1016/j.foodchem.2004.01.033.

Yu, Lu Gang, Jeremy D. Milton, David G. Fernig, and Jonathan M. Rhodes. 2001. “Opposite Effects on Human Colon Cancer Cell Proliferation of Two Dietary Thomse-Friedenreich Antigen-Binding Lectins.” *Journal of Cellular Physiology* 186 (2): 282–287. https://doi.org/10.1002/1097-4652(200102)186:2 < 282::AID-JCP1028 > 3.0.CO;2-2.

Zhang, Hua, Chien-An A. Hu, Jennifer Kovacs-Nolan, and Yoshinori Mine. 2015. “Bioactive Dietary Peptides and Amino Acids in Inflammatory Bowel Disease.” *Amino Acids* 47 (10): 2127–2141. https://doi.org/10.1007/s00726-014-1886-9.

Zhuang, Yu, Huirong Wu, Xiangxiang Wang, Jieyu He, Shanping He, and Yulong Yin. 2019. “Resveratrol Attenuates Oxidative Stress-Induced Intestinal Barrier Injury through PI3K/Akt-Mediated Nrf2 Signaling Pathway.” *Oxidative Medicine and Cellular Longevity*: 1–14. https://doi.org/10.1155/2019/7591840.

Chapter 10

Development of Delivery Systems of Bioactive Compounds Using Chia Seed By-Products

Luciana Magdalena Julio, Vanesa Yanet Ixtaina, and Mabel Cristina Tomás

CONTENTS

10.1 INTRODUCTION

In recent years there has been a considerable increase in awareness about the impact of diet composition on the health benefits by consumers (Saguansri and Augustin 2010). Thus, they are interested in functional foods in terms of their direct health benefits, the reduction of certain disease risks, or their promotion of well-being (Bigliardi and Galati 2013). Among the food components of great interest are the omega-3 fatty acids (omega-3 FAs) due to their deficiency in Western diets characterized by a high omega-6 and low omega-3 FA intake (Simopoulos 2002; Lane and Derbyshire 2015). Many research works deal with

DOI: 10.1201/9781003087618-10

the benefit of omega-3 FAs at an appropriate level on human health (Adkins and Kelley 2010; Lopez-Huertas 2010; Kaushik et al. 2015). The polyunsaturated fatty acids (PUFAs), such as omega-3, contribute to promote brain development (Kitajka et al. 2002; Derbyshire 2018), reducing the incidence of cardiovascular diseases (Elagizi et al. 2018; Rimm et al. 2018), controlling inflammation (Laye et al. 2018; Macedo Rogero and Calder 2018), suppressing tumor carcinogenesis (D'Eliseo and Velotti 2016; Manson et al. 2019), and managing obesity (Simopoulos 2016; Albracht-Schulte et al. 2018). However, to encourage their consumption, it is necessary to consider the preservation of these sensitive compounds due to their high susceptibility to the occurrence of lipid oxidation.

On the other hand, it has been notable growth in interest by both consumers and researchers toward ancient Latin American crops like chia, among others. Chia seeds (*Salvia hispanica* L.) are vegetable sources with the highest content known of α-linolenic acid (omega-3 $\cong$ 65%) and 20% of omega-6 (linoleic acid), which are both essential fatty acids. Also, their composition includes proteins, dietary fiber, phytosterols, and natural antioxidants. In contrast, this crop does not present natural toxic substances like cyanogenic glycosides or gluten proteins (Ixtaina 2010). Chia is considered a food by the US Food and Drug Administration (US Food and Drug Administration 2009); also, the European Parliament and European Council have approved it as a novel food (European Commission 2009).

The food enrichment in their compounds represents a multidisciplinary challenge that involves different sectors such as food science and technology, nutrition, and academia (Jacobsen 2010). It is relevant to consider that some of the main nutraceutical and functional food components need to be delivered into different matrices, trying to overcome the limitations to their current utilization within the food industry (McClements et al. 2009; Wang et al. 2020). As with drug delivery systems in the pharmaceutical field, more attention has been paid to delivery systems in the food industry (Lacatusu et al. 2013). New approaches include food-based strategies as alternatives to improve the stability of omega-3 FAs and contribute to consumer's health (Lane and Derbyshire 2015). Noticeably, there is a growing demand for studies dealing with functional emulsions and microparticles as omega-6 and omega-3 FAs delivery systems (Julio et al. 2015; Valoppi et al. 2019).

It is important to remark that an edible delivery system plays different roles: providing an efficient method to encapsulate an appreciable amount of the functional components being easily incorporated into food systems; protecting sensitive compounds from chemical degradation like lipid oxidation; releasing these components at a specific site of action, at a controlled rate in response to a specific chemical environment; making compatible with a certain food matrix and resistant to the environmental stresses during its production, storage, transport, and consumption (McClements et al. 2009). In this context, different lipid-based structured delivery systems have been obtained using emulsion technology and/or the microencapsulation of bioactive compounds (Chen et al. 2006; Duncan 2011).

This chapter reviews the design of food vehicles and delivery systems to improve the stability of bioactive and functional omega-3 FAs from chia seeds. Also, the chapter discusses the application of chia by-products in the different delivery systems, being useful to the further development of functional foods.

10.2 CHIA SEED BY-PRODUCTS

Chia seeds are a source of bioactive compounds present in the different by-products obtained from their processing, which are shown in Figure 10.1.

10.2.1 Oil

Chia is one of the significant oilseed sources of polyunsaturated fatty acids. These seeds and those from sacha inchi possess ~85% of essential fatty acids,

Figure 10.1 Chia (*Salvia hispanica* L.) seed by-products.

followed by safflower, flaxseed, and sunflower with 75%, 72%, and 67%, respectively. The difference between them is that chia oil contains a higher amount of α-linolenic acid than sacha inchi. The oil content of chia seeds ranges between 25% and 38%, presenting the highest percentage of α-linolenic acid known to date (62%–65%), as well as a high linoleic acid concentration (Ayerza 1995; Ixtaina et al. 2011a).

Different countries market the chia seed oil as crude or virgin. The most common process used for chia seed oil extraction is cold-pressing, an alternative to conventional pressing. In this process, the raw material is not subjected to heating before pressing, and the temperature is controlled during the extraction. These practices allow the retention of a higher quantity of interesting phytochemical compounds, such as some natural antioxidants (Ixtaina et al. 2011a). However, one disadvantage associated with this extraction method is its low oil yield. For this reason, some other alternatives have been studied, such as the use of CO_2-supercritical extraction (Ixtaina et al. 2010, 2011a, 2011b; Uribe et al. 2011).

The α-linolenic acid of the chia seed oil is present in most of the twelve triacylglycerol molecular species identified, trilinolenin being the most important of these compounds. In addition to its content of essential fatty acids, the existence of other components in this oil, such as pigments and antioxidants, makes it even more interesting for its application in different industrial fields. It has been reported the presence of carotenoids (0.5–1.2 mg/kg), mainly of β-carotene, as well as tocopherols (~240 and 430 mg/kg) and polyphenolic compounds (6×10^{-6} to 2.1×10^{-5} mol/kg). Among tocopherols, γ-tocopherol stands out due to its higher concentration (>85%), followed by δ- and α-tocopherol. Besides, the chlorogenic and caffeic acids, myricetin, quercetin, and kaempferol are the polyphenols reported in the scientific literature (Taga et al. 1984; Ixtaina et al. 2011a, 2012; da Silva Marineli et al. 2014).

Despite the presence of antioxidant compounds, the chia seed oil is highly unstable, caused mainly by the high level of PUFAs.

10.2.2 Proteins

While the growing demand for chia seed is especially encouraged by the high content of omega-3 FAs of its oil, it is further a source of proteins with potential health and nutrition contributions. The protein content of chia seed is about 19%–23% on a dry weight basis, which is superior to those from major grains such as wheat (14%), corn (14%), rice (8.5%), oats (15.3%), and barley (9.2%), and greatly dependent upon environmental and agronomic factors (Sandoval-Oliveros and Paredes-López 2013).

An important characteristic of this seed is related to the absence of gluten, so it can be used to gluten-free foods to be ingested by celiac consumers.

These vegetable proteins are recovered mainly from the residual chia meal obtained after the oil extraction process. The chia seed meal can be a valuable source of protein that would comprise over 17%–24% of this macronutrient (Sandoval-Oliveros and Paredes-López 2013; Timilsena et al. 2016a). Several research works have driven the extraction and characterization of protein-rich-fractions, protein concentrates, and protein isolates from the defatted chia meal by applying dry fractionation or chemical processes (Olivos-Lugo et al. 2010; Vázquez-Ovando et al. 2010; Sandoval-Oliveros and Paredes-López 2013; Timilsena et al. 2016a; López et al. 2018; Coelho and Salas-Mellado 2018; Julio et al. 2019b).

It is considered that chia proteins have high nutritional quality because of their significant levels of digestibility (~78.9%) and the content of all the essential amino acids for human nutrition (Sandoval-Oliveros and Paredes-López 2013). The content of essential amino acids in chia seeds is higher than those found in soybean and sunflower oilseeds. Regarding this, leucine, isoleucine, and valine essential compounds comprise 42.2%–42.9% of the total present in chia seeds, lysine being the limiting essential amino acid with the lowest coverage of requirement (Olivos-Lugo et al. 2010). Chia seed also contains considerable amounts of non-essential amino acids, such as sulfur amino acids and aspartic and glutamic acids; the latter two are of interest in the diet due to their key role in hormonal regulation and stimulation of the immune system (Olivos-Lugo et al. 2010).

Based on solubility criteria, several storage proteins have been isolated and characterized, resulting in a decreasing concentration order as follows: globulins, albumins, glutelins, and prolamins as the most abundant fractions (Vázquez-Ovando et al. 2010; Sandoval-Oliveros and Paredes-López 2013; Timilsena et al. 2016a; Coelho and Salas-Mellado 2018; Julio et al. 2019b).

10.2.3 Soluble Fiber (Mucilage)

The chia mucilage (CHM) is located in the outer three layers of the nutlets (commonly named "seeds"; Muñoz et al. 2012). When chia seeds are soaked in water, they exude a clear mucilaginous gel, which remains tightly bonded to the seed (Muñoz et al. 2012; Capitani et al. 2013). Its content is about 5%–6% of the seeds and is part of the soluble fiber. It is a tetrasaccharide with 4-O-methyl-α-D-glu coronopyranosyl residues as branches of β-D-xylopyranosyl on the main chain, with a molecular weight of 800–2000 kDa (Lin et al. 1994).

CHM possesses interesting functional properties from a technological and physiological point of view. Like other plant gums, it is of interest to the food industry because of its different applications such as texture modifiers, gelling agents, thickeners, stabilizers, emulsifiers, bulking agents, encapsulants, syneresis inhibitors, and film/coating agents, in addition to its constituting soluble dietary fiber (Capitani et al. 2013; Timilsena et al. 2016b).

Several researchers have extracted CHM with different compositions according to the employed method. Capitani et al. (2015) obtained CHM by two methods (MI and MII), whose carbohydrate contents were 67.68 (MI) and 82.61% d.b. (MII), and protein concentrations of 18.85 (MI) and 6.79% d.b. (MII). Timilsena et al. (2016b) purified chia crude mucilage, achieving a gum with 93.8% of carbohydrates and a small amount of protein (2.6%). These authors reported that CHM is a hydrophilic heteropolysaccharide mainly composed of xylose and glucose, uronic acids, and lower amounts of arabinose and galactose. CHM presents anionic character at pH values above 1.8 due to the high uronic acid amount. Besides it is highly soluble at different temperatures and pH levels and presents high viscosity, even at low concentrations, due to the intermolecular associations between the 4-O-methyl glucuronic acid substituents. CHM is also stable at temperatures up to 244 °C (Capitani et al. 2013, 2015; Timilsena et al. 2015, 2016b). Regarding rheology, it presents a shear-thinning or pseudoplastic behavior (Segura-Campos et al. 2014).

10.3 APPLICATION OF CHIA BY-PRODUCTS IN DELIVERY SYSTEMS

Because of the high susceptibility against lipid oxidation of omega-3 PUFAs rich oils such as chia seed oil, their preservation is one of the main challenges when developing enriched foods in these bioactive compounds, in addition to sensory characteristics. Considering this, it is crucial to evaluate the optimal way to add them into foods either as neat oils or through a delivery system.

Properly designed delivery systems could encapsulate and protect PUFAs against the oxidation both before and after incorporation into foods, preserving their health benefits. Regarding this, among the different systems developed to minimize PUFAs oxidation, emulsification and microencapsulation are the most promising techniques (Figure 10.2).

10.3.1 Emulsions

Oil-in-water (O/W) emulsions are widely used systems to limit the oxidation of bioactive lipophilic compounds like PUFAs, offering advantages such as ease of preparation, relatively high stability, high bioavailability, and low cost (McClements and Decker 2000). They can be applied not only in the formulation of liquid food systems, such as sauces, beverages, dressing, soups, and desserts, but also in the development of powder products (microparticles) from the drying of these emulsions (Hogan et al. 2001; Figure 10.3).

Conventional O/W emulsions, the most used system by the food industry, consist of oil droplets dispersed within a continuous aqueous phase in which

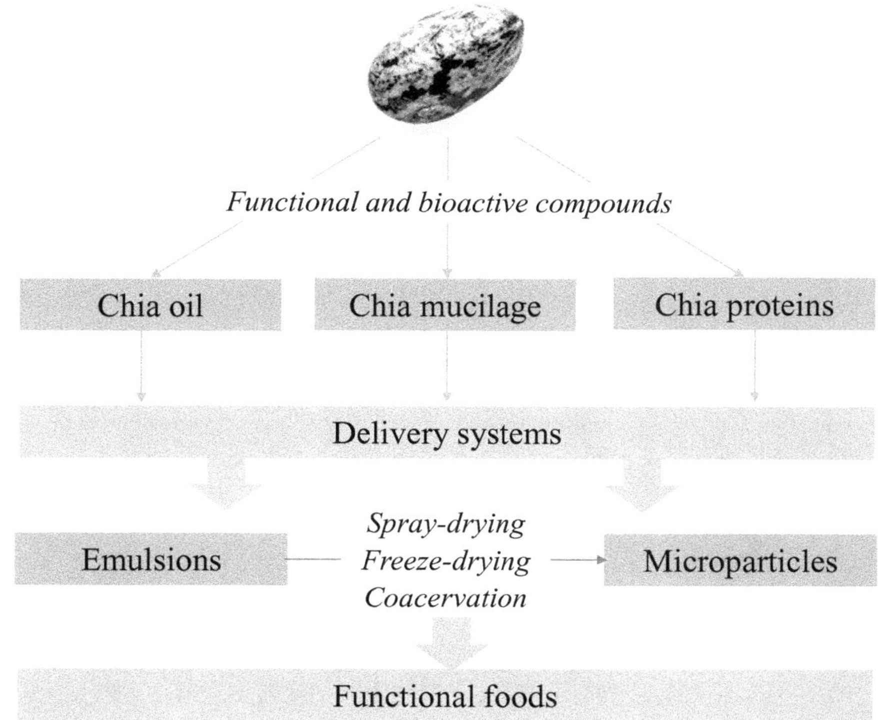

Figure 10.2 Delivery systems of chia seed bioactive compounds.

emulsifier agents adsorb to the freshly formed interface of oil droplets created during the homogenization process. They are thermodynamically unfavorable colloidal systems that tend to divide into two phases over time, undergoing creaming, flocculation, coalescence, and Ostwald ripening. The extent of the stability of the emulsions is highly dependent on their rheology, superficial charge, and droplet characteristics. This latest, in turn, is determined by the homogenization conditions (e.g., intensity and duration of energy input) and system composition (e.g., the type and concentration of emulsifier used or the viscosity ratio between disperse and continuous phases; Julio et al. 2015). Therefore, emulsion composition, emulsification conditions, and structure design are key factors to develop protective and stable systems.

Different emulsions have been applied for chia seed oil entrapment, including conventional, multilayer, and multiple (Table 10.1). Regarding the formulation of emulsions, a series of studies have investigated the effect of different

TABLE 10.1 CHIA OIL EMULSIFICATION USING DIFFERENT PROCESSES AND EMULSIFIERS/STABILIZER AGENTS

Type of emulsion	Emulsifier/ stabilizer	Emulsification process	References
Conventional O/W emulsion	NaCas and Lac	Ultraturrax, High-pressure homogenization	Ixtaina et al. (2015)
Conventional O/W emulsion	Chia PRF, CHM, Lac, Mx, NaCas	Ultraturrax, High-pressure homogenization	González et al. (2016)
Conventional O/W emulsion	NaCas, CHM	Ultraturrax, Ultrasonic processor	Capitani et al. (2018)
Double (W_1/O/W_2) emulsion	mesquite gum, Mx and WPC	Ultraturrax	Timilsena et al. (2016c)
Mono and bilayer emulsions	Modified sunflower lecithins, chitosan, Mx	Ultraturrax, High-pressure homogenization	Martinez et al. (2015)
Mono and double-layer emulsions	WPC and pectin	Ultraturrax, ultrasound sonication	Rodea-González et al. (2012), Rodea-González et al. (2012), Rodea-González et al. (2012)
O/W nanoemulsions	Tween 80 and Span80, NaCas, and sucrose monopalmitate	Spontaneous emulsification, microfluidization	Teng et al. (2018)

Abbreviations: CHM, chia mucilage; Lac, lactose; Mx, maltodextrin; NaCas, sodium caseinate; PRF, protein rich fraction; WPC, whey protein concentrate.

aqueous phase compositions (emulsifying agent, thickening agent) and homogenization process conditions on emulsions systems to protect and deliver chia seed oil. Among the emulsifying agents used to stabilize emulsions containing this oil, some proteins (sodium caseinate, whey, and soybean proteins, phospholipids (modified sunflower lecithins), and small-molecule synthetic surfactants like Tween 80 and Span 80 are found. Overall, different polysaccharides such as lactose, maltodextrin, pectin, carrageenan, and mesquite gum were also added to retard the destabilization process and/or act as wall material in the emulsion drying.

As mentioned earlier, depending on the method and process applied to create an emulsion, it is possible to design diverse structured systems. Carrillo-Navas (2012) developed viscoelastic and stable $W_1/O/W_2$ double emulsions containing chia oil through a two-stage emulsification procedure using a low-energy homogenizer (Ultraturrax) and different biopolymer ratios. However, these rotor-stator devices are typically used in combination with a high-energy method such as ultrasound, high-pressure homogenization, or microfluidization to achieve smaller oil droplets and thus enhance global stability. Considering this, other authors have emulsified chia oil by applying a pre-emulsification with Ultraturrax followed by a droplet reduction using sonication (Capitani et al. 2018; Vélez-Erazo et al. 2018). In this sense, Vélez-Erazo et al. 2018 reported a greater effect caused by the power amplitude than the time to produce simple and double layer sonicated emulsions. Similarly, other authors prepared a series of emulsion-based systems with chia seed oil through a homogenization process that consisted of the first step with Ultraturrax and a second one by high-pressure homogenization (Julio et al. 2015, 2016, 2018). On the other hand, Teng et al. (2017) obtained different nanoemulsions using Tween 80 and Span 80 through spontaneous emulsification process and microfluidization, and using sodium caseinate or sucrose monopalmitate. These authors reported that all nanoemulsions remained stable for at least 2 weeks at 4 °C or ambient temperature, recording the highest transparency and the smallest droplet diameter (~47 nm) systems with sucrose monopalmitate.

Regarding the oxidative stability of PUFAs, the *layer-by-layer* (LBL) deposition technique has proved to be an effective strategy to protect and enhance the oxidative stability of sensible oils into an emulsion. This type of sophisticated emulsions consists of multiple layers of emulsifying agents/biopolymer with opposite charges. Julio et al. (2018) have demonstrated the efficient limitation of chia oil oxidation by the LBL technique application into the emulsion structure design. Emulsions stabilized with chitosan-modified sunflower lecithin bilayers showed higher oxidative stability in terms of hydroperoxide values and ω-3 fatty acid content than monolayer ones and neat chia oil stored at the same conditions, 4 °C for 50 days. Because of the greater thickness and the cationic nature of the oil droplets, the interfacial double-layer may constitute a barrier against the penetration and the diffusion of pro-oxidant agents such as transition metals (Julio et al. 2018). Furthermore, bilayer emulsions showed better physical stability than the monolayer ones with no significant alterations during the total storage period (Julio et al. 2018).

Different authors have conducted studies to produce effective delivery systems for bioactive compounds from chia seeds, including several chia by-products. Capitani et al. (2018) studied the addition of different concentrations of chia mucilage (CHM; 0–0.8% wt/wt) and sodium caseinate (NaCas; 0.1–5.0% wt/wt) to conventional O/W emulsions containing chia seed oil. These

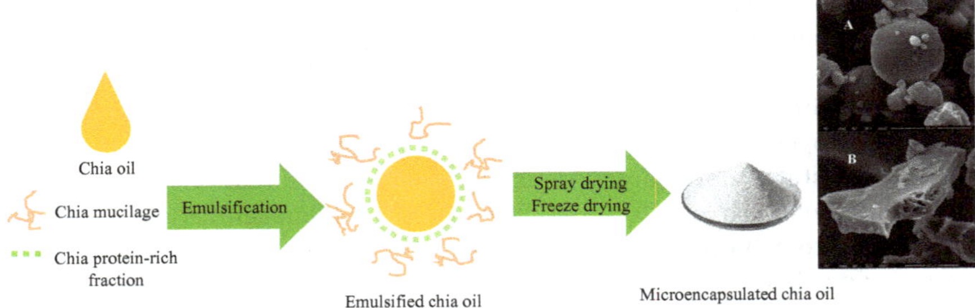

Figure 10.3 Emulsions and microparticles based-systems of bioactive compounds with chia seed by-products. SEM micrographs of microencapsulated chia oil by (A) spray-drying and (B) freeze-drying.

researchers found smaller droplet mean diameter $D_{4,3}$ and greater physical stability over the storage period in systems prepared with high concentrations of CHM (0.8%) and NaCas (2.0%–5.0%). They concluded that the addition of chia mucilage to O/W emulsions confers more stability to the emulsions.

Julio et al. (2016) have developed chia O/W emulsions using CHM and a protein-rich fraction (PRF) from chia seeds (Figure 10.3). These researchers applied different protein-carbohydrate combinations (NaCas-Lac, NaCas-Mx, PRF-Mx) and CHM (0.2% wt/wt) in the aqueous phase of chia O/W emulsions. The use of NaCas-Lac or NaCas-Mx led to emulsions with high physical stability, small droplet size, and pseudoplastic behavior. In contrast, those with chia PRF presented higher droplet mean diameters, Newtonian behavior, and limited global stability. In the case of CHM, it showed an effective role as a thickener, mainly affecting the viscosity of the emulsions, slowing down the movement of the oil droplets, and thus improving the stability of these systems. Thus, the potential application of these novel ingredients (CHM and PRF) into the development of functional foods (emulsion-based liquid or emulsions dried to be converted in their powdered form) is a key issue to be considered.

10.3.2 Microparticles

Microencapsulation is a strategy to deliver bioactive compounds from chia seeds, which can constitute either the core or the wall material of the particles (Table 10.2). In general, the chia seed oil is directly encapsulated alone or with the addition of other bioactive compounds, such as antioxidants (Martinez et al. 2015; Rodriguez et al. 2019; Fırtın et al. 2020). Concerning the chia by-products as wall material, different works have included both the CHM and the PRF as encapsulating agents (Timilsena et al. 2016a, 2016c, 2017; Us-Medina et al. 2018).

TABLE 10.2 MICROENCAPSULATION OF CHIA OIL USING DIFFERENT PROCESSES AND WALL MATERIALS

Wall material/emulsifiers	Microencapsulation process	References
WPC, gum Arabic, mesquite gum	Spray-drying O/W emulsions	Rodea-González et al. (2012)
NaCas and Lac	Spray-drying O/W emulsions	Ixtaina et al. (2015)
HMC and Mx	Spray-drying O/W emulsions	Martinez et al. (2015)
WPC, pectin, and Mx	Spray-drying multilayer emulsions obtained by LBL	Noello et al. (2016)
Isolated soy protein, Mx with/without MRPs	Spray-drying or freeze-drying O/W emulsions	González et al. (2016)
Chia seed protein isolate and/or CHM	Complex coacervation with transglutaminase and spray- or freeze-drying	Timilsena et al. (2016c, 2019)
CHM, alginate, and Tween 80	External ionic gelation	Us-Medina et al. (2017)
NaCas and Lac with/without MRPs	Freeze-drying O/W emulsions	Copado et al. (2017)
Stearic acid and Tween 80	O/W emulsion obtained by hot homogenization technique followed by freeze-drying	Souza et al. (2017)
NaCas, Lac, Mx, CHM, chia PRF	Spray-drying O/W emulsions	Us-Medina et al. (2018)
Carnauba wax, NaCas	Dispersion followed by freeze-drying	Guimarães-Inácio et al. (2018)
NaCas and Lac with/without MRPs	Spray-drying O/W emulsions	Copado et al. (2019)
Modified sunflower lecithins, chitosan, and Mx	Freeze-drying multilayer emulsions obtained by LBL	Julio et al. (2019a)
WPI, Mx, and Arabic gum	Spray-drying O/W emulsions	Alcântara et al. (2019)
NaCas and Lac	Freeze-drying O/W emulsions	Rodriguez et al. (2019)
Arabic gum and Mx	Freeze-drying O/W emulsions	Fırtın et al. (2020)

Abbreviations: NaCas, sodium caseinate; Lac, lactose; MRPs, Maillard reaction products; CHM, chia mucilage; HMC, hydroxypropylmethylcellulose; Mx, maltodextrin; PRF, protein-rich fraction; WPC, whey protein concentrate; WPI, whey protein isolate.

Several researchers have studied different methods to microencapsulate chia oil to protect it from oxidative deterioration. Most of them have applied an emulsification process followed by spray-drying or freeze-drying to obtain the powdered oil (Table 10.2). The wall materials used in these systems include proteins (whey protein concentrate, sodium caseinate, isolated soy protein, chia protein-rich fraction) commonly combined with carbohydrates (maltodextrin, lactose), gums (Arabic or mesquite gum, CHM), lipids (carnauba wax, stearic acid, native or modified lecithins) or cellulose materials (hydroxypropylmethylcellulose). On the whole, the particles obtained by spray-drying showed spherical shapes and varied sizes, with some depressions formed by shrinkages, whereas those obtained by freeze-drying are like sheets with irregular geometry and compact structure (Figure 10.3).

The microencapsulation efficiencies reported by the different authors present a wide range of variation (40%–99%), depending mainly on the emulsification process and the wall material. It is known that the operating conditions during the stages of emulsification or drying affect the physicochemical characteristics of the particles. Thereby, the smaller the particle sizes of the parent emulsions, the higher the microencapsulation efficiencies. Regardless of the wall material, emulsions prepared only with a low-energy homogenizer (Ultraturrax; Rodea-González et al. 2012; Martinez et al. 2015; González et al. 2016) presented a larger particle size than those performed with high-energy emulsification processes, such as sonication (Alcântara et al. 2019) or high-pressure homogenization (Ixtaina et al. 2015; Us-Medina et al. 2018; Copado et al. 2019). Thus, the differences in the particle size of the parent emulsions have influenced the microencapsulation efficiency values.

About the spray-drying, the most common drying inlet/outlet temperatures used by the different researchers were 130–135/80 °C (Rodea-González et al. 2012; Ixtaina et al. 2015; González et al. 2016) or 160–170/90 °C (Ixtaina et al. 2015; Martinez et al. 2015; Us-Medina et al. 2018; Copado et al. 2017), mainly affecting the oxidative stability of the microencapsulated chia oil. In this sense, when the drying temperature was the highest, the oxidative stability of the oil was the lowest. Most of the research works found that the microencapsulation process was efficient in protecting chia seed oil. However, Martinez et al. (2015) reported that the spray-drying microencapsulation carried out at a drying temperature of 163 °C using maltodextrin in combination with hydroxypropylmethylcellulose as wall material was not effective in protecting the chia oil during the storage. These authors attributed these results to the chemical damage caused by the temperature in the chia oil during drying, showing that in this case, the matrix of the wall could not protect the lipid core during the obtaining of the microcapsules.

Different strategies have been carried out to confer spray-dried microencapsulated chia oil additional protection against lipid oxidation. One of them is the study carried out by Copado et al. (2019) that reported that the Maillard reaction products generated from different heat treatments of sodium caseinate–lactose mixture as a wall material presented a higher level of antioxidant activity than non-thermally treated systems. The addition of antioxidant compounds either into the aqueous or the oily phase of the O/W emulsions before drying constituted another alternative in this sense. Rodriguez et al. (2019) added different natural antioxidants such as extracts from chamomile and rosemary, and essential oils from *Origanum vulgare, Origanum x majoricum*, and *Mentha spicata* to chia seed oil microencapsulated by the freeze-drying. These authors reported that most of these antioxidants further increased the oxidative stability of microencapsulated chia oil, presenting the best effect, the blend of chamomile and rosemary extracts. A different strategy consists of the development of multilayered systems. They are obtained by the layer-by-layer technique, which involves the electrostatic deposition of a biopolymer with an opposite charge on the surface of the oil droplet coated with the emulsifier (Noello et al. 2016; Julio et al. 2019a). These powder systems have been more efficient than the monolayer ones to protect chia oil against lipid oxidation (Julio et al. 2019a).

The hot homogenization technique (Souza et al. 2017), the complex coacervation (Timilsena et al. 2016c), and the ionic gelation (Us-Medina et al. 2017) are other methods used to microencapsulate chia oil.

Souza et al. (2017) carried out an encapsulation process of chia oil in stearic acid particles. The applied process consisted of obtaining dispersion by sonication at 75 °C and subsequent freeze-drying. These authors reported a microencapsulation efficiency of 95% and an increase in the oxidative stability of the encapsulated oil concerning the bulk one.

Timilsena et al. (2016c) produced particles of chia oil using chia seed protein isolate (CHPI), mucilage (CHM), and CHPI-CHM as encapsulating agents to obtain complex coacervates. These authors recorded higher microencapsulation efficiency (~94%) and oxidative stability for CHPI-CHM complex coacervates than those produced by CHMG or CHPI individually. Us-Medina et al. (2017) used the combination of CHM with alginate as wall materials and calcium chloride as a cross-linking agent in the encapsulation of chia oil by external ionic gelation, obtaining spherical capsules with a uniform diameter of ~1.8 cm, a microencapsulation efficiency of ~83%, and high oxidative stability.

Chia by-products application has also been studied to microencapsulate bioactive compounds from other sources, including essential oil from lemon (Cortés-Camargo et al. 2019) and oregano (Hernández-Nava et al. 2020) as well as natural beet dye (Antigo et al. 2020) and probiotic bacteria *Lactobacillus plantarum* and *Bifidobacterium infantis* (Bustamante et al. 2017). Most of these research works selected CHM as encapsulating material in combination with proteins or other gums. Also, in the microencapsulation of probiotic bacteria, the soluble protein extracted from chia was used. According to Cortés-Camargo et al. (2019), the combination of CHM with mesquite gum contributed to improve the oxidative stability and delay the release rate of microencapsulated lemon essential oil. Similarly, Hernández-Nava et al. (2020) proposed the blend of gelatin with CHM as an alternative system for complex coacervation instead of the most frequently applied (gelatin and gum Arabic), since a high microencapsulation efficiency of oregano essential oil was obtained with gelatin/CHM. Finally, Bustamante et al. (2017) reported that CHM and soluble proteins from chia seeds used as encapsulating agents with maltodextrin improved probiotic survival during spray-drying and its viability during storage.

10.4 CONCLUSION

The notable growth in interest of consumers and researchers toward ancient Latin American crops, such as chia, has driven the revalue of its consumption. The potential use of this crop as a source of ingredients of novel functional foods because of its high content of bioactive compounds (essential lipids like omega-3 fatty acids, proteins, dietary fiber, natural antioxidants) have led to optimizing the production of valuable by-products (oil, protein-rich fraction, mucilage).

An overview of the strategies developed through the design of different delivery systems to preserve the chia seed oil from the adverse influence of the environment with the application of their by-products by diverse methodologies has been provided. Various types of food-based emulsions (conventional, multilayers, and multiples) and spray-dried or freeze-dried microparticles of chia seed oil were described and characterized in terms of their physicochemical and functional properties, among others.

Furthermore, chia proteins and/ or mucilage have been successfully used as emulsifying, stabilizing, thickening, and encapsulating agents to supply their own bioactive compounds as well as those from other sources.

This information is relevant key for the food industry and researchers to be applied as knowledge in the development of functional foods (i.e., food products enriched with omega-3 fatty acids) containing bioactive compounds from chia seed with an improvement in their shelf life and suitable sensory attributes.

Further studies will also be required to evaluate the various aspects related to the bioactivity and bioaccessibility of these functional foods.

ACKNOWLEDGMENTS

This material is based upon work supported by Agencia Nacional de Promoción Científica y Tecnológica (ANPCyT PICT 2016-0323, 2019-1775), Project X907 Universidad Nacional de La Plata (UNLP) Argentina and Project CYTED 119RT0567, Spain.

REFERENCES

Adkins, Yuriko, and Darshan S. Kelley. 2010. "Mechanisms Underlying the Cardioprotective Effects of Omega-3 Polyunsaturated Fatty Acids." *The Journal of Nutritional Biochemistry* 21 (9): 781–792. https://doi.org/10.1016/j.jnutbio.2009.12.004.

Albracht-Schulte, Kembra, Nisha Kalupahana, Latha Ramalingam, Shu Wang, Shaik Rahman, Jacalyn Robert-McComb, and Naima Moustaid-Moussa, N. 2018. "Omega-3 Fatty Acids in Obesity and Metabolic Syndrome: A Mechanistic Update." *Journal of Nutritional Biochemistry* 58: 1–16. https://doi.org/10.1016/j.jnutbio.2018.02.012.

Alcântara, Maristela Alves, Anderson Eduardo Alcântara de Lima, Ana Luiza Mattos Braga, Renata Valeriano Tonon, Melicia Cintia Galdeano, Mariana da Costa Mattos, Ana Iraidy Santa Brígida, Raul Rosenhaim, Nataly Albuquerque dos Santos, and Angela Maria Tribuzy de Magalhães Cordeiro. 2019. "Influence of the Emulsion Homogenization Method on the Stability of Chia Oil Microencapsulated by Spray Drying." *Powder Technology* 354: 877–885. https://doi.org/10.1016/j.powtec.2019.06.026.

Antigo, Jéssica Loraine D., Ana Paula Stafussa, Rita de Cassia Bergamasco, and Grasiele Scaramal Madrona. 2020. "Chia Seed Mucilage as a Potential Encapsulating Agent of a Natural Food Dye." *Journal of Food Engineering*: 110101. https://doi.org/10.1016/j.jfoodeng.2020.110101.

Ayerza, Ricardo. 1995. "Oil Content and Fatty Acid Composition of Chia (*Salvia hispanica* L.) from Five Northwestern Locations in Argentina." *Journal of the American Oil Chemists' Society* 72: 1079–1081.

Bigliardi, Barbara, and Francesco Galati. 2013. "Innovation Trends in the Food Industry: The Case of Functional Foods." *Trends in Food Science and Technology* 31: 118–129. https://doi.org/10.1016/j.tifs.2013.03.006.

Bustamante, Mariela B., Dave Oomah, Mónica Rubilar, and Carolina Shene. 2017. "Effective *Lactobacillus plantarum* and *Bifidobacterium infantis* Encapsulation with Chia Seed (*Salvia hispanica* L.) and Flaxseed (*Linum usitatissimum* L.) Mucilage and Soluble Protein by Spray Drying." *Food Chemistry* 216: 97–105. https://doi.org/10.1016/j.foodchem.2016.08.019.

Capitani, Marianela I., Vanesa Y. Ixtaina, Susana M. Nolasco, and Mabel C. Tomás. 2013. "Microstructure, Chemical Composition and Mucilage Exudation of Chia (*Salvia hispanica* L.) Nutlets from Argentina." *Journal of the Science of Food and Agriculture* 93 (15): 3856–3862. https://doi.org/10.1002/jsfa.6327.

Capitani, Marianela I., Luis Jorge Corzo-Rios, Luis Chel-Guerrero, David Betancur-Ancona, Susana Maria Nolasco, and Mabel Cristina Tomás. 2015. "Rheological Properties of Aqueous Dispersions of Chia (*Salvia hispanica* L.) Mucilage." *Journal of Food Engineering* 149: 70–77. https://doi.org/10.1016/j.jfoodeng.2014.09.043.

Capitani, Marianela I., Mukthar Sandoval-Peraza, Luis A. Chel-Guerrero, David A. Betancur-Ancona, Susana M. Nolasco, and Mabel C. Tomás. 2018. "Functional Chia Oil-in-Water Emulsions Stabilized with Chia Mucilage and Sodium Caseinate." *Journal of the American Oil Chemists' Society* 95 (9): 1213–1221. https://doi.org/10.1002/aocs.12082.

Carrillo-Navas, Hector, Cruz-Olivares, Julian, Varela-Guerrero, Victor, Alamilla-Beltrán, Liliana, Vernon-Carter, Eduardo J., and Pérez-Alonso, César. 2012.

"Rheological properties of a double emulsion nutraceutical system incorporating chia essential oil and ascorbic acid stabilized by carbohydrate polymer–protein blends." *Carbohydrate Polymers* 87(2): 1231–1235.

Chen, Hongda, JochenWeiss, and Fereidoon Shahidi. 2006. "Nanotechnology in Nutraceuticals and Functional Foods." *Food Technology* 60: 30–36.

Coelho, Michele S., and Myriam M. Salas-Mellado. 2018. "How Extraction Method Affects the Physicochemical and Functional Properties of Chia Proteins." *LWT-Food Science and Technology* 96: 26–33. https://doi.org/10.1016/j.lwt.2018.05.010.

Copado, Claudia N., Bernd W. K. Diehl, Vanesa Y. Ixtaina, and Mabel C. Tomás. 2017. "Application of Maillard Reaction Products on Chia Seed Oil Microcapsules with Different Core/Wall Ratios." *LWT-Food Science and Technology* 86: 408–417. https://doi.org/10.1016/j.lwt.2017.08.010.

Copado, Claudia N., Bernd W. K. Diehl, Vanesa Y. Ixtaina, and Mabel C. Tomás. 2019. "Microencapsulation with Maillard Reaction Products to Improve the Oxidative Stability of Chia Oil." *Inform* 30 (7): 14–17.

Cortés-Camargo, Stefani, Pedro Estanislao Acuña-Avila, María Eva Rodríguez-Huezo, Angélica Román-Guerrero, Victor Varela-Guerrero, and César Pérez-Alonso. 2019. "Effect of Chia Mucilage Addition on Oxidation and Release Kinetics of Lemon Essential Oil Microencapsulated Using Mesquite Gum—Chia Mucilage Mixtures." *Food Research International* 116: 1010–1019. https://doi.org/10.1016/j.foodres.2018.09.040.

da Silva Marineli, Rafaela, Érica Aguiar Moraes, Sabrina Alves Lenquiste, Adriana Teixeira Godoy, Marcos Nogueira Eberlin, and Mário Roberto Maróstica, Jr. 2014. "Chemical Characterization and Antioxidant Potential of Chilean Chia Seeds and Oil (*Salvia hispanica* L.)." *LWT-Food Science and Technology* 59 (2): 1304–1310. https://doi.org/10.1016/j.lwt.2014.04.014.

D'Eliseo, Donatella, and Francesca Velotti. 2016. "Omega-3 Fatty Acids and Cancer Cell Cytotoxicity: Implications for Multi-Targeted Cancer Therapy." *Journal of Clinical Medicine* 5 (2): 15. https://doi.org/10.3390/jcm5020015.

Derbyshire, Emma. 2018. "Brain Health Across the Lifespan: A Systematic Review on the Role of Omega-3 Fatty Acid Supplements." *Nutrients* 10 (8): 1094. https://doi.org/10.3390/nu10081094.

Duncan, Timothy V. 2011. "Applications of Nanotechnology in Food Packaging and Food Safety: Barrier Materials, Antimicrobials and Sensors." *Journal of Colloid and Interface Science* 363: 1–24. https://doi.org/10.1016/j.jcis.2011.07.017.

Elagizi, Andrew, Carl J. Lavie, Keri Marshall, James J. DiNicolantonio, James H. O'Keefe, and Richard V. Milani. 2018. "Omega-3 Polyunsaturated Fatty Acids and Cardiovascular Health: A Comprehensive Review." *Progress in Cardiovascular Diseases* 61 (1): 76–85. https://doi.org/10.1016/j.pcad. 2018.03.006.

European Commission. 2009. "Authorizing the Placing on the Market of Chia Seed (*Salvia hispanica*) as Novel Food Ingredient Under Regulation (EC) N° 258/97 of European Parliament of the Council." *Official Journal of the European Union, C* 7645. https://doi.org/10.1002/jsfa.9294.

Fırtın, Burcu, Hande Yenipazar, Ayşe Saygün, and Neşe Şahin-Yeşilçubuk. 2020. "Encapsulation of Chia Seed Oil with Curcumin and Investigation of Release Behavior & Antioxidant Properties of Microcapsules During in Vitro Digestion

Studies." *LWT-Food Science and Technology* 134: 109947. https://doi.org/10.1016/j.lwt.2020.109947.

González, Agustín, Marcela L. Martínez, Alejandro J. Paredes, Alberto E. León, and Pablo D. Ribotta. 2016. "Study of the Preparation Process and Variation of Wall Components in Chia (*Salvia hispanica* L.) Oil Microencapsulation." *Powder Technology* 301: 868–875. https://doi.org/10.1016/j.powtec.2016.07.026.

Guimarães-Inácio, Alexandre, Cristhian Rafael Lopes Francisco, Valquíria Maeda Rojas, Roberta de Souza Leone, Patrícia Valderrama, Evandro Bona, Fernanda Vitória Leimann, Ailey Aparecia Coelho Tanamati, and Odinei Hess Gonçalves. 2018. "Evaluation of the Oxidative Stability of Chia Oil-Loaded Microparticles by Thermal, Spectroscopic and Chemometric Methods." *LWT-Food Science and Technology* 87: 498–506. https://doi.org/10.1016/j.lwt.2017.09.031.

Hernández-Nava, Ruth, Aurelio López-Malo, Enrique Palou, Nelly Ramírez-Corona, and María Teresa Jiménez-Munguía. 2020. "Encapsulation of Oregano Essential Oil (*Origanum vulgare*) by Complex Coacervation Between Gelatin and Chia Mucilage and Its Properties After Spray Drying." *Food Hydrocolloids*: 106077. https://doi.org/10.1111/1750-3841.14605.

Hogan, Sean A., Brian F. McNamee, Dolores O'Riordan, and Michael O'Sullivan. 2001. "Microencapsulating Properties of Whey Protein Concentrate." *Journal of Food Science* 66 (5): 675–680. https://doi.org/10.1111/j.1365-2621.2001.tb04620.x.

Ixtaina, Vanesa Y. 2010. "Caracterización de la semilla y el aceite de chía (*Salvia hispanica* L.). Aplicaciones en tecnología de alimentos." Doctoral thesis. Facultad de Ciencias Exactas, Universidad Nacional de La Plata (FCE-UNLP), La Plata.

Ixtaina, Vanesa Y., Luciana M. Julio, Jorge R. Wagner, Susana M. Nolasco, and Mabel C. Tomás. 2015. "Physicochemical Characterization and Stability of Chia Oil Microencapsulated with Sodium Caseinate and Lactose by Spray-Drying." *Powder Technology* 271: 26–34. https://doi.org/10.1016/j.powtec.2014.11.006.

Ixtaina, Vanesa Y., Marcela L. Martínez, Viviana Spotorno, Carmen M. Mateo, Damián M. Maestri, Bernd W. K. Diehl, Susana M. Nolasco, and Mabel C. Tomás. 2011a. "Characterization of Chia Seed Oils Obtained by Pressing and Solvent Extraction." *Journal of Food Composition and Analysis* 24 (2): 166–174. https://doi.org/10.1016/j.jfca.2010.08.006.

Ixtaina, Vanesa Y., Facundo Mattea, Damián A. Cardarelli, Miguel A. Mattea, Susana M. Nolasco, and Mabel C. Tomás. 2011b. "Supercritical Carbon Dioxide Extraction and Characterization of Argentinean Chia Seed Oil." *Journal of the American Oil Chemists' Society* 88 (2): 289–298. https://doi.org/10.1007/s11746-010-1670-2.

Ixtaina, Vanesa Y., Susana M. Nolasco, and Mabel C. Tomás. 2012. "Oxidative Stability of Chia (*Salvia hispanica* L.) Seed Oil: Effect of Antioxidants and Storage Conditions." *Journal of the American Oil Chemists' Society* 89 (6): 1077–1090. https://doi.org/10.1007/s11746-011-1990-x.

Ixtaina, Vanesa Y., Andrea Vega, Susana M. Nolasco, Mabel C. Tomás, Miquel Gimeno, Eduardo Bárzana, and Alberto Tecante. 2010. "Supercritical Carbon Dioxide Extraction of Oil from Mexican Chia Seed (*Salvia hispanica* L.): Characterization and Process Optimization." *The Journal of Supercritical Fluids* 55 (1): 192–199. https://doi.org/10.1016/j.supflu.2010.06.003.

Jacobsen, Charlotte. 2010. "Enrichment of Foods with Omega-3 Fatty Acids: A Multidisciplinary Challenge." *Annals of the New York Academy of Sciences*: 141–150. https://doi: 10.1111/j.1749-6632.2009.05263.x.

Julio, Luciana M., Claudia N. Copado, Rosana Crespo, Bernd W. K. Diehl, Vanesa Y. Ixtaina, and Mabel C. Tomás. 2019a. "Design of Microparticles of Chia Seed Oil by Using the Electrostatic Layer-by-Layer Deposition Technique." *Powder Technology* 345: 750–757. https://doi.org/10.1016/j.powtec.2019.01.047.

Julio, Luciana M., Claudia N. Copado, Bernd W. K. Diehl, Vanesa Y. Ixtaina, and Mabel C. Tomás. 2018. "Chia Bilayer Emulsions with Modified Sunflower Lecithins and Chitosan as Delivery Systems of Omega-3 Fatty Acids." *LWT-Food Science and Technology* 89: 581–590. https://doi.org/10.1016/j.lwt.2017.11.

Julio, Luciana M., Vanesa Y. Ixtaina, Mariela Fernández, Rosa Torres Sánchez, Jorge R.Wagner, Susana M. Nolasco, and Mabel C. Tomás. 2015. "Chia Seed Oil-in-Water Emulsions as Potential Delivery Systems of Omega-3 Fatty Acids." *Journal of Food Engineering* 162: 48–55. https://doi.org/10.1016/j.jfoodeng.2015.04.005.

Julio, Luciana M., Vanesa Y. Ixtaina, Mariela Fernández, Rosa M. Torres Sánchez, Susana M. Nolasco, and Mabel C. Tomás. 2016. "Development and Characterization of Functional O/W Emulsions with Chia Seed (*Salvia hispanica* L.) By-Products." *Journal of Food Science and Technology* 53 (8): 3206–3214. https://doi.org/10.1007/s13197-016-2295-8.

Julio, Luciana M., Jorge C. Ruiz-Ruiz, Mabel C. Tomás, and Maira R. Segura-Campos. 2019b. "Chia (*Salvia hispanica*) Protein Fractions: Characterization and Emulsifying Properties." *Journal of Food Measurement and Characterization* 13 (4): 3318–3328. https://doi.org/10.1007/s11694-019-00254-w.

Kaushik, Pratibha, Kim Dowling, Colin J. Barrow, and Benu Adhikari. 2015. "Microencapsulation of Omega-3 Fatty Acids: A Review of Microencapsulation and Characterization Methods." *Journal of Functional Foods* 19: 868–881. https://doi.org/10.1016/j.jff.2014.06.029.

Kitajka, Klára, Lázló G. Puskas, Agnes Zvara, Lázló Hackler, Jr., Gwendolyn Barcelo-Coblijn, Young K. Yeo, and Tibor Farkas, T. 2002. "The Role of n-3 Polyunsaturated Fatty Acids in Brain: Modulation of Rat Brain Gene Expression by Dietary n-3 Fatty Acids." *Proceedings of the National Academy of Sciences of the United States of America* 99 (5): 2619–2624. https://doi.org/10.1073/pnas.042698699.

Lacatusu, Ioana, Elena Mitrea, Nicoleta Badea, Raluca Stan, Ovidiu Oprea, and Aurelia Meghea. 2013. "Lipid Nanoparticles Based on Omega-3 Fatty Acids as Effective Carriers for Lutein Delivery: Preparation and in Vitro Characterization Studies." *Journal of Functional Foods* 5: 1260–1269. https://doi.org/10.1002/jsfa.9294.

Lane, Katie E., and Emma Derbyshire. 2015. "Omega-3 Fatty Acids—A Review of Existing and Innovative Delivery Methods." *Critical Reviews in Food Science and Nutrition*. http://doi.org/10.1080/10408398.2014.994699.

Laye, Sophie, Agnés Nadjar, Corinne Joffre, and Richard P. Bazinet. 2018. "Antiinflammatory Effects of Omega-3 Fatty Acids in the Brain: Physiological Mechanisms and Relevance to Pharmacology." *Pharmacological Reviews* 70 (1): 12–38. https://doi.org/10.1124/pr.117.014092.

Lin, Kuei-Ying, James R. Daniel, and Roy L. Whistler. 1994. "Structure of Chia Seed Polysaccharide Exudate." *Carbohydrate Polymers* 23 (1): 13–18. https://doi.org/10.1016/0144-8617(94)90085-X.

López, Débora N., Romina Ingrassia, Pablo Busti, Jorge Wagner, Valeria Boeris, and Darío Spelzini. 2018. "Effects of Extraction pH of Chia Protein Isolates on Functional Properties." *LWT-Food Science and Technology* 97: 523–529. https://doi.org/10.1016/j.lwt.2018.07.036.

Lopez-Huertas, Eduardo. 2010. "Health Effects of Oleic Acid and Long Chain Omega-3 Fatty Acids (EPA and DHA) Enriched Milks: A Review of Intervention Studies." *Pharmacological Research* 61 (3): 200–207. https://doi.org/10.1016/j.phrs.2009.10.007.

Macedo Rogero, Marcelo, and Philipe C. Calder. 2018. "Obesity, Inflammation, Toll-Like Receptor 4 and Fatty Acids." *Nutrients* 10 (4): 432. https://doi.org/10.3390/nu10040432.

Manson, Jo Ann E., Nancy R. Cook, I-Min Lee, William Christen, Shari S. Bassuk, Samia Mora, Heike Gibson, Christine M. Albert, David Gordon, Trisha Copeland, Denise D'Agostino, Georgina Friedenberg, Claire Ridge, Vadim Bubes, Edward L. Giovannucci, Walter C. Willett, and Julie E. Buring. 2019. "Marine n-3 Fatty Acids and Prevention of Cardiovascular Disease and Cancer." *New England Journal of Medicine* 380 (1): 23–32. https://doi.org/10.1056/NEJMoa1811403.

Martinez, Marcela Lilian, María Isabel Curti, Paola Roccia, Juan Manuel Llabot, Maria Cecilia Penci, Romina Mariana Bodoira, and Pablo Daniel Ribotta. 2015. "Oxidative Stability of Walnut (*Juglans regia* L.) and Chia (*Salvia hispanica* L.) Oils Microencapsulated by Spray Drying." *Powder Technology* 270: 271–277. https://doi.org/10.1016/j.powtec.2014.10.031.

McClements, David J., and Eric A. Decker. 2000. "Lipid Oxidation in Oil-in-Water Emulsions: Impact of Molecular Environment on Chemical Reactions in Heterogeneous Food Systems." *Journal of Food Science* 65 (8): 1270–1282. https://doi.org/10.1111/j.1365-2621.2000.tb10596.x.

McClements, David J., Eric A. Decker, Yeonhwa Park, and Jochen Weiss. 2009. "Structural Design Principles for Delivery of Bioactive Components in Nutraceuticals and Functional Foods." *Critical Reviews in Food Science and Nutrition* 49: 577–606. https://doi.org/10.1080/10408390902841529.

Muñoz, Loreto A., Ángel Cobos, Olga Diaz, and José M. Aguilera. 2012. "Chia Seeds: Microstructure, Mucilage Extraction and Hydration." *Journal of Food Engineering* 108 (1): 216–224. https://doi.org/10.1016/j.jfoodeng.2011.06.037.

Noello, Carla, Águeda G. S. Carvalho, Vanesa Martins da Silva, and Míriam D. Hubinger. 2016. "Spray Dried Microparticles of Chia Oil Using Emulsion Stabilized by Whey Protein Concentrate and Pectin by Electrostatic Deposition." *Food Research International* 89: 549–557. https://doi.org/10.1016/j.foodres.2016.09.003.

Olivos-Lugo, Blanca L., María Á. Valdivia-López, and Alberto Tecante. 2010. "Thermal and Physicochemical Properties and Nutritional Value of the Protein Fraction of Mexican Chia Seed (*Salvia hispanica* L.)." *Food Science and Technology International* 16 (1): 89–96. https://doi.org/10.1177/1082013209353087.

Rimm, Eric B., J. Lawrence, Stephanie E. Appel, Luc Djousse Chiuve, Mary B. Engler, Penny M. Kris-Etherton, Dariush Mozaffarian, David S. Siscovick, and Alice H. Lichtenstein. 2018. "Seafood Long-Chain n-3 Polyunsaturated Fatty Acids and Cardiovascular Disease: A Science Advisory from the American Heart Association." *Circulation* 138 (1): 35–47. https://doi.org/10.1161/CIR.0000000000000574.

Rodea-González, Dulce Anahi, Julian Cruz-Olivares, Angélica Román-Guerrero, María Eva Rodríguez-Huezo, Eduardo Jaime Vernon-Carter, and César Pérez-Alonso. 2012. "Spray-Dried Encapsulation of Chia Essential Oil (*Salvia hispanica* L.) in Whey Protein Concentrate-Polysaccharide Matrices." *Journal of Food Engineering* 111 (1): 102–109. https://doi.org/10.1016/j.jfoodeng.2012.01.020.

Rodriguez, Erica S., Luciana M. Julio, Cynthia Henning, Bernd W. K. Diehl, Mabel C. Tomás, and Vanesa Y. Ixtaina. 2019. "Effect of Natural Antioxidants on the Physicochemical Properties and Stability of Freeze-Dried Microencapsulated Chia Seed Oil." *Journal of the Science of Food and Agriculture* 99 (4): 1682–1690. https://doi.org/10.1002/jsfa.9355.

Saguansri, Luz, and Mary Ann Augustin 2010. "Microencapsulation in Functional Food Product Development." In *Functional Food Product Development*. Oxford: Wiley Blackwell, Chapter 1, 3–19. https://doi.org/10.1002/jsfa.9294.

Sandoval-Oliveros, María R., and Octavio Paredes-López. 2013. "Isolation and Characterization of Proteins from Chia Seeds (*Salvia hispanica* L.)." *Journal of Agricultural and Food Chemistry* 61 (1): 193–201. https://doi.org/10.1021/jf3034978.

Segura-Campos, Maira Rubi, Norma Ciau-Solís, Gabriel Rosado-Rubio, Luis Chel-Guerrero, and David Betancur-Ancona. 2014. "Chemical and Functional Properties of Chia Seed (*Salvia hispanica* L.) Gum." *International Journal of Food Science* 3: 1–5. https://doi.org/10.1155/2014/241053.

Simopoulos, Artemis P. 2002. "Omega-3 Fatty Acids in Inflammation and Autoimmune Diseases." *Journal of the American College of Nutrition* 21 (6): 495–505. https://doi.org/10.1080/07315724.2002.10719248.

Simopoulos, Artemis P. 2016. "An Increase in the Omega-6/Omega-3 Fatty Acid Ratio Increases the Risk for Obesity." *Nutrients* 8 (3): 128. https://doi.org/10.3390/nu8030128.

Souza, Mateus F., Cristhian R. L. Francisco, Jorge L. Sanchez, Alexandre Guimarães-Inácio, Patrícia Valderrama, Evandro Bona, Augusto A. C. Tanamati, Fernanda V. Leimann, and Odinei H. Gonçalves. 2017. "Fatty Acids Profile of Chia Oil-Loaded Lipid Microparticles." *Brazilian Journal of Chemical Engineering* 34 (3): 659–669. https://doi.org/10.1590/0104-6632.20170343s20150669.

Taga, María Silvia, Evelyn E. Miller, and Dan E. Pratt. 1984. "Chia Seeds as a Source of Natural Lipid Antioxidants." *Journal of the American Oil Chemists' Society* 61 (5): 928–931. https://doi.org/10.1007/BF02542169.

Teng, Jing, Xiaoqian Hu, Mingfu Wang, and Ningping Tao. 2018. "Fabrication of Chia (*Salvia hispanica* L.) Seed Oil Nanoemulsions Using Different Emulsifiers." *Journal of Food Processing and Preservation* 42 (1): 13416. https://doi.org/10.1111/jfpp.13416.

Timilsena, Yakindra Prasad, Raju Adhikari, Stefan Kasapis, and Benu Adhikari. 2015. "Rheological and Microstructural Properties of the Chia Seed Polysaccharide." *International Journal of Biological Macromolecules* 81: 991–999. https://doi.org/10.1016/j.ijbiomac.2015.09.040.

Timilsena, Yakindra Prasad, Bo Wang, Raju Adhikari, and Benu Adhikari. 2016a. "Preparation and Characterization of Chia Seed Protein Isolate—Chia Seed Gum Complex Coacervates." *Food Hydrocolloids* 52: 554–563. https://doi.org/10.1016/j.foodhyd.2015.07.033.

Timilsena, Yakindra P., Raju Adhikari, Stefan Kasapis, and Benu Adhikari. 2016b. "Molecular and Functional Characteristics of Purified Gum from Australian

Chia Seeds." *Carbohydrate Polymers* 136: 128–136. https://doi.org/10.1016/j.foodchem.2016.06.017.

Timilsena, Yakindra Prasad, Raju Adhikari, Colin J Barrow, and Benu Adhikari. 2016c. "Microencapsulation of Chia Seed Oil Using Chia Seed Protein Isolate—Chia Seed Gum Complex Coacervates." *International Journal of Biological Macromolecules* 91: 347–357. https://doi.org/10.1016/j.ijbiomac.2016.05.058.

Timilsena, Yakindra Prasad, Raju Adhikari, Colin J. Barrow, and Benu Adhikari. 2017. "Digestion Behavior of Chia Seed Oil Encapsulated in Chia Seed Protein-Gum Complex Coacervates." *Food Hydrocolloids* 66: 71–81. https://doi.org/10.1016/j.foodhyd.2016.12.017.

Timilsena, Yakindra Prasad, Jitraporn Vongsvivut, Mark J. Tobin, Raju Adhikari, Colin Barrow, and Benu Adhikari. 2019. "Investigation of Oil Distribution in Spray-Dried Chia Seed Oil Microcapsules Using Synchrotron-FTIR Microspectroscopy." *Food Chemistry* 275: 457–466. https://doi.org/10.1016/j.foodchem.2018.09.043.

Uribe, José Antonio Rocha, Jorge Iván Novelo Pérez, Henry Castillo Kauil, Gabriel Rosado Rubio, and Carlos Guillermo Alcocer. 2011. "Extraction of Oil from Chia Seeds with Supercritical CO_2." *The Journal of Supercritical Fluids* 56 (2): 174–178. https://doi.org/10.1016/j.supflu.2010.12.007.

US Food and Drug Administration. 2009. "US Food and Drug Administration Home Page." www.fda.gov/. Accessed July 3, 2017.

Us-Medina, Ulil, Luciana M. Julio, Maira R. Segura-Campos, Vanesa Y. Ixtaina, and Mabel C. Tomás. 2018. "Development and Characterization of Spray-Dried Chia Oil Microcapsules Using by-Products from Chia as Wall Material." *Powder Technology* 334: 1–8. https://doi.org/10.1016/j.powtec.2018.04.060.

Us-Medina, Ulil, Jorge Carlos Ruiz-Ruiz, Patricia Quintana-Owen, and Maira Rubi Segura-Campos. 2017. "*Salvia hispanica* Mucilage-Alginate Properties and Performance as an Encapsulation Matrix for Chia Seed Oil." *Journal of Food Processing and Preservation* 41 (6): e13270. https://doi.org/10.1111/jfpp.13270.

Valoppi, Fabio, Ndegwa Allén Maina, Roberta Miglioli Maria, Petri O. Kilpeläinen, and Kirsi S. Mikkonen 2019. "Spruce Galactoglucomannan-Stabilized Emulsions as Essential Fatty Acid Delivery Systems for Functionalized Drinkable Yogurt and Oat-Based Beverage." *European Food Research and Technology* 245: 1387–1398. https://doi.org/10.1007/s00217-019-03273-5.

Vázquez-Ovando, José A., José G. Rosado-Rubio, Luis A. Chel-Guerrero, and David A. Betancur-Ancona. 2010. "Dry Processing of Chia (*Salvia hispanica* L.) Flour: Chemical Characterization of Fiber and Protein." *CyTA—Journal of Food* 8 (2): 117–127. https://doi.org/10.1080/19476330903223580.

Vélez-Erazo, Eliana M., Larissa Consoli, and Miriam Dupas Hubinger. 2018. "Mono and Double-Layer Emulsions of Chia Oil Produced with Ultrasound Mediation." *Food and Bioproducts Processing* 112: 108–118. https://doi.org/10.1016/j.fbp.2018.09.007.

Wang Chenxi, Cuixia Sun, Wei Lu, Khalid Gul, Analucia Mata and Yapeng Fang. 2020. "Emulsion Structure Design for Improving the Oxidative Stability of Polyunsaturated Fatty Acids." *Comprehensive Reviews in Food Science and Food Safety*: 1–17. https://doi.org/10.1111/1541-4337.1262.

Chapter 11

New Approaches about Nutraceutical Aspects of Dietary Fiber From Chia Seeds as a Functional Ingredient

Loreto A. Muñoz

CONTENTS

11.1 INTRODUCTION

In recent years, dietary fiber, both soluble and insoluble, has been positioned as a functional and nutraceutical ingredient primarily because of the physiological effects produced at the level of the gastrointestinal tract. The consumption of fiber, specifically a high intake, has been associated with a reduction in the risk of developing a wide range of noncommunicable diseases, particularly diabetes, overweight, obesity, and cardiovascular disorders (Anderson et al. 2009; European Heart Network 2011; Cho and Almeida

DOI: 10.1201/9781003087618-11

2012). The wide range of health benefits and physiological effects of dietary fiber can be attributed to many factors such as rheological versus physiological properties in the gastrointestinal tract; the function of fiber within the food matrix; the biochemical and physicochemical characteristics of each type of dietary fiber; the effect of dietary fiber on large bowel microbiota diversity and its metabolic activity producing by-products of fermentation; and the ability to expand the fecal bulk, decreasing the transit time in the colon (Brownlee 2011; Mackie, Bajka et al. 2016; Goff et al. 2018); however, the main effects produced by fiber have been associated with its viscous and gel-forming properties (Brownlee 2014).

In this sense, chia seed is an important source of omega-3 fatty acids, proteins, and carbohydrates, of which the last ranges from 35–45 g/100 g seed (USDA 2019). Of this amount of carbohydrates, approximately 30–40 g/100 g seed corresponds to dietary fiber, which includes soluble and insoluble fiber; for this reason, chia has become an attractive raw material to produce functional ingredients that can later be incorporated into foods, helping to prevent some chronic diseases (Ayerza and Coates 2011; Muñoz et al. 2013). Many of the health effects produced by the consumption of chia seed are related to the dietary fiber content, as in the case of the studies by Vuksan et al. (2010) and Vuksan et al. (2017), who associated the reduction in cholesterol and postprandial glycemia with the viscous behavior exhibited by this fiber. In addition, in recent years, colonic health has been more associated with the maintenance of overall health and the reduction/prevention of the risk of chronic diseases by producing changes in the diet and lifestyle (Wong et al. 2006). Certainly, dietary fiber and other dietary components can affect the colon environment, influencing the gut microbiota by altering bacterial fermentation and enhancing the production of different metabolites. The major metabolites produced by the microbiota are short-chain fatty acids such as acetic, butyric, and propionic acids, which are the main end products produced by fermentation of dietary fibers by anaerobic intestinal microbiota (Lahner and Annibale 2017). The colony size and distribution of the different species of bacteria in the colon determine the metabolic profile of the microbiota, which has a potential physiologic effect on health, promoting well-being and wellness (Sawicki et al. 2017). In general terms, these short-chain fatty acids may play an important role in energy homeostasis, immune function, weight regulation, improved insulin sensitivity and host microbiome signaling, among others (Nicholson et al. 2012; Myhrstad et al. 2020).

This chapter examines the relationship between intake and health benefits of dietary fiber from chia seed and the mechanisms involved during digestion and explores the potential functionality to improve the glycemic and lipid profile as well the growth and metabolic activity of the human gut microbiota.

11.2 CHIA SEED AS SOURCE OF DIETARY FIBER

Chia seed has been recognized as a natural source of dietary fiber by several authors (Ullah et al. 2016; de Falco et al. 2017; Brütsch et al. 2019; Capitani et al. 2015). Its uses have been studied in diverse food matrices, such as bread, pasta, cookies, meat products, and others, with promising functional and technological results (Menga et al. 2017; Miranda-Ramos et al. 2020; Fernandes and Salas-Mellado 2017; Rani et al. 2021). The dietary fiber content has been analyzed by several authors; for example, da Silva et al. (2017) analyzed Brazilian seeds and reported a range of 33.37–37.78 g per 100 g of seeds of total dietary fiber (TDF), of which 2.89–3.88 g/100 g corresponds to the soluble portion and 30.47–33.30 to insoluble fiber. In addition, Reyes-Caudillo et al. (2008) analyzed chia seeds from Mexico (Jalisco and Sinaloa) and reported 41.33 g/100 of TDF with 6.84 and 34.9 g/100 g of soluble and insoluble fiber for Jaliscan seeds and 41.33 g/100 of TDF with 6.16 and 32.87 g/100 g of soluble and insoluble fiber for Sinaloan seeds, respectively. In general terms, chia seed contains approximately 30 to 40 g of TDF, of which approximately 7%–15% corresponds to soluble fiber and is part of the mucilage (Kulczyński et al. 2019). For instance, chia seed exceeds the dietary fiber content compared to that of quinoa, oat bran, buckwheat, and flaxseed, among others.

Furthermore, the mucilage from this seed has been analyzed and characterized by several authors. Goh et al. (2016) determined that it was composed of 95% nonstarch polysaccharide, of which 35% corresponds to the neutral soluble fraction and 65% corresponds to the negatively charged insoluble fraction. Both fractions, when hydrated, produce fibrous transparent microgels with irregular shapes, also described by Muñoz et al. (2012), such as continuous transparent capsule formed by filamentous branch structures localized in cellular structures called mucilaginous cells that are part of the seed coat. In Figure 11.1, images of the mucilage from chia seed can be seen. Figure 11.1A shows the fibers of safranin-dyed mucilage forming a filamentous structure; the strands that come from the mucilaginous cells have different thicknesses and form a branched tree-like structure. Figure 11.1B displays a scanning electron microscopy (SEM) image in which the fibers or strands forming the mucilage create a compact three-dimensional network. In Figure 11.1C, spray-dried dehydrated mucilage is shown; this is a white powder when observed by the naked eye, but through SEM, a flake-like structure is seen. During hydration, the water enters the flakes, allowing swelling and formation of the 3D network. In addition, Timilsena et al. (2016) described the mucilage as white powder and determined a composition of 93.8% carbohydrates consisting of xylose, glucose, arabinose, galactose, and glucuronic and galacturonic acids.

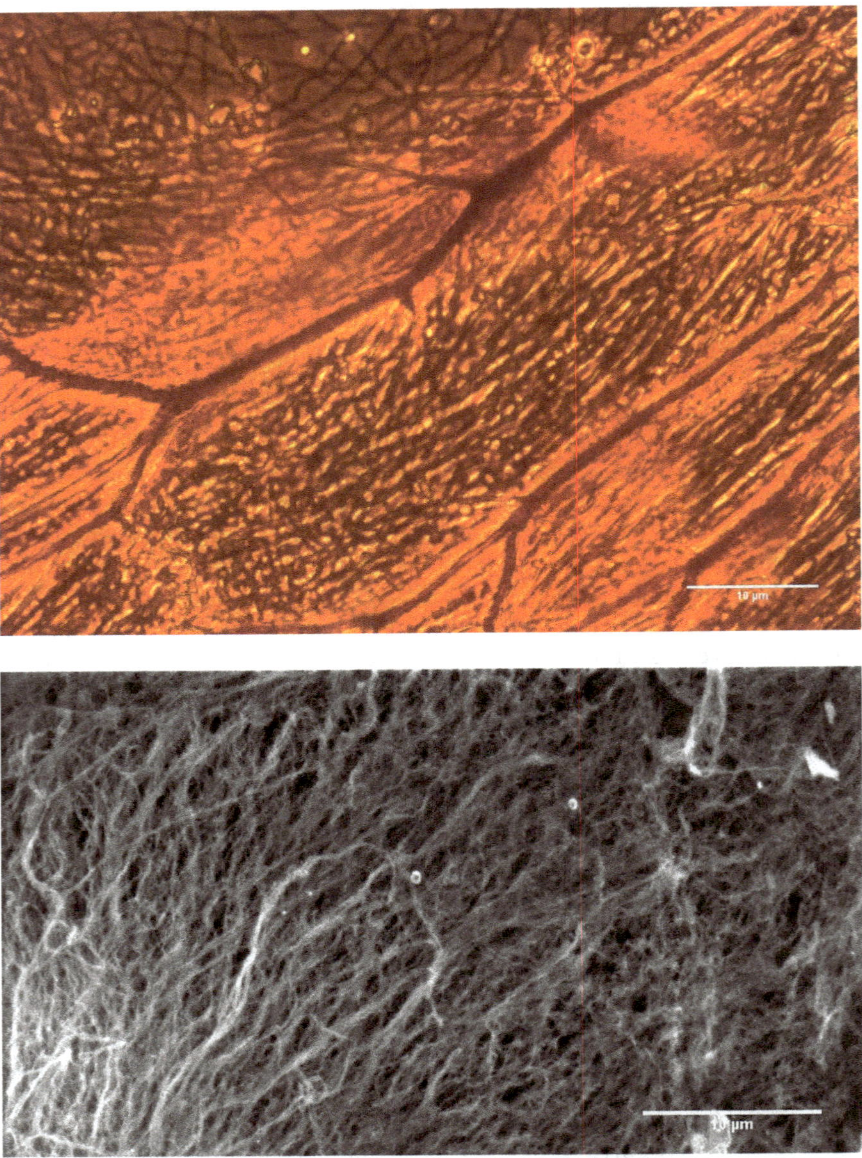

Figure 11.1 Mucilage from chia seed. (A) Mucilage dyed with safranin (optical image). (B) Mucilage 3D structure (SEM image). (C) Mucilage powder after spray-drying (SEM image).

Figure 11.1 (Continued)

11.3 FUNCTIONAL PROPERTIES

Many of the functional properties of dietary fiber are associated with its potential physiological effects. For instance, some dietary fiber, especially soluble, such as mucilage from chia seeds, has the ability to modify the physical properties of digestive contents (Vera et al. 2019; Capuano 2017). In this sense, physical properties such as solubility, swelling behavior, water absorption and adsorption capacity, viscosity, and the structure of the polysaccharide (linear or branched), among others, could influence the behavior of dietary fiber in the gastrointestinal tract (Tan et al. 2017; Mackie, Bajka et al. 2016). In this context, the soluble fiber from chia seeds shows high solubility, 85% to 98%, within a wide range of temperatures (30–80 °C) and pH values; for example, at pH 9, it has 97% solubility, while at pH 1.8, it reaches 85 to 88% in the same temperature range (Timilsena et al. 2016). The high solubility in water at temperatures close to human body temperature and physiologic pH is an indication of its behavior in the gastrointestinal tract. This property has a strong influence on functionality because it determines whether the fiber is totally, partially, or not dissolved (Guillon and Champ 2000). Furthermore, mucilage exhibits a great capacity to absorb and to retain water, at 27 g water/g mucilage (Muñoz et al. 2012). This property relates the structure of the polysaccharide with the way it interacts with water, including the swelling capacity, the volume occupied after hydration and the amount of water retained under specified conditions (Capuano 2017). This property has also been correlated with the water binding capacity of the digesta and the formation of gels in the stomach, which may slow gastric emptying and increase stomach distension or fullness (Tan et al. 2017).

The structure of this polysaccharide has not been fully elucidated; however, recently it has been described by Phan and Burton (2018) as a novel tetrasaccharide usually composed of pectic, noncellulosic, and cellulosic polysaccharides. In particular, Xing et al. (2017) isolated mucilage from chia seeds by using solid-phase state extraction and determined that the polysaccharide contains two nonreducing ends corresponding to terminal glucopyranose and terminal galactopyranose and possesses a molecular mass of 504 Da. The oligosaccharide was identified as planteose by 2D NMR (nuclear magnetic resonance) analysis, which is the major oligosaccharide found in psyllium seeds. These characteristics also make it a potential prebiotic.

In addition, this soluble fiber produces highly viscous aqueous dispersions at low concentrations, which was observed by Timilsena et al. (2015) in solutions with 0.02% (w/v) mucilage, where the mucilage exhibited pseudoplastic behavior. Similar behavior was reported by Capitani et al. (2015), who studied mucilage in a range of 0.25% to 1%, and all the dispersions exhibited non-Newtonian shear-thinning behavior at 25 °C. Despite these findings, it is precisely the viscous property that has been associated with main health effects (Brownlee 2014).

In this sense, in the study by Vera et al. (2019), where xanthan gum, pectin, guar gum, and mucilage from chia seed were analyzed comparatively during *in vitro* digestion, the steady flow behavior of chia mucilage at different concentrations during *in vitro* digestion shows a behavior similar to xanthan gum in the same conditions. That is, the apparent viscosity showed a slow decrease along with its digestion, but the mucilage from chia seeds has the ability to retain more viscosity than the other soluble fiber.

11.4 REGULATIONS AND NUTRITIONAL RECOMMENDATIONS OF DIETARY FIBER INTAKE AND CHIA SEEDS

According to American Dietetic Association, FDA/WHO and EFSA, dietary reference intake recommends a consumption of 14 g of dietary fiber per 1000 kcal or 25 g for adults (Marlett et al. 2002; American Dietetic Association 2008; EFSA

Panel on Dietetic Products 2010; FDA 2020). In this sense, chia seed contains between 30 and 40 g of dietary fiber per 100 g, which is equivalent to 100% of the daily recommendation. Chia seeds and by-products in the United States (Food and Drug Administration [FDA]), Latin America, Australia, and other countries are considered food, therefore making these products exempt from regulation (Cassiday 2017). However, the European Union (EU) recognized chia seeds as a novel food in 2009, allowing their incorporation into bread at up to 5% (EFSA 2009).

In March 2019, in light of the increasing dietary intake of chia seed in different authorized uses, the European Commission, the EFSA Panel on Nutrition, Novel Foods and Food Allergens (NDA) pronounced the safety of chia seeds as a novel food for extended uses pursuant to Regulation EU 2015/2283, concluding that the use of chia seeds whole and ground is safe under the assessed conditions and allowing its incorporation into baked products, breakfast cereals, and other items in 10% of whole seed and 5% whole or ground into bread products (EFSA Panel on Nutrition et al. 2019). In addition, in May of that same year, the EFSA pronounced the safety of chia seed powders such as partially defatted powders obtained by extrusion, concluding that these powders are safe under the assessed conditions of use (EFSA 2019).

11.5 GUT MICROBIOTA, CHIA SEED, AND ITS DIETARY FIBER

Many factors determine the human colon microbiota composition, but one of the most relevant is the diet. The reason is that microorganisms in the colon obtain energy for their growth by metabolizing complex dietary compounds from the diet, and in particular, dietary fibers can be used as substrates (Hamaker and Tuncil 2014; De Filippo et al. 2010). In the same way, diet, and in particular those diets high in fruit/legume/seed fiber, seems to be critical to influence the composition and metabolic activity of microbiota, determining the levels of short-chain fatty acids (SCFAs), butyric, acetic, and propionic acids, important components for gut health (Simpson and Campbell 2015).

In the last 20 years, evidence about the changes in the microbiota that fiber consumption can produce has increased. According to Gibson et al. (2004), for dietary fiber to have a "prebiotic effect," it must meet certain criteria, for example, resist gastric acidity, resist hydrolysis by digestive enzymes, be fermented by the intestinal microbiota and stimulate the growth and/or activity of intestinal bacteria to produce beneficial products.

In general terms, dietary fiber, especially soluble fiber, increases colonic fermentation by the intestinal microbiota, promoting an increase in the production of SCFAs (Tan et al. 2016). In this context, several studies have demonstrated that an increase in SCFAs levels plays a key role in regulating host metabolism, enhancing glucose homeostasis, lipid metabolism, and weight control management and preventing inflammation, minimizing the risk of developing metabolic diseases (Myhrstad et al. 2020).

In a recent work carried out by Tamargo et al. (2018), soluble fiber from chia seeds was assessed for the first time to analyze the relationship between apparent viscosity, growth and metabolic activity of the human gut microbiota by using a dynamic gastrointestinal model, simgi®. The mucilage was studied at three concentrations (0.3%, 0.5%, and 0.8%). The results showed that the differences in viscosity observed in the first stages of digestion disappeared when the digesta reached the descending colon and were dependent on gastrointestinal tract shear but were not affected by the microbiota. In terms of microbial activity, the fermentation of soluble fiber from chia seeds in the large bowel (ascending, transverse, and descending) influenced the microbiota composition and bacterial metabolism, showing a significant increase in SCFAs, mainly at 0.5% and 0.8% of chia mucilage concentrations. Figure 11.2 shows that the highest production of SCFAs was in the descending colon at the three concentrations of chia mucilage. In addition, when the 0.5% concentration was used, the highest production of acetic, propionic, and butyric acid was observed in the ascending, transverse, and descending colon.

Additionally, all the concentrations of mucilage led to significant differences in all bacterial groups, but the most significant changes in total aerobes and anaerobes were observed with 0.3% mucilage in the transverse and descending colon, while changes in *Lactobacillus* species were recorded at 0.8% chia mucilage in all compartments of the large bowel.

In conclusion, the concentration of mucilage affects the growth of some representative intestinal bacterial groups in a different way, but a concentration dependence was not observed; therefore, further studies are necessary to determine the specific role of mucilage from chia seeds (Tamargo et al. 2018). In agreement with Duncan et al. (2009), the changes in pH produced by the incorporation of dietary fiber from chia seed could modify the human colonic microbiota, lowering the value of this parameter. This decrease in pH could inhibit the growth of gram-negative Enterobacteriaceae, including pathogens such as *Salmonella* spp. and *Escherichia coli* (Simpson and Campbell 2015).

In addition, in a recent study by Pereira da Silva et al. (2019), soluble extracts from chia seeds were assessed to compare the effects of intra-amniotic administration on Fe and zinc status, intestinal morphology, and intestinal bacterial populations in an animal model (*Gallus gallus*). The results demonstrated that the administration of chia seed soluble extracts upregulated the expression of proteins related to zinc metabolism; moreover, the administration of chia soluble extracts increased the *Bifidobacterium, Lactobacillus*, and *Clostridium*

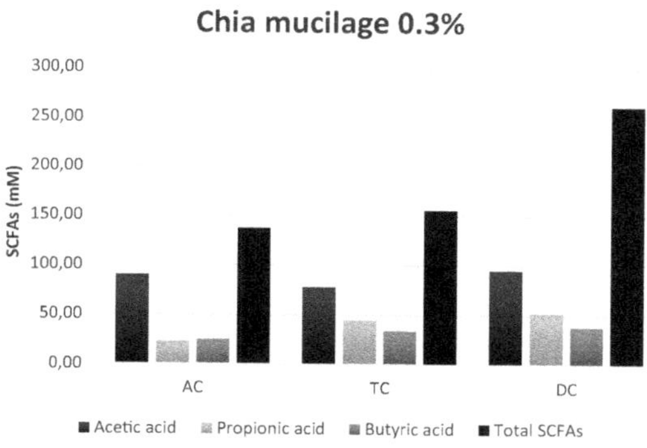

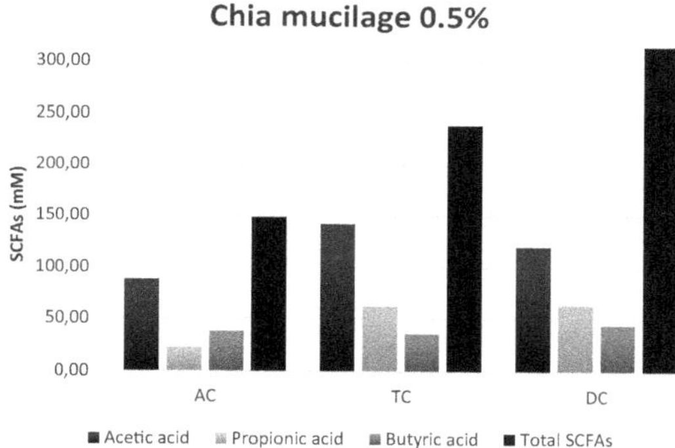

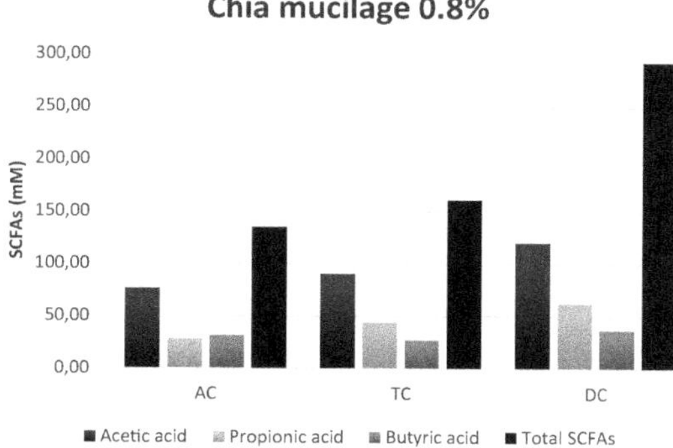

Figure 11.2 Effect of soluble fiber from chia seeds on the production of SCFAs in the large bowel.

Abbreviations: AC, ascending colon; TC, transverse colon; DC, descending colon.

Source: Adapted from Tamargo et al. (2018).

populations in the cecum content of animals that received 0.5% of the soluble extract. The authors highlight that the increase in lactobacilli and bifidobacteria abundance due to the consumption of dietary fiber may further contribute directly or indirectly to the increased bioavailability of iron and zinc in vulnerable populations.

In the same way, the effect of the pre-Hispanic Mexican diet, including chia seeds as a dietary fiber source, on metabolic and cognitive abnormalities and gut microbiota dysbiosis caused by a sucrose-enriched high-fat diet in rats was analyzed by Avila-Nava et al. (2017). The authors used a diet based mainly on corn, beans, tomato, cactus, pumpkin, and chia seeds in dehydrated form and examined the effect of this diet in a sucrose-enriched high-fat animal model. The results showed decreased glucose intolerance, body weight gain, serum and liver triglycerides, and leptin. The diet was also able to modulate the microbiota, increasing the relative abundance of bifidobacteria to a higher extent than the control diet and lactobacilli, which is indicative that the diet contains some foods with probiotic effects. This fact was associated mainly with the presence of cactus, chia, and pumpkin seeds. Along the same line, the effect of a functional-based dietary intervention on fecal microbiota and biochemical parameters in patients with type 2 diabetes was analyzed by Medina-Vera et al. (2019). The diet used for the experiments consisted of high-fiber, polyphenol-rich, and vegetable protein functional foods comprising 14 g of dehydrated cactus, 4 g of chia seeds, 30 g of soy protein, and 4 g of inulin. This intervention was accompanied by an eating plan according to the patients' usual diet as recommended by the US National Institutes of Health (NHLBI 1998).

On the other hand, other ancestral seeds such as quinoa and amaranth have been shown to have prebiotic effects, increasing the population of certain bacterial groups and the amount of SCFAs (Gullon et al. 2016).

11.6 HYPOGLYCEMIC AND HYPOCHOLESTEROLEMIC EFFECTS OF CHIA SEED SOLUBLE FIBER

One of the physiological responses associated with the consumption of dietary fiber is the ability to modulate the postprandial glycemic response (Cameron-Smith et al. 2007; Brennan 2005; Goff et al. 2018). The behavior of glucose in transit through the intestine can be affected by the flow of the digesta, which influences how glucose can reach the epithelium via diffusion in laminar or turbulent flow (Takahashi et al. 2012). For instance, when dietary fiber is a part of the digestive content, the flow behavior can be modified, causing a decrease in the glucose absorption rate. In this sense, in an investigation of Tamargo et al. (2018) mucilage from chia seeds was used in a food model to analyze how the bioaccessibility of glucose was modified during digestion in the dynamic gastrointestinal model simgi®. The experiments were performed with 0.75% and 0.95% soluble fiber from chia seeds supplemented with 1 g of glucose, and the percentage of glucose bioaccessibility was determined in terms of glucose diffusion across a dialysis membrane. The results showed 53.4% reduction in glucose calculated in terms of the glucose dialysis retardation index (GDRI) when 0.95% of mucilage was added to the food model, which was significantly higher than that of psyllium with 38.81% GDRI, fiber-rich orange pomace with 19.25% GDRI and cellulose with 8.69% (Huang et al. 2019). In addition, the total reduction in glucose bioaccessibility was over 66%, which was significantly high (Figure 11.3). These findings suggest a directly proportional relationship between the concentration of soluble fiber from chia seed and the reduction in glucose bioaccessibility; however, more studies are necessary to corroborate these claims. One of the mechanisms proposed to explain this phenomenon has been described by Zheng et al. (2019) and Capuano (2017), such as the increase in viscosity in the digesta producing changes in the flow behavior from turbulence to laminar modifying the glucose transport and delaying the absorption by the small intestine.

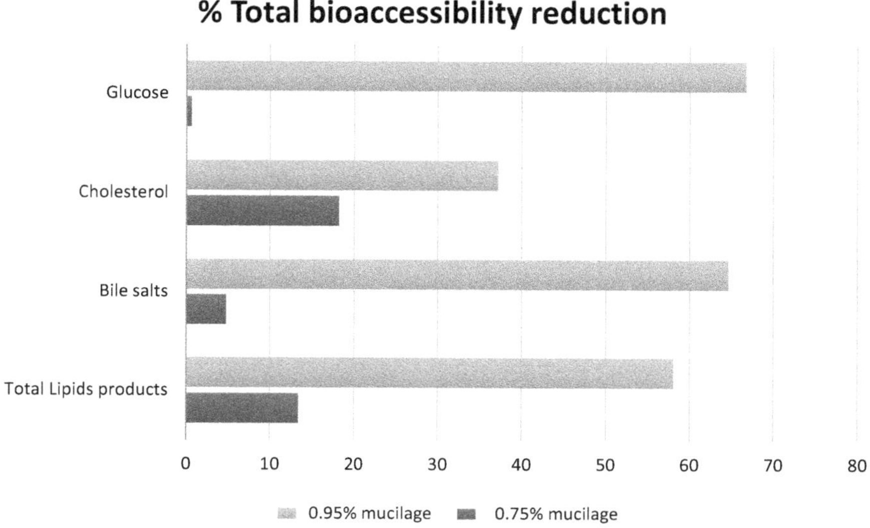

Figure 11.3 Reduction in bioaccessibility of glucose, cholesterol, and lipid products. *Source*: Adapted from Tamargo et al. (2020).

However, this behavior has also been associated with the ability of the fibers to entrap endogenous compounds, leading to a lower diffusion of nutrients, as previously described by Mackie, Rigby et al. (2016). In the case of soluble fiber from chia seeds, because of the ability to form viscous solutions at low concentrations, this gel-forming 3D network structure creates a physical barrier between carbohydrates and digestive enzymes, decreasing the conversion of carbohydrates into sugars (Goh et al. 2016).

Otherwise, some soluble fibers have the ability to lower serum cholesterol concentrations, but not all soluble fibers can produce this effect; only highly viscous fibers have exhibited viscosity-dependent health benefits (McRorie and McKeown 2017). Dietary fiber can interact with bile acids, preventing their reabsorption and promoting their excretion into the colon, and the primary mechanism described is that highly viscous fiber lowers serum cholesterol by trapping and eliminating bile via the stool (Naumann et al. 2019; McRorie and McKeown 2017). In this case, in a recent study by Tamargo et al. (2020), the use of mucilage from chia seeds showed promising results in terms of the bioaccessibility of lipid products, cholesterol, and bile salts. In the investigation, the same concentrations of mucilage described above were used (0.75% and 0.95%), and 100 mg of cholesterol and 1 g of refined olive oil were added to build the food model. The samples were subjected to *in vitro* dynamic digestion, and the bioaccessibility was analyzed in comparison with a control system without soluble fiber from chia seeds. The reduction in the bioaccessibility of total lipid products (free fatty acids and monoglycerides) was close to 60%, and cholesterol and bile salt bioaccessibility was 37% and 65%, respectively (Figure 11.3). These results also suggest that mucilage can limit the solubility of bile salts, which is essential for the aqueous dispersion of dietary lipids and cholesterol; this phenomenon explains the reduction in the bioaccessibility of lipid products and cholesterol (Tamargo et al. 2020).

The mechanism by which dietary fiber lowers cholesterol was proposed by Mackie, Bajka, et al. (2016), such as the disruption of the normal recycling of bile, possibly by sequestering the rate of absorption as a result of entanglement with intestinal mucus. In addition, Zacherl et al. (2011) suggest binding and excretion of bile acids in the small intestine by water-soluble and water-insoluble fiber as the main mechanism for cholesterol-lowering effects.

The effect of chia seeds on carbohydrate and lipid metabolism has also been studied with whole seeds. In the research of Da Silva et al. (2016), the influence on glucose and lipid homeostasis and the integrity of the liver and intestinal morphology of Wistar rats by supplying whole seeds and heat-treated flour (90 °C, 20 min) was evaluated. The authors confirmed the hypoglycemic effect and found that the animals receiving whole chia seeds or flour with or without treatment had improved lipid and glycemic profiles. Furthermore, Miranda-Ramos et al. (2020) used whole chia seed and its flour as ingredients to enrich bread nutritionally. The authors found that bread supplemented with chia seeds and by-products had a lower glycemic index than the control. Additionally, in a randomized controlled trial by Ho et al. (2013), whole and ground chia seeds at doses of 7, 15, and 24 g were incorporated into white bread and later consumed by heathy volunteers. It was found that the consumption of bread with both chia and by-products using 24 g produced lower postprandial glycemia than that in volunteers who consumed bread without chia products. In this case, the authors suggest that the hypoglycemic effect is produced by the high content of dietary fiber.

These findings suggest that the consumption of dietary fiber from chia seeds not only benefits glucose and lipid metabolism but also has potential use in functional food development to help prevent chronic and metabolic diseases.

11.7 CONCLUSIONS

In consideration of the findings described above, dietary fiber from chia seeds can be considered a nutraceutical and functional ingredient since it shows a wide range of functionality, among which the effect on human microbiota and its metabolic activity, its action in reducing the bioaccessibility of glucose in the digesta, and the substantial reduction in bile salts, cholesterol, and total lipids available for absorption by the small intestine stand out.

The use of this type of fiber provides a range of benefits; therefore, dietary fiber from chia seed could be used in the development of functional food as a strategy to reduce the risk of and prevent some noncommunicable diseases. However, more studies are necessary to corroborate these health effects and to expand the knowledge of the future uses of this fiber in the prevention of diseases.

ACKNOWLEDGMENTS

This work was carried out with the financial support of FONDECYT Project 1201489 from the National Agency for Research and Development (ANID), Chile and CYTED Program, Project 119RT0567, Spain.

REFERENCES

American Dietetic Association. 2008. "Position of the American Dietetic Association: Health Implications of Dietary Fiber." *Journal of the American Dietetic Association* 108 (10): 1716–1731. https://doi.org/10.1016/j.jada.2008.08.007.

Anderson, James W., Pat Baird, Richard H. Davis, Jr., Stefanie Ferreri, Mary Knudtson, Ashraf Koraym, Valerie Waters, and Christine L. Williams 2009. "Health Benefits of Dietary Fiber." *Nutrition Reviews* 67 (4): 188–205. https://doi.org/10.1111/j.1753-4887.2009.00189.x.

Avila-Nava, Azalia, Lilia G. Noriega, Armando R. Tovar, Omar Granados, Claudia Perez-Cruz, José Pedraza-Chaverri, and Nimbe Torres. 2017. "Food Combination Based on a Pre-Hispanic Mexican Diet Decreases Metabolic and Cognitive Abnormalities and Gut Microbiota Dysbiosis Caused by a Sucrose-Enriched High-Fat Diet in Rats." *Molecular Nutrition & Food Research* 61 (1): 1501023. https://doi.org/10.1002/mnfr.201501023.

Ayerza, Ricardo, and Wayne Coates. 2011. "Protein Content, Oil Content and Fatty Acid Profiles as Potential Criteria to Determine the Origin of Commercially Grown Chia (*Salvia hispanica* L.)." *Industrial Crops and Products* 34 (2): 1366–1371. http://doi.org/10.1016/j.indcrop.2010.12.007.

Brennan, Charles S. 2005. "Dietary Fibre, Glycaemic Response, and Diabetes." *Molecular Nutrition & Food Research* 49 (6): 560–570. https://doi.org/10.1002/mnfr.200500025.

Brownlee, Iain. 2011. "The Physiological Roles of Dietary Fibre." *Food Hydrocolloids* 25 (2): 238–250. https://doi.org/10.1016/j.foodhyd.2009.11.013.

Brownlee, Iain. 2014. "The Impact of Dietary Fibre Intake on the Physiology and Health of the Stomach and Upper Gastrointestinal Tract." *Bioactive Carbohydrates and Dietary Fibre* 4 (2): 155–169. http://doi.org/10.1016/j.bcdf.2014.09.005.

Brütsch, Linda, Fiona J. Stringer, Simon Kuster, Erich J. Windhab, and Peter Fischer. 2019. "Chia Seed Mucilage—a Vegan Thickener: Isolation, Tailoring Viscoelasticity and Rehydration." *Food & Function* 10 (8): 4854–4860. https://doi.org/10.1039/C8FO00173A.

Cameron-Smith, D., G. R. Collier, and K. O'Dea. 2007. "Effect of Soluble Dietary Fibre on the Viscosity of Gastrointestinal Contents and the Acute Glycaemic Response in the Rat." *British Journal of Nutrition* 71 (4): 563–571. https://doi.org/10.1079/BJN19940163.

Capitani, M. I., L. J. Corzo-Rios, L. A. Chel-Guerrero, D. A. Betancur-Ancona, S. M. Nolasco, and M. C. Tomás. 2015. "Rheological Properties of Aqueous Dispersions of Chia (*Salvia hispanica* L.) Mucilage." *Journal of Food Engineering* 149: 70–77. http://doi.org/10.1016/j.jfoodeng.2014.09.043.

Capuano, Edoardo. 2017. "The Behavior of Dietary Fiber in the Gastrointestinal Tract Determines Its Physiological Effect." *Critical Reviews in Food Science and Nutrition* 57 (16): 3543–3564. https://doi.org/10.1080/10408398.2016.1180501.

Cassiday, Laura. 2017. "Chia: Superfood or Superfad?" *Inform Magazine from AOCS.*

Cho, Susan S., and Nelson Almeida. 2012. *Dietary Fiber and Health.* Boca Raton, FL: CRC Press, 1st edition. https://doi.org/10.1201/b12156.

Da Silva, Bárbara Pereira, Pamella Cristine Anunciação, Jessika Camila Da Silva Matyelka, Lucia Della, Mattos Ceres, Hércia Stampini Duarte Martino, and Helena Maria Pinheiro-Santána. 2017. "Chemical Composition of Brazilian Chia Seeds Grown in Different Places." *Food Chemistry* 221: 1709–1716. http://doi.org/10.1016/j.foodchem.2016.10.115.

Da Silva, Bárbara Pereira, Desirrê Morais Dias, Moreira De Castro, Toledo Maria Eliza, Celi Lopes Renata, Sérgio Luis Pinto Da Matta, Ceres Mattos Della Lucia, Hércia Stampini Duarte Martino, and Helena Maria Pinheiro-Santána. 2016. "Chia Seed Shows Good Protein Quality, Hypoglycemic Effect and Improves the Lipid Profile and Liver and Intestinal Morphology of Wistar Rats." *Plant Foods for Human Nutrition* 71 (3): 225–230. https://doi.org/10.1007/s11130-016-0543-8.

De Falco, Bruna, Mariana Amato, and Virginia Lanzotti. 2017. "Chia Seeds Products: An Overview." *Phytochemistry Reviews* 16 (4): 745–760. https://doi.org/10.1007/s11101-017-9511-7.

De Filippo, Carlotta, Duccio Cavalieri, Monica Di Paola, Matteo Ramazzotti, Jean Baptiste Poullet, Sebastien Massart, Silvia Collini, Giuseppe Pieraccini, and Paolo Lionetti. 2010. "Impact of Diet in Shaping Gut Microbiota Revealed by a Comparative Study in Children from Europe and Rural Africa." *Proceedings of the National Academy of Sciences* 107 (33): 14691. https://doi.org/10.1073/pnas.1005963107.

Duncan, S. H., P. Louis, J. M. Thomson, and H. J. Flint. 2009. "The Role of pH in Determining the Species Composition of the Human Colonic Microbiota." *Environmental Microbiology*11 (8): 2112–2222. https://doi.org/10.1111/j.1462-2920.2009.01931.x.

EFSA. 2009. "Opinion on the Safety of 'Chia Seeds (Salvia hispanica L.) and Ground Whole Chia Seeds' as a Food Ingredient." *EFSA Journal* 7 (4): 996. https://doi.org/10.2903/j.efsa.2009.996.

EFSA. 2019. "Safety of Chia Seeds (Salvia hispanica L.) Powders, as Novel Foods, Pursuant to Regulation (EU) 2015/2283." *EFSA Journal* 17 (6): e05716. https://doi.org/10.2903/j.efsa.2019.5716.

EFSA Panel on Dietetic Products, Nutrition, and Allergies (NDA). 2010. "Scientific Opinion on Dietary Reference Values for Carbohydrates and Dietary Fibre." *EFSA Journal* 8 (3): 1462. https://doi.org/doi:10.2903/j.efsa.2010.1462.

EFSA Panel on Nutrition, Novel Foods, Allergens, Food, Dominique Turck, Jacqueline Castenmiller, Stefaan De Henauw, Karen Ildico Hirsch-Ernst, John Kearney, Alexandre Maciuk, Inge Mangelsdorf, Harry J. Mcardle, Androniki Naska, Carmen Pelaez, Kristina Pentieva, Alfonso Siani, Frank Thies, Sophia Tsabouri, Marco Vinceti, Francesco Cubadda, Karl-Heinz Engel, Thomas Frenzel, Marina Heinonen, Rosangela Marchelli, Monika Neuhäuser-Berthold, Annette Pöting, Morten Poulsen, Yolanda Sanz, Josef Rudolf Schlatter, Henk Van Loveren, Wolfgang Gelbmann, Leonard Matijević, Patricia Romero, and Helle Katrine Knutsen. 2019. "Safety of Chia Seeds (Salvia hispanica L.) as a Novel Food for Extended Uses Pursuant to Regulation (EU) 2015/2283." *EFSA Journal* 17 (4): e05657. https://doi.org/10.2903/j.efsa.2019.5657.

European Heart Network. 2011. *Diet, Physical Activity and Cardiovascular Disease Prevention in Europe*. Brussels: EHN.

Fernandes, Sibele Santos, and Myriam De Las Mercedes Salas-Mellado. 2017. "Addition of Chia Seed Mucilage for Reduction of Fat Content in Bread and Cakes." *Food Chemistry* 227 (Suppl. C): 237–244. https://doi.org/10.1016/j.foodchem.2017.01.075.

Food and Drug Administration FDA. 2020. "Nutrition Facts Label: Dietary Fiber." https://www.fda.gov/food/food-labeling-nutrition.

Gibson, G. R., H. M. Probert, J. V. Loo, R. A. Rastall, and M. B. Roberfroid. 2004. "Dietary Modulation of the Human Colonic Microbiota: Updating the Concept of Prebiotics." *Nutrition Research Review* 17 (2): 259–275. https://doi.org/10.1079/nrr200479.

Goff, H. Douglas, Nikolay Repin, Hrvoje Fabek, Dalia El Khoury, and Michael J. Gidley. 2018. "Dietary Fibre for Glycaemia Control: Towards a Mechanistic Understanding." *Bioactive Carbohydrates and Dietary Fibre* 14: 39–53. https://doi.org/10.1016/j.bcdf.2017.07.005.

Goh, Kelvin Kim Tha, Lara Matia-Merino, Jie Hong Chiang, Ruisong Quek, Stephanie Jun Bing Soh, and Roger G. Lentle. 2016. "The Physico-Chemical Properties of Chia Seed Polysaccharide and Its Microgel Dispersion Rheology." *Carbohydrate Polymers* 149: 297–307. http://doi.org/10.1016/j.carbpol.2016.04.126.

Guillon, F., and M. Champ. 2000. "Structural and Physical Properties of Dietary Fibres, and Consequences of Processing on Human Physiology." *Food Research International* 33 (3): 233–245. https://doi.org/10.1016/S0963-9969(00)00038-7.

Gullon, Beatriz, Patricia Gullon, Freni K. Tavaria, and Remedios Yanez. 2016. "Assessment of the Prebiotic Effect of Quinoa and Amaranth in the Human Intestinal Ecosystem." *Food & Function* 7 (9): 3782–3788. https://doi.org/10.1039/C6FO00924G.

Hamaker, Bruce R., and Yunus E. Tuncil. 2014. "A Perspective on the Complexity of Dietary Fiber Structures and Their Potential Effect on the Gut Microbiota." *Journal of Molecular Biology* 426 (23): 3838–3850. https://doi.org/10.1016/j.jmb.2014.07.028.

Ho, H., A. S. Lee, E. Jovanovski, A. L. Jenkins, R. Desouza, and V. Vuksan. 2013. "Effect of Whole and Ground Salba Seeds (Salvia Hispanica L.) on Postprandial Glycemia in Healthy Volunteers: A Randomized Controlled, Dose-Response Trial." *European Journal of Clinical Nutrition* 67 (7): 786–788. https://doi.org/10.1038/ejcn.2013.103.

Huang, Ya-Ling, Ya-Sheng Ma, Yung-Hsiang Tsai, and Sam K. C. Chang. 2019. "In Vitro Hypoglycemic, Cholesterol-Lowering and Fermentation Capacities of Fiber-Rich Orange Pomace as Affected by Extrusion." *International Journal of Biological Macromolecules* 124: 796–801. https://doi.org/10.1016/j.ijbiomac.2018.11.249.

Kulczyński, Bartosz, Joanna Kobus-Cisowska, Maciej Taczanowski, Dominik Kmiecik, and Anna Gramza-Michałowska. 2019. "The Chemical Composition and Nutritional Value of Chia Seeds-Current State of Knowledge." *Nutrients* 11 (6): 1242. https://doi.org/10.3390/nu11061242.

Lahner, E., and B. Annibale. 2017. "Dietary Fiber and Gut Microbiota Modulation—A Possible Switch from Disease to Health for Subjects with Diverticular Disease?" *JSM Nutritional Disorders* 1 (1): 1001.

Mackie, Alan, Balazs Bajka, and Neil Rigby. 2016. "Roles for Dietary Fibre in the Upper GI Tract: The Importance of Viscosity." *Food Research International* 88: 234–238. http://dx.doi.org/10.1016/j.foodres.2015.11.011.

Mackie, Alan, Neil Rigby, Pascale Harvey, and Balazs Bajka. 2016. "Increasing Dietary Oat Fibre Decreases the Permeability of Intestinal Mucus." *Journal of Functional Foods* 26: 418–427. http://doi.org/10.1016/j.jff.2016.08.018.

Marlett, Judith A., Michael I. McBurney, and Joanne L. Slavin. 2002. "Position of the American Dietetic Association: Health Implications of Dietary Fiber." *Journal of the American Dietetic Association* 102 (7): 993–1000. https://doi.org/10.1016/S0002-8223(02)90228-2.

McRorie, Johnson W., and Nicola M. Mckeown. 2017. "Understanding the Physics of Functional Fibers in the Gastrointestinal Tract: An Evidence-Based Approach to Resolving Enduring Misconceptions About Insoluble and Soluble Fiber." *Journal of the Academy of Nutrition and Dietetics* 117 (2): 251–264. https://doi.org/10.1016/j.jand.2016.09.021.

Medina-Vera, I., M. Sanchez-Tapia, L. Noriega-López, O. Granados-Portillo, M. Guevara-Cruz, A. Flores-López, A. Avila-Nava, M. L. Fernández, A. R. Tovar, and N. Torres. 2019. "A Dietary Intervention with Functional Foods Reduces Metabolic Endotoxaemia and Attenuates Biochemical Abnormalities by Modifying Faecal Microbiota in People with Type 2 Diabetes." *Diabetes and Metabolism* 45 (2): 122–131. https://doi.org/10.1016/j.diabet.2018.09.004.

Menga, Valeria, Mariana Amato, Tim D. Phillips, Donato Angelino, Federico Morreale, and Clara Fares. 2017. "Gluten-Free Pasta Incorporating Chia (Salvia hispanica L.) as Thickening Agent: An Approach to Naturally Improve the Nutritional Profile and the in Vitro Carbohydrate Digestibility." *Food Chemistry* 221: 1954–1961. http://doi.org/10.1016/j.foodchem.2016.11.151.

Miranda-Ramos, Karla, Ma Carmen Millán-Linares, and Claudia Monika Haros. 2020. "Effect of Chia as Breadmaking Ingredient on Nutritional Quality, Mineral Availability, and Glycemic Index of Bread." *Foods (Basel, Switzerland)* 9 (5): 663. https://doi.org/10.3390/foods9050663.

Muñoz, L. A., A. Cobos, O. Diaz, and J. M. Aguilera. 2012. "Chia Seeds: Microstructure, Mucilage Extraction and Hydration." *Journal of Food Engineering* 108 (1): 216–224. http://doi.org/10.1016/j.jfoodeng.2011.06.037.

Muñoz, L. A., Angel Cobos, Olga Diaz, and José Miguel Aguilera. 2013. "Chia Seed (*Salvia hispanica*): An Ancient Grain and a New Functional Food." *Food Reviews International* 29 (4): 394–408. https://doi.org/10.1080/87559129.2013.818014.

Myhrstad, Mari C. W., Hege Tunsjø, Colin Charnock, and Vibeke H. Telle-Hansen. 2020. "Dietary Fiber, Gut Microbiota, and Metabolic Regulation—Current Status in Human Randomized Trials." *Nutrients* 12 (3): 859. https://doi.org/10.3390/nu12030859.

Naumann, Susanne, Ute Schweiggert-Weisz, Julia Eglmeier, Dirk Haller, and Peter Eisner. 2019. "In Vitro Interactions of Dietary Fibre Enriched Food Ingredients with Primary and Secondary Bile Acids." *Nutrients* 11 (6): 1424. https://doi.org/10.3390/nu11061424.

NHLBI. 1998. *Obesity Education Initiative Expert Panel on the Identification, Evaluation, and Treatment of Obesity in Adults (US). Clinical Guidelines on the Identification, Evaluation, and Treatment of Overweight and Obesity in Adults.* The Evidence Report. Bethesda, MD: National Heart, Lung, and Blood Institute.

Nicholson, Jeremy K., Elaine Holmes, James Kinross, Remy Burcelin, Glenn Gibson, Wei Jia, and Sven Pettersson. 2012. "Host-Gut Microbiota Metabolic Interactions." *Science* 336 (6086): 1262–1267. https://doi.org/10.1126/science.1223813.

Pereira Da Silva, Bárbara, Nikolai Kolba, Hércia Stampini Duarte Martino, Jonathan Hart, and Elad Tako. 2019. "Soluble Extracts from Chia Seed (Salvia hispanica L.) Affect Brush Border Membrane Functionality, Morphology and Intestinal Bacterial Populations In Vivo (*Gallus gallus*)." *Nutrients* 11 (10): 2457. https://doi.org/10.3390/nu11102457.

Phan, J. L., and R. A. Burton. 2018. "New Insights into the Composition and Structure of Seed Mucilage." *Annual Plant Reviews Online* 1 (1): 1–41. https://doi.org/10.1002/9781119312994.apr0606.

Rani, Reetu, Surender Kumar, and Sanjay Yadav. 2021. "Pumpkin and Chia Seed as Dietary Fibre Source in Meat Products: A Review." *The Pharma Innovation Journal* 10 (1): 477–485.

Reyes-Caudillo, E., A. Tecante, and M. A. Valdivia-López. 2008. "Dietary Fibre Content and Antioxidant Activity of Phenolic Compounds Present in Mexican Chia (*Salvia hispanica* L.) Seeds." *Food Chemistry* 107 (2): 656–663. https://doi.org/10.1016/j.foodchem.2007.08.062.

Sawicki, Caleigh M., Kara A. Livingston, Martin, Obin, Susan B. Roberts, Mei Chung, and Nicola M. McKeown. 2017. "Dietary Fiber and the Human Gut Microbiota: Application of Evidence Mapping Methodology." *Nutrients* 9 (2): 125. https://doi.org/10.3390/nu9020125.

Simpson, H. L., and B. J. Campbell. 2015. "Review Article: Dietary Fibre—Microbiota Interactions." *Alimentary Pharmacology & Therapeutics* 42 (2): 158–179. https://doi.org/10.1111/apt.13248.

Takahashi, Toru, Noborikawa, Mari, Oda, Sen-Ichi, Maruyama, Satomi, Koda, Tomoko, Tokunaga, Miki, Naoi, Mitsuko, and Kitamori, Kazunari 2012. "Digesta Viscosity and Glucose Behavior in the Small Intestine Lumen." In Susan S. Cho and Nelson Almeida (Eds.), *Dietary Fibre and Health*. Boca Raton, FL: CRC Press, 185–193.

Tamargo, Alba, Carolina Cueva, Laura Laguna, M. Victoria Moreno-Arribas, and Loreto A. Muñoz. 2018. "Understanding the Impact of Chia Seed Mucilage on Human Gut Microbiota by Using the Dynamic Gastrointestinal Model Simgi®." *Journal of Functional Foods* 50: 104–111. https://doi.org/10.1016/j.jff.2018.09.028.

Tamargo, Alba, Diana Martin, Joaquín Navarro Del Hierro, M. Victoria Moreno-Arribas, and Loreto A. Muñoz. 2020. "Intake of Soluble Fibre from Chia Seed Reduces Bioaccessibility of Lipids, Cholesterol and Glucose in the Dynamic Gastrointestinal Model Simgi®." *Food Research International* 137: 109364. https://doi.org/10.1016/j.foodres.2020.109364.

Tan, Chengquan, Hongkui Wei, Xichen Zhao, Chuanhui Xu, and Jian Peng. 2017. "Effects of Dietary Fibers with High Water-Binding Capacity and Swelling Capacity on Gastrointestinal Functions, Food Intake and Body Weight in Male Rats." *Food and Nutrition Research* 61 (1): 1308118. https://doi.org/10.1080/16546628.2017.1308118.

Tan, Chengquan, Hongkui Wei, Xichen Zhao, Chuanhui Xu, Yuanfei Zhou, and Jian Peng. 2016. "Soluble Fiber with High Water-Binding Capacity, Swelling Capacity, and Fermentability Reduces Food Intake by Promoting Satiety Rather Than Satiation in Rats." *Nutrients* 8 (10): 615. https://doi.org/10.3390/nu8100615.

Timilsena, Yakindra Prasad, Raju Adhikari, Stefan Kasapis, and Benu Adhikari. 2015. "Rheological and Microstructural Properties of the Chia Seed Polysaccharide." *International Journal of Biological Macromolecules* 81: 991–999. http://doi.org/10.1016/j.ijbiomac.2015.09.040.

Timilsena, Yakindra Prasad, Raju Adhikari, Stefan Kasapis, and Benu Adhikari. 2016. "Molecular and Functional Characteristics of Purified Gum from Australian Chia Seeds." *Carbohydrate Polymers* 136: 128–136. http://doi.org/10.1016/j.carbpol.2015.09.035.

Ullah, Rahman, M. Nadeem, A. Khalique, M. Imran, S. Mehmood, A. Javid, and J. Hussain. 2016. "Nutritional and Therapeutic Perspectives of Chia (Salvia hispanica L.): A Review." *Journal of Food Science and Technology* 53 (4): 1750–1758. https://doi.org/10.1007/s13197-015-1967-0.

USDA. 2019. "Agricultural Research Service: Food Data Central." www.fdc.nal.usda.gov.

Vera, Natalia, Laura Laguna, Liliana Zura, Luis Puente, and Loreto A. Muñoz. 2019. "Evaluation of the Physical Changes of Different Soluble Fibres Produced During an in Vitro Digestion." *Journal of Functional Foods* 62: 103518. https://doi.org/10.1016/j.jff.2019.103518.

Vuksan, V., L. Choleva, E. Jovanovski, A. L. Jenkins, F. Au-Yeung, A. G. Dias, H. V. T. Ho, A. Zurbau, and L. Duvnjak. 2017. "Comparison of Flax (*Linum usitatissimum*) and Salba-Chia (*Salvia hispanica* L.) Seeds on Postprandial Glycemia and Satiety in Healthy Individuals: A Randomized, Controlled, Crossover Study." *European Journal of Clinical Nutrition* 71 (2): 234–238. https://doi.org/10.1038/ejcn.2016.148.

Vuksan, V., A. L. Jenkins, A. G. Dias, A. S. Lee, E. Jovanovski, A. L. Rogovik, and A. Hanna. 2010. "Reduction in Postprandial Glucose Excursion and Prolongation of Satiety: Possible Explanation of the Long-Term Effects of Whole Grain Salba (*Salvia hispanica* L.)." *European Journal of Clinical Nutrition* 64 (4): 436–438. https://doi.org/10.1038/ejcn.2009.159.

Wong, Julia M. W., Russell De Souza, Cyril W. C. Kendall, Azadeh Emam, and David J. A. Jenkins. 2006. "Colonic Health: Fermentation and Short Chain Fatty Acids." *Journal of Clinical Gastroenterology* 40 (3): 235–243. https://doi.org/10.1097/00004836-200603000-00015.

Xing, Xiaohui, Yves S. Y. Hsieh, Kuok Yap, Main E. Ang, Jelle Lahnstein, Matthew R. Tucker, Rachel A. Burton, and Vincent Bulone. 2017. "Isolation and Structural Elucidation by 2D NMR of Planteose, a Major Oligosaccharide in the Mucilage of Chia (*Salvia hispanica* L.) Seeds." *Carbohydrate Polymers* 175: 231–240. https://doi.org/10.1016/j.carbpol.2017.07.059.

Zacherl, Christian, Peter Eisner, and Karl-Heinz Engel. 2011. "In Vitro Model to Correlate Viscosity and Bile Acid-Binding Capacity of Digested Water-Soluble and Insoluble Dietary Fibres." *Food Chemistry* 126 (2): 423–428. https://doi.org/10.1016/j.foodchem.2010.10.113.

Zheng, Yafeng, Qi Wang, Juqing Huang, Dongya Fang, Weijing Zhuang, Xianliang Luo, Xiaobo Zou, Baodong Zheng, and Hui Cao. 2019. "Hypoglycemic Effect of Dietary Fibers from Bamboo Shoot Shell: An in Vitro and in Vivo Study." *Food and Chemical Toxicology* 127: 120–126. https://doi.org/10.1016/j.fct.2019.03.008.

Chapter 12

Chia Proteins as a Source of Bioactive Peptides to Enhance Human Health Benefits

Juliana Cotabarren, Adriana Mabel Rosso, Walter David Obregón, and Mónica Graciela Parisi

CONTENTS

12.1 INTRODUCTION

Salvia hispanica L., commonly called *chia*, is a well-known annual herbaceous plant that belongs to the Lamiaceae family and it is native of Mesoamerica, exhibiting the greatest genetic diversity in the slope of the Pacific Ocean from

DOI: 10.1201/9781003087618-12

central Mexico to northern Guatemala (Kulczyński et al. 2019; Cahill 2004). Chia seed is known to have been used as staple food for the pre-Columbian populations of Mesoamerica around the year 3500 BC, being cultivated between 2600 BC and 900 BC in the Valley of Mexico by the Teotihuacan and Toltec civilizations. It was also one of the main components of the Aztec diet along with quinoa, amaranth, corn, and some varieties of beans (Peláez et al. 2019; Ixtaina et al. 2011). The Mayan and Aztec civilizations used it as a medicine and a food supplement to obtain the necessary energy, endurance, and strength in extreme conditions, as well as in offerings to the gods during religious ceremonies (Muñoz et al. 2013).

Chia seeds are small, oval-shaped, and flat, measuring between 2.0 to 2.5 mm in length, 1.2 to 1.5 mm width and 0.8 to 1.0 mm thickness, available in different shades of colors from dark brown to black, and sometimes gray or white. Chia is an oilseed crop that grows in arid or semiarid climates (tropical and subtropical conditions) and is not tolerant to frost. Regarding the edaphic conditions in which it develops, chia plant can grow in a wide range of well-drained clay and sandy soils, with reasonable salt and acid tolerance (Bochicchio et al. 2015; USDA 2011), producing 500–600 kg seed/acre, but under appropriate agronomic conditions a yield of 2500 kg/acre has also been reported (Ullah et al. 2015). However, a low nitrogen content can be a limiting factor to obtain good yields (Ayerza and Coates 2011). Peiretti and Gai (2009) have studied the crude protein content in vegetative parts of the plant and demonstrated the potential of chia seeds in feeding. Peiretti (2010) has also reported the use of whole chia plants as functional forage. Recently, Jamshidi et al. (2019) have reported that the crude protein in chia biomass increases with manure fertilization compared to other organic fertilizers, also with a narrow spacing between rows and a variation in the content of plant protein according to the pH of the soil and inoculation with arbuscular mycorrhiza.

Chia is currently grown in Argentina, Mexico, Bolivia, Guatemala, Ecuador and Australia, among other countries, due to its healthy properties and new popularity as a nutritional supplement and as a functional food.

This chapter focuses on the potential nutraceutical properties of proteins from chia seeds as a source of bioactive peptides with benefits on human health.

12.2 NUTRITIONAL COMPOSITION OF CHIA SEEDS

Chia seed is an attractive but underutilized pseudo-cereal considered a super food, whose consumption is increasing due to its high content of healthy ω-3 (α-linolenic acid, eicosapentaenoic acid, and docosahexaenoic acid), and ω-6 polyunsaturated fatty acids (linoleic acid and arachidonic acid), with low content of saturated fatty acids (Timilsena et al. 2017), as well as a high dietary fiber content, being higher than dried fruits, cereals, or nuts (Kulczyński et al. 2019). An important property of chia is due to the presence of the mucilage, a polysaccharide that exudes when the seeds are immersed in water, producing gels with excellent functional properties such as water and oil retention capacity, viscosity, and potential emulsifying activity with potential application as a thickener, emulsifier, and stabilizer for frozen foods (Ixtaina et al. 2008; Yedida et al. 2020). Chia seeds are rich in vitamins (vitamin E and complex B vitamins like riboflavin, niacin, and thiamine), carotenoids, minerals (phosphorus, manganese, calcium, potassium, and sodium) and also contain considerable amounts of natural antioxidants like chlorogenic and caffeic acids, along with trace amounts of myricetin, quercetin, kaempferol, and flavanols, capable of preventing the oxidation of lipids, proteins, and DNA (Ali et al. 2012). Reyes-Caudillo et al. (2008) reported the presence of phenolic compounds such as rosmarinic acid and daidzein in chia seeds, finding a relationship between the content of phenolic compounds, flavonoids, and the *in vitro* potential antioxidant of chia seed extracts.

Chia seeds are known to contain a high concentration of proteins (19%–23%) of good nutritional quality and digestibility, and a good balance of essential amino acids being higher than most of grains, especially due to the cysteine and methionine content (Bethapudi et al. 2019; Grancieri et al. 2019b; Muñoz et al. 2013; Sandoval-Oliveros and Paredes-López 2013; USDA 2011). Another important characteristic is the absence of gluten, which makes it highly valued for celiac patients (Grancieri et al. 2019b). All the characteristics could produce health benefits in the reduction of cardiovascular diseases, obesity, the regulation of intestinal transit, cholesterol and triglyceride levels, and the prevention of metabolic diseases such as type 2 diabetes and also some types of cancer. Although little is known about the nutritional and therapeutic potential of chia seed as a functional ingredient, this knowledge offers great prospects for the food, medical, pharmaceutical, cosmetic, and nutraceutical industries (Miranda-Ramos et al. 2020; Ullah et al. 2015; Sandoval-Oliveros and Paredes-López 2013; Skov et al. 1999).

12.3 CHARACTERIZATION OF CHIA SEED PROTEIN FRACTIONS

Chia seed proteins can be classified based on different criteria such as function and differential solvent solubility, among others. Storage proteins are those that are present to supply intermediate nitrogen compounds for biosynthesis in a metabolically active stage of seed development (López et al. 2018; Kačmárová et al. 2016). The chia protein profile showed that globulins are the majority protein fraction followed by albumins, glutelins, and prolamins, the first two being found in higher quantity than the others (Hernández-Pérez et al. 2020; Orona-Tamayo et al. 2015; Sandoval-Oliveros and Paredes-López 2013). In contrast, Segura-Campos et al. (2013b) reported that globulins were the major fraction followed by glutelins, albumins, and prolamins. Regarding the variety in amino acids, chia proteins contain nine essential amino acids in appreciable quantity (isoleucine, leucine, lysine, methionine, phenylalanine, threonine, tryptophan, histidine, and valine) and with a good balance of essential and non-essential amino acids, lysine being the limiting amino acid with the lowest coverage of requirement (Salazar-Vega et al. 2020; Grancieri et al. 2019b).

Globulins constitute approximately 52%–54% of the total protein. Some studies have reported that edible seeds generally contain two main types of storage proteins that differ in size: the first group includes proteins with sedimentation coefficients around 11S, which are referred to as "legumin like" or 11S globulins; and the second group of proteins has sedimentation coefficients around 7S, which are classified as "vicilin-like," or 7S globulins with molecular masses between 15–50 kDa. There is also another type of proteins to a lesser extent, 2S-like proteins. Mass spectrometry analysis identified four major globulin peptides as belonging to the 11S type, and two of them as 7S proteins (Orona-Tamayo et al. 2017).

Electrophoretic experiments showed four bands of globulins in the range of 104–628 kDa under native conditions, but eight protein bands have molecular masses of approximately 10, 15, 18, 24, 28, 33, 50, and 60 kDa under denaturing conditions. Albumin (16%–18%) presented low-intensity bands with molecular masses between 10 and 250 kDa, but with apparently intensive bands that range from 12, 25, 28, 35, 60, and 68 kDa. Prolamins showed a different electrophoretic pattern, with only three sharp bands between 12 and 25 kDa. Finally, chia glutelins showed main bands of 25, 28, 35, 60, and 68 kDa, with similarity to the main bands of albumins (Hrnčič et al. 2020; Orona-Tamayo et al. 2017).

Thermal stability using Differential Scanning Calorimetry (DSC) showed that albumins and globulins have better thermal stability than prolamins and glutelins, since denaturation temperatures were 103.6 °C, 104.7 °C, 85.6 °C, and 91.3 °C, respectively—higher than those from other plant proteins. These values indicated that their structures were mainly stabilized by hydrophobic interactions leading to an increase in enthalpy change of denaturation (ΔHd) with significant energy absorption (Ullah et al. 2015; Sanchez Del Angel et al. 2003). This makes these protein fractions suitable for food products undergoing high heat treatment.

12.4 PRODUCTION OF BIOACTIVE PEPTIDES (BAPs) FROM CHIA SEEDS

The study of dietary proteins has increasingly become a research topic in food technology. Numerous dietary proteins of animal and plant origin have been exploited as sources of BAPs. The field of BAPs is highly promising in functional foods, and there is growing interest from consumers, industry, and scientists. Consumers see the possibility of improving their health or preventing disease through healthy eating. The industry perceives this field as an opportunity to expand its market and diversify their offer of products made with high added value. Finally, scientists appreciate this field as an area where new challenges are constantly being posed in the search for new BAPs, but also in the verification through *in vivo* tests of the beneficial effects on health of the new peptides (Montesano et al. 2020).

BAPs may be natural compounds obtained from food or part of a protein (small amino acid sequences of 2–15 amino acids), which are inactive in the precursor molecule but acquire biological activity after release during gastrointestinal digestion or by enzymatic hydrolysis during processing of food or are transported to the active site (Karaś 2019). They can also be produced by microorganisms in the fermentation process. To obtain biopeptides with specific activity, proteases with broad specificity of action are used for proteolysis. They are extracted from plant tissues (e.g., ficin, papain, bromelain), animal tissues (e.g., pepsin, chymotrypsin, trypsin), and microbial cells (e.g., proteinase K, pronase, collagenase, subtilisin A, Alcalase, Flavourzyme). Processing conditions, as hydrolysis time, degree of hydrolysis of the proteins, kind of enzyme, enzyme-substrate ratios, and pretreatment of the protein prior to hydrolysis can influence the bioactive properties of the peptides. Peptide properties can also be influenced by net charge, hydrophobicity, and the size of the peptide, which are factors that affect their absorption across the enterocytes (Udenigwe and Aluko 2012).

Although the production of BAPs could be achieved through different processes (production of endogenous peptides, production by enzymatic hydrolysis, or production by microbial fermentation), in the particular case of chia, the production of BAPs reported to date is limited to enzymatic hydrolysis with commercial proteases (Table 12.1). BAPs from chia obtained under simulated gastrointestinal digestion by sequential hydrolysis with pepsin and pancreatin were reported by different authors (Chan-Zapata et al. 2019; Grancieri et al. 2019a; Orona-Tamayo et al. 2015; Sosa-Crespo et al. 2018), being able to evaluate the formation of BAPs after normal consumption of dietary proteins (Udenigwe and Aluko 2012). BAPs were also obtained using microbial proteases such as Alcalase and Flavourzyme (Aguilar-Toalá et al. 2020; López-García et al. 2019; Urbizo-Reyes et al. 2019; Silveira Coelho et al. 2018, 2019; Sosa-Crespo et al. 2018; Chim-Chi et al. 2017; Segura-Campos et al. 2013a-2013b), which act by interacting with peptide bonds through a serine residue located in the active site (Ottesen and Svendsen 1970). Finally, BAPs obtained by enzymatic hydrolysis with papain (Cotabarren et al. 2019), pepsin, trypsin,

TABLE 12.1 BIOACTIVE PEPTIDES (BAPS) FROM CHIA GENERATED BY ENZYMATIC HYDROLYSIS

Origin	Enzyme	pH	Time (min)	Temp (°C)	Fractionation method	Fractions	Reference
Seeds	Alcalase + Flavourzyme	7.0 8.0	60 90	50	UF (1, 3, 5, and 10 kDa cut-off)	>10 kDa 5–10 kDa 3–5 kDa 1–3 kDa <1 kDa	Segura-Campos et al. (2013a)
Seeds	Alcalase + Flavourzyme	7.0 8.0	60 150	50	–	–	Segura-Campos et al. (2013b)
Seeds	Pepsin + Trypsin + Pancreatin	2.0 7.5 7.5	180 180 180	37	Alkaline solubilization	Globulin, albumin, prolamin, glutelin	Orona-Tamayo et al. (2015)
Seeds	Alcalase + Flavourzyme	8.0 7.0	45 45	50	UF (1, 3, 5, and 10 kDa cut-off)	>10 kDa 5–10 kDa 3–5 kDa 1–3 kDa <1 kDa	Sosa-Crespo et al. (2018)
	Pepsin + Pancreatin	2.0 7.5	45 45	37			
Seeds	Pepsin + Pancreatin	2.0 7.5	45 45	37	UF (1, 3, 5, and 10 kDa cut-off)	>10 kDa 5–10 kDa 3–5 kDa 1–3 kDa <1 kDa	Chan-Zapata et al. (2019)
Seeds	Pepsin + Pancreatin	2.0 7.5	120 120	37	Alkaline solubilization	Globulin, albumin, prolamin, glutelin	Grancieri et al. (2019a)
Seeds	Alcalase	8.0	180	50	–	–	San Pablo-Osorio et al. (2019)
	Pepsin	2.0	180	37	SEC (Superdex 30 Peptide)	3–14 kDa	
	Trypsin	8.0	180	37	–	–	
	Chymotrypsin	8.0	180	37	–	–	

Seeds	Alcalase	8.0	60	MW	–	–	Urbizo-Reyes et al. (2019)
	Alcalase + Flavourzyme	8.0	45 45	MW			
Seeds	Alcalase	8.0	60	MW	UF (3 and 10 kDa cut-off)	3–10 kDa, <3 kDa	Aguilar-Toalá et al. (2020)
	Alcalase + Flavourzyme	8.0	45 45	MW			
Defatted flour	Alcalase + Flavourzyme	8.0	120 240	50	–	–	Chim-Chi et al. (2017)
Defatted flour	Alcalase	7.0	240	50	–	–	Silveira Coelho et al. (2018)
	Flavourzyme	8.0	240	50			
	Alcalase + Flavourzyme	7.0 8.0	60 180	50			
Defatted flour	Alcalase	7.0	240	50	UF (3 and 10 kDa cut-off)	>10 kDa, 3–10 kDa, <3 kDa	Silveira Coelho et al. (2019)
	Flavourzyme	8.0	240	50			
	Alcalase + Flavourzyme	7.0 8.0	60 180	50			
Expeller	Papain	7.0	120	45	SEC (Sephacryl S-100 HR)	3–15 kDa <3 kDa	Cotabarren et al. (2019)
Expeller	Alcalase	7.0	60	50	–	–	López-García et al. (2019)
	Flavourzyme	8.0	150	50			
	Alcalase + Flavourzyme	7.0 8.0	60 150	50			

Abbreviations: MW, microwave; UF, ultrafiltration; SEC, size exclusion chromatography.

or chymotrypsin (San Pablo-Osorio et al. 2019) were reported. Therefore, these commercial proteases could represent an alternative for the industrial production of BAPs.

After enzymatic hydrolysis, BAPs are subjected to purification procedures with subsequent centrifugation and/or washing to remove residual reagents, as well as secondary reaction products. Subsequently, peptides are separated by ultrafiltration techniques and size-exclusion chromatography. Finally, BAPs are characterized by their molecular mass, amino acid content, and biological activity, and identified by proteomic techniques (Figure 12.1).

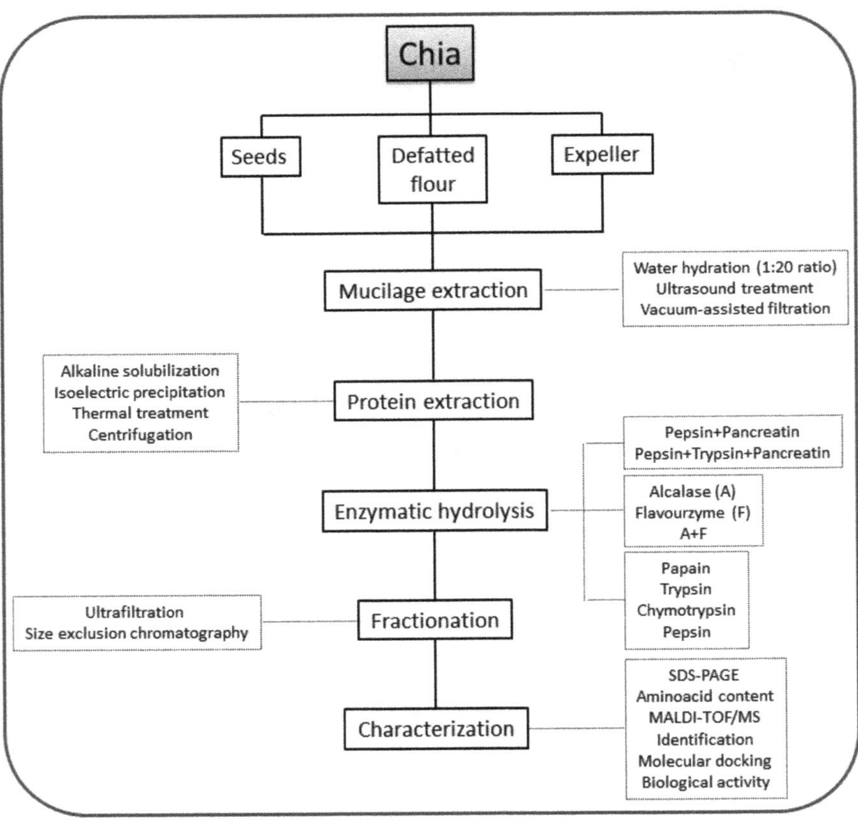

Figure 12.1 Production of bioactive peptides (BAPs) from chia seeds.

12.5 BIOLOGICAL ACTIVITIES OF BAPs FROM CHIA

Many biological activities were reported for chia BAPs. Table 12.2 summarizes the key findings from each study.

TABLE 12.2 BIOLOGICAL ACTIVITIES OF BIOACTIVE PEPTIDES (BAPs) FROM CHIA SEEDS

Bioactive peptide fractions	Biological activity method	Key findings	Reference
Antioxidant activity			
Total hydrolysate	ABTS	TE = 4.49 mM per mg	Segura-Campos et al. (2013b)
Globulin	ABTS	IC50 = 91.9 μg/mL	Orona-Tamayo et al. (2015)
Albumin		IC50 = 89.3 μg/mL	
Prolamin		IC50 = 161.5 μg/mL	
Glutelin		IC50 = 184.7 μg/mL	
Globulin	DPPH	IC50 = 74.7 μg/mL	
Albumin		IC50 = 124.4 μg/mL	
Prolamin		IC50 = 242.8 μg/mL	
Glutelin		IC50 = 238.4 μg/mL	
Globulin	Fe	IC50 = 1.1 mg/mL	
Albumin		IC50 = 1.4 mg/mL	
Prolamin		IC50 = 0.94 mg/mL	
Glutelin		IC50 = 1.2 mg/mL	
Total hydrolysate	β-carotene	CAA = 79.79%	Chim-Chi et al. (2017)
	FRAP	77.76%	
	Fe	86.89%	
	Cu	38.67%	
3–15 kDa	ABTS	IC50 = 25.1 μg/mL	Cotabarren et al. (2019)
<3 kDa		IC50 = 31.6 μg/mL	
3–15 kDa	DPPH	IC50 = 398.1 μg/mL	
<3 kDa		IC50 = 316.2 μg/mL	
Globulin	DPPH	TE = 125 μM per mg/mL	Grancieri et al. (2019a)
Albumin		TE = 160 μM per mg/mL	

(*Continued*)

TABLE 12.2 (CONTINUED)

Bioactive peptide fractions	Biological activity method	Key findings	Reference
Prolamin		TE = 75 μM per mg/mL	
Glutelin		TE = 130 μM per mg/mL	
Globulin	O_2	AAE = 125 μM per mg/mL	
Albumin		AAE = 125 μM per mg/mL	
Prolamin		AAE = 70 μM per mg/mL	
Glutelin		AAE = 175 μM per mg/mL	
Globulin	NO	TE = 12 μM per mg/mL	
Albumin		TE = 25 μM per mg/mL	
Prolamin		TE = 13 μM per mg/mL	
Glutelin		TE = 14 μM per mg/mL	
Globulin	H_2O_2	TE = 75 μM per mg/mL	
Albumin		TE = 10 μM per mg/mL	
Prolamin		TE = 75 μM per mg/mL	
Glutelin		TE = 40 μM per mg/mL	
Total hydrolysate	ABTS	TE = 12.5 mM per mg	López-García et al. (2019)
	DPPH	77.5%	
>10 kDa	ABTS	TE = 3835.2 μM per g	Silveira Coelho et al. (2019)
3–10 kDa		TE = 701 μM per g	
<3 kDa		TE = 754.2 μM per g	
>10 kDa	DPPH	IC50 = 71.2 mg/g	
3–10 kDa		IC50 = 64.3 mg/g	
<3 kDa		IC50 = 68.3 mg/g	
>10 kDa	FRAP	319.6 μmol/g	
3–10 kDa		350.1 μmol/g	
<3 kDa		220.8 μmol/g	
Total hydrolysate	ABTS	TE = 506.07 μmol/mg	Urbizo-Reyes et al. (2019)
Total hydrolysate	DPPH	TE = 178 μmol/mg	
Total hydrolysate	Fe	76.8%	
Total hydrolysate	ORAC	TE = 1122.7 μmol/mg	
Total hydrolysate	CAA	94.7% per mg	

Bioactive peptide fractions	Biological activity method	Key findings	Reference
Antihypertensive activity			
>10 kDa	Inhibition of ACE-I	53.8%	Segura-Campos et al. (2013a)
5–10 kDa		59.1%	
3–5 kDa		58.8%	
1–3 kDa		64.5%	
<1 kDa		69.3%	
Total hydrolysate	Inhibition of ACE-I	IC50 = 8.86 μg/mL	Segura-Campos et al. (2013b)
Globulin	Inhibition of ACE-I	IC50 = 339 μg/mL	Orona-Tamayo et al. (2015)
Albumin		IC50 = 377 μg/mL	
Prolamin		IC50 = 700 μg/mL	
Glutelin		IC50 = 700 μg/mL	
3–14 kDa	Inhibition of ACE-I	IC50 = 0.130 mg/mL	San Pablo-Osorio et al. (2019)
Total hydrolysate	Inhibition of ACE-I	IC50 = 0.55 mg/mL	Urbizo-Reyes et al. (2019)
Hypocholesterolemic activity			
Total hydrolysate	Inhibition of HMG-CoA reductase	80%	Silveira Coelho et al. (2018)
Antimicrobial activity			
Total hydrolysate	Agar disk diffusion	No antimicrobial activity was observed on *Escherichia coli, Salmonella typhi, Shigella flexneri, Klebsiella pneumonia, Staphylococcus aureus, Bacillus subtilis, and Streptococcus agalactae* for 50 mg/mL	Segura-Campos et al. (2013b)
Total hydrolysate	Agar disk diffusion	*S. aureus* presented an MIC of 0.73 mg/mL	Silveira Coelho et al. (2018)

(*Continued*)

TABLE 12.2 (CONTINUED)

Bioactive peptide fractions	Biological activity method	Key findings	Reference
3–10 kDa	OD	42% inhibition for *E. coli*, 62.3% inhibition for *S. typhi*, 67.2% inhibition for *Listeria monocytogenes*, and 28.9% inhibition for *Listeria innocua*	Aguilar-Toalá et al. (2020)
<3 kDa		61.9% inhibition for *E. coli*, 65% inhibition for *S. typhi*, 60.9% inhibition for *L. monocytogenes*, and 51.7% inhibition for *L. innocua*	
Anti-inflammatory activity			
>10 kDa	Inhibition of H2O2, IL-1β, IL-6, and TNF-α	25%, 25%, 30%, and 25% respectively	Chan-Zapata et al. (2019)
5–10 kDa		10%, 25%, 50%, and 40% respectively	
3–5 kDa		25%, 25%, 60%, and 40% respectively	
1–3 kDa		25%, 50%, 70%, and 50% respectively	
<1 kDa		25% for all assays	
Globulin	Inhibition of 5-LOX	AAE = 13 μM per mg/mL	Grancieri et al. (2019a)
Albumin		AAE = 4 μM per mg/mL	
Prolamin		AAE = 12 μM per mg/mL	
Glutelin		AAE = 12 μM per mg/mL	
Globulin	Inhibition of COX-1	IC50 = 66.7 μg/mL	
Albumin		IC50 = 9.8 μg/mL	
Prolamin		IC50 = 1030 μg/mL	
Glutelin		IC50 = 9.9 μg/mL	

Bioactive peptide fractions	Biological activity method	Key findings	Reference
Globulin	Inhibition of COX-2	IC50 = 27.7 μg/mL	
Albumin		IC50 = 13.8 μg/mL	
Prolamin		IC50 = 1087 μg/mL	
Glutelin		IC50 = 23.2 μg/mL	
Globulin	Inhibition of iNOS	IC50 = 7.7 μg/mL	
Albumin		IC50 = 0.1 μg/mL	
Prolamin		IC50 = 1120 μg/mL	
Glutelin		IC50 = 24.9 μg/mL	
Antidiabetic activity			
>10 kDa–AF	Inhibition of α-amylase and α-glucosidase	39.9% for α-amylase	Sosa-Crespo et al. (2018)
5–10 kDa–AF		28.8% for α-amylase	
3–5 kDa–AF		23.4% and 28.4%, respectively	
1–3 kDa–AF		18.3% and 33.7%, respectively	
<1 kDa–AF		8.2% and 28.5%, respectively	
>10 kDa–GIS	Inhibition of α-amylase and α-glucosidase	85.6% and 78%, respectively	
5–10 kDa–GIS		79.2% and 83.6%, respectively	
3–5 kDa–GIS		12.2% and 77.8%, respectively	
1–3 kDa–GIS		96.9% for α-glucosidase	
<1 kDa–GIS		56% for α-glucosidase	
Total hydrolysate	Inhibition of DPP-IV	IC50 = 2.1 mg/mL	Urbizo-Reyes et al. (2019)
Anticoagulant activity			
3–10 kDa	Inhibition of aPTT	80 mg/mL produce complete inhibition	Valicenti et al. (2019)
3–10 kDa	Inhibition of PT	IC50 = 50 mg/mL	

(*Continued*)

TABLE 12.2 (CONTINUED)

Abbreviations: AF, Alcalase-Flavourzyme; GIS, gastrointestinal simulation; ABTS, 2,2′-azino-bis(3-ethylbenzothiazoline-6-sulfonic acid) diammonium salt; DPPH, 1,1-diphenyl-2-picrylhydrazyl; Fe, ferrous ion chelating activity; FRAP, antioxidant power of iron reduction; Cu, copper chelating capacity; O_2, superoxide radical (O_2^-) scavenging activity; NO, nitric oxide (NO) activity; H_2O_2, hydrogen peroxide capacity; ORAC, oxygen radical absorbance capacity; CAA, cellular antioxidant activity; ACE-I, angiotensin-I-converting enzyme; HMG-CoA reductase, 3-hydroxy-3-methylglutaryl coenzyme A reductase; OD, optical density; IL-1β, interleukin 1β; IL-6, interleukin 6; TNF-α, tumor necrosis factor α; 5-LOX, 5-lipooxygenase; COX-1, cyclooxygenase-1; COX-2, cyclooxygenase-2; iNOS, inducible nitric oxide synthase; DPP-IV, dipeptidyl peptidase IV; aPTT, time of activated partial thromboplastin; PT, prothrombin time; TE, Trolox equivalent; IC50, concentration needed to inhibit 50% activity; AAE, ascorbic acid equivalent; MIC, minimal inhibitory concentration.

12.5.1 Antioxidant Activity

Since antioxidants can protect the human body against oxidation by free radicals and reactive oxygen species (ROS), including singlet oxygen, several studies have shown that hydrogen peroxide, superoxide anion and hydroxyl radicals, have positive effects on human health. Reactive oxygen species and free radicals can modify macromolecules such as DNA, proteins, and lipids, causing severe health disorders such as inflammatory diseases, aging, diabetes, neurodegenerative, cardiovascular disease, and cancer (Ngo et al. 2011; Butterfield et al. 2006). Many synthetic antioxidants including butylated hydroxytoluene (BHT), butylated hydroxyanisole (BHA), tert-butylhydroquinone (TBHQ), and propyl gallate (PG) have been used to delay the lipid peroxidation process in foods (Ngo et al. 2012). The antioxidant activity of these synthetic antioxidants is stronger than that found in natural compounds such as α-tocopherol and ascorbic acid, but they are strictly regulated due to their potential health hazards. The interest in the development and use of natural antioxidants as an alternative to synthetic products has grown steadily. Chakrabarti et al. (2014) focused their studies on the biological importance of antioxidant peptides and found that BAPs could improve the actual nutraceutical and functional framework of foods by enhancing biological defense mechanisms against inflammatory diseases due to oxidative stress. Furthermore, Cicero et al. (2017) emphasized the beneficial impact of antioxidant peptides on heart-related diseases.

The chia flour by-product (CF), a protein-rich fraction (PRF), and chia protein concentrates (CPC1 and CPC2) were hydrolyzed with Flavourzyme (F), Alcalase (A), and a sequential system (AF). The hydrolysates were then ultrafiltered, obtaining F1 (>10 kDa), F2 (3–10 kDa), and F3 (<3 kDa) fractions, in which the antioxidant capacity was measured. Peptides corresponding to CPC1-F3 and CPC2-F3 fractions displayed the best antioxidant properties, although good results were also obtained with the CFF2 and PRF-F2 fractions. Flavourzyme-generated hydrolysates were the best antioxidants, followed by those prepared with Alcalase. CPC2-AF F2 (500 μg/g), CF-AF F3, CF-A F3, and PRF-A F3 fractions at concentrations of 250 and 500 μg/g showed the best action against the lipid oxidation in a model system of meat, indicating the possible efficacy of these peptides as dietary antioxidants. Furthermore, CF-AF F3, CPC2-F F2, CPC1-A F2, and CPC1-F F2 significantly increased the survival of *Saccharomyces cerevisiae* when exposed to hydrogen peroxide (Silveira Coelho et al. 2019).

Chim-Chi et al. (2017) have recently described the antioxidant activity, reducing power of iron and the chelation (iron and copper) of chia hydrolysates generated by sequential hydrolysis with Alcalase and Flavourzyme, demonstrating that antioxidants act through mechanisms such as free radical's entrapment, chelation of the pro-oxidant metal, inhibition of reactive oxygen species, and reduction of hydroperoxides. Segura-Campos et al. (2013b) also observed that the antioxidant activity of the hydrolysates produced by sequential hydrolysis with Alcalase-Flavourzyme, expressed as TEAC values (mmol/L per mg protein), decreased as DH increased. The highest TEAC value was observed for the hydrolysate produced at 90 minutes (7.31 mmol/L per mg protein), followed by those produced at 120 minutes (4.66 mmol/L per mg protein) and 150 minutes (4.49 mmol/L per mg protein). In another study, sequential hydrolysis with microwave treatment showed a higher *in vitro* antioxidant activity compared to the traditional water bath treatment, establishing a positive correlation between antioxidant assays and cellular antioxidant activity, and showing that 2,2′-azino-bis (3-ethylbenzothiazoline-6-sulfonic acid) diammonium salt (ABTS) and 2,2-diphenyl-1-picrylhydrazyl (DPPH) were the most efficient in predicting cellular antioxidant activity (Urbizo-Reyes et al. 2019).

Peptides enzymatically released from chia albumin and globulin fractions after simulated gastrointestinal digestion exhibited the highest antiradical capacity against ABTS and DPPH assay, while peptides released from the prolamin fraction exhibited the highest ability to chelate ferrous ion (Grancieri et al. 2019a; Orona-Tamayo et al. 2015). Cotabarren et al. (2019) reported potential antioxidant peptides with molecular mass lower than 15 kDa produced from the expeller, a by-product obtained after the extruded and pressing process of chia seed for oil production, by digestion with papain. These peptides showed a potent radical scavenging effect against DPPH (IC50 = 316.2 μg/mL) and ABTS radicals (IC50 = 31.6 μg/mL) compared to undigested samples. López-García et al. 2019 reported that 60 min Alcalase–150 min Flavourzyme hydrolysates obtained from the chia seed by-product also presented antioxidant activity by DPPH (77.5%) and ABTS (12.5 mM TE/mg).

12.5.2 Antihypertensive Activity

Cardiovascular disease (CVD) affects both the heart and blood vessels, being the leading cause of death worldwide. It is considered the main risk factor for CVD, high blood pressure, or hypertension, which consists of a sustained increase in blood pressure levels. Angiotensin I-converting enzyme (ACE, dipeptidyl carboxypeptidase, EC 3.4.15.1) is a multifunctional enzyme containing zinc and found in different tissues (Bougatef et al. 2010). Through the renin-angiotensin system, ACE plays an important physiological role in the regulation of blood pressure by converting angiotensin I to the potent vasoconstrictor angiotensin II, and thus inactivating the vasodilator bradykinin. ACE inhibition mainly produces a hypotensive effect, but it can also influence the regulatory systems involved in immune defense mechanisms and nervous system activity. Pharmacological therapy usually used to control hypertension consists of ACE inhibitors, such as captopril, enalapril, alacepril, and lisinopril; β-blockers, such as carvedilol and pindolol; renin inhibitors, such as aliskiren and zankiren; and angiotensin II type I receptor (AT1) blockers, such as losartan and valsartan (Ramalingam et al. 2017). However, the use of antihypertensive drugs can cause side effects, such as headache, cough, taste disorders, and rashes, among others. In this sense, the use of natural products with the potential to block ACE without known side effects is a promising alternative as adjuvants for the control of blood pressure (Jemil et al. 2016; Girgih et al. 2011).

A protein-rich fraction from chia seeds was hydrolyzed with an Alcalase-Flavourzyme sequential system showing a higher ACE-I inhibitory activity after 60 min Alcalase–150 min Flavourzyme (IC50 = 8.86 mg protein/mL), followed by those produced at 120 min Flavourzyme (IC50 = 20.76 mg/mL) and at 90 min Flavourzyme (IC50 = 44.01 mg/mL; Segura-Campos et al. 2013b). In a later study, the hydrolysate was ultrafiltered through four membranes with different molecular weight cut-offs (1 kDa, 3 kDa, 5 kDa, and 10 kDa) and the ACE-I inhibitory activity was quantified in both the hydrolysate and the ultrafiltrates. The hydrolysate was extensively digested (DH = 51.64%) with an ACE-inhibitory activity of 58.46%. In addition, it was found that the inhibition ranged from 53.84% to 69.31% in the five ultrafiltered fractions being highest in the <1 kDa fraction (69.31%; Segura-Campos et al. 2013a). Recently, Urbizo-Reyes et al. (2019) evaluated the ACE-I inhibitory activity on microwave-assisted hydrolysates and the peptides displayed a 57% inhibition with a 46% DH. Nevertheless, an influence of microwave energy in peptide conformation is not clear for ACE inhibition.

Peptides obtained by simulated gastrointestinal digestion of chia seeds were analyzed at different concentrations to evaluate the antihypertensive action against the ACE enzyme. Albumin and globulin peptides showed similar inhibition patterns against ACE enzymes with EC50 values of 377 and 339 μg/mL, respectively. Bioactive peptides from prolamin and glutelin resulted in a similar inhibitory effect against ACE enzyme (IC50 of 700 μg/mL; Orona-Tamayo et al. 2015).

12.5.3 Hypocholesterolemic Activity

BAPs presenting antibacterial and hypocholesterolemic effects possess potential applications in food quality and safety and human health. Cholesterol is an important molecule in living cells, where approximately 70% of the total cholesterol in the human body is produced by endogenous biosynthesis, and the remainder is provided by the diet (Grundy and Veja 1988). To treat hypercholesterolemia, 3-hydroxy-3-methylglutaryl coenzyme A reductase (HMGCoA reductase) has been a major target for developing drugs such as statins. The efficacy of these drugs in controlling blood cholesterol levels has been well recognized, however, they cause severe adverse effects, such as distal muscle weakness, headache, and acute renal failure (Rader 2003).

To date, the hypocholesterolemic activity of chia proteins has been poorly reported. Silveira Coelho et al. (2018) reported the effects of the method and the enzyme used to produce hydrolysates on the inhibition of HMG-CoA reductase; and observed that the CF-AF peptide showed the highest inhibitory activity at the highest concentration evaluated (3 mg/mL), while PRF-F peptide showed the highest activity at the lowest concentration (1 mg/mL).

12.5.4 Antimicrobial Activity

The shelf life and safety of foods are the major concern in the food industry as they are affected by the incidence of pathogenic bacteria and spoilage that can contaminate food. A broad spectrum of methods has been used to prevent bacteria contamination in food, including the use of synthetic and natural antibacterial agents. However, the *Codex Alimentarius* regulates the use of synthetic agents due to the antimicrobial resistance, potential negative impact on the environment and their effect on human health. Therefore, novel antibacterial agents from natural sources are highly required (Osman et al. 2016).

The rapid emergence of pathogens resistant to antibacterial drugs is also a serious public health problem, and great efforts have been focused on the development of new classes of antimicrobial agents. Antimicrobial peptides are often the first line of defense against invasion by pathogens, playing an important role in innate immunity (Park et al. 2004).

Despite a great variety of sequences and structures, antimicrobial peptides have certain common features, such as size and the presence of hydrophobic amino acid. Its main destructive mechanism of action causes a rapid death of microbial pathogens by disrupting the microbial cell membrane. Many antimicrobial peptides not only exhibit a potent activity against a wide range of microbes, including the more well-known gram-negative and gram-positive bacteria, yeast, and fungi, but also have antimicrobial activity against enveloped viruses such as HIV, herpes simplex virus, and influenza A virus, thus making them promising agents in the development of novel therapeutic agents aimed at overcoming the challenge of bacterial resistance (Rydlo et al. 2006).

There are few studies on the antimicrobial activity of protein hydrolysates from chia seeds, with contradictory results. Segura-Campos et al. (2013b) verified that none of the chia seed protein hydrolysates generated by sequential hydrolysis with Alcalase-Flavourzyme (50 mg protein/mL) exhibited antibacterial activity even at the lowest concentration evaluated (1×10^3 UFC/mL). In a later study, Silveira Coelho et al. (2018) reported that defatted chia flour hydrolysates exhibited antimicrobial activity against *Staphylococcus aureus*, with a lower MIC after sequential hydrolysis compared to Alcalase hydrolysates. Recently, Aguilar-Toalá et al. (2020) reported that peptides generated by microwave-assisted hydrolysis with sequential enzymes (Alcalase-Flavourzyme) showed antimicrobial activity toward *Escherichia coli*, *Salmonella enterica*, and *Listeria monocytogenes*. The peptide fraction of molecular mass <3 kDa showed higher antimicrobial activity than both the hydrolysate and the peptide fraction between 3–10 kDa, by increasing the membrane permeability of *E. coli* (71.49% crystal violet uptake) and *L. monocytogenes* (80.10% crystal violet uptake). These peptides caused a significant extension in the lag phase, decreases in the maximum growth, and growth rate in the bacteria and promoted multiple indentations (transmembrane tunnels), membrane wrinkling, and pronounced deformations in the integrity of the bacterial cell membranes.

12.5.5 Anti-Inflammatory Activity

Inflammation occurs in response to infection or injury, and has been universally associated with diabetes, cancer and cardiovascular or neurodegenerative diseases (Okin and Medzhitov 2012). The first cellular signs of inflammation include infiltration of monocytes, dendritic cells, and neutrophils at the site of injury (Nathan and Ding 2010). The inflammatory response begins with the neutrophils, which destroy and remove foreign particles and damaged tissue. Later, macrophages appear in the site of inflammation and continue the process of phagocytosis, playing an important role in inflammatory diseases (Velnar et al. 2009). Receptors as Toll-like receptor (TLR) and lipoxygenase (LOX) can be activated, sending signals for activation of cellular molecules, such as nuclear factor kappa B (NF-κB; Morettini et al. 2015), and promoting the activation of the inducible nitric oxide synthase enzyme (iNOS) and cyclooxygenase (COX-2), which will produce nitric oxide (NO) and prostaglandins (PGE2), respectively (Khodabandehloo et al. 2016). When macrophages are stimulated, these activated cells release cytokines involved in the regulation of the inflammation process, such as interleukin 1β (IL-1β), interleukin 6 (IL-6), interleukin 10 (IL-10), and tumor necrosis factor α (TNF-α), which mediate intercellular communication and orchestrate a variety of processes ranging from the cellular proliferation to metabolism, chemotaxis, and inhibition of other cytokines (Arango Duque and Descoteaux 2014; Lawrence and Fong 2010; Zhang and An 2007). This process can be deepened by reactive oxygen species (ROS) as superoxide and hydrogen peroxide (Mittal et al. 2014).

There are many different types of drugs with relevant clinical applications for the treatment of inflammatory disorders; however, these drugs are associated with undesirable side effects in some patients and a high cost of treatment. Therefore, BAPs may be a good source for the possible development in drugs for treating inflammation (Gautam and Jachak 2009).

Some protein derivatives isolated from *S. hispanica* L. seeds obtained by sequential hydrolysis with pepsin and pancreatin exhibited a decrease in the cellular release of NO and H_2O_2 in a concentration dependent manner, decreasing the values of pro-inflammatory cytokines (IL-1β, IL-6, and TNF-α) and increasing the levels of an anti-inflammatory cytokine (IL-10). The 1–3 kDa fraction showed the highest inhibitory effect at the maximum concentration evaluated, suggesting that the low-molecular-weight peptides contained in this fraction possess important anti-inflammatory properties at high doses (Chan-Zapata et al. 2019). Grancieri et al. 2019a found that peptides released from albumin, globulin, prolamin and glutelin fractions showed interaction with cyclooxygenase-2 (COX-2), p65-nuclear factor kappa B, lipoxygenase-1 (LOX-1) and toll-like receptor 4 after simulated gastrointestinal digestion, and that the inhibition of 5-LOX, COX-1–2, and inducible nitric oxide synthase (iNOS) enzymes, demonstrating the potential beneficial health effects of the consumption of chia seed proteins as part of the diet.

12.5.6 Antidiabetic Activity

Various therapeutic alternatives have been applied in hyperglycemia disorders, with the main purpose of reducing postprandial hyperglycemia and therefore, microvascular complications associated with the dysregulation of blood glucose levels (Holman et al. 2008). In this context, the inhibition of the α-glucosidase enzyme—responsible for catalyzing the release of glucose from complex carbohydrates—has been effective in reducing blood glucose levels (Johnson et al. 2011). An example is diabetes mellitus, which is characterized by hyperglycemia associated with disorders in carbohydrate, lipid, and protein metabolism (Schmidt and Hickey 2009). Current treatments that involve the use of hypoglycemic drugs—such as acarbose, voglibose and miglitol—although effective, have serious gastrointestinal side effects. Thus, dipeptidyl peptidase-IV (DPP-IV) inhibitors offer novel opportunities for treatments. Several DPP-IV inhibitory drugs, known as gliptins, have IC50 values in the nanomolar range (Sortino et al. 2013). Natural alternatives to DPP-IV inhibitory drugs may exist in the diet, including DPP-IV inhibitory peptides released by enzymatic hydrolysis of dietary proteins (Nongonierma and FitzGerald 2015). Microwave-assisted hydrolysis with Alcalase and Flavourzyme increased DPP-IV inhibition as a function of DH, suggesting that higher hydrolysis will result in improved inhibitions. DPP-IV inhibition was higher with sequential hydrolysis with Alcalase and Flavourzyme (IC50 = 2.1 mg/mL) compared to Alcalase (IC50 = 3.6 mg/mL), suggesting the benefit of applying sequential hydrolysis of enzymes (Urbizo-Reyes et al. 2019).

Sosa-Crespo et al. 2018 also evaluated the α-amylase inhibitory activity of BAPs generated by sequential Alcalase-Flavourzyme and pepsin-pancreatin hydrolysis. For the 45 min Alcalase–45 min Flavourzyme hydrolysates, the highest inhibition (39.90%) was observed with the >10 kDa fraction; however, this result was lower than that obtained by acarbose (93.42%) at a concentration of 2 mg/mL. The lowest inhibition (8.29%) was found with the <1 kDa fraction of this dual sequential system. Data showed that there was no significant difference ($p < 0.05$) in the fractions generated by the pepsin-pancreatin system with the highest molecular mass (>10 and 5–10 kDa), these being the ones with the highest activity (85.61% and 79.19%, respectively) on α-amylase. This suggests that the highest inhibition is achieved with peptides whose polypeptide chains are longer. For the sequential hydrolysis with Alcalase-Flavourzyme, a higher inhibition of α-glucosidase (33.75%) was observed with the 1–3 kDa fraction at a concentration of 20 mg/mL. For the pepsin-pancreatin system, the results showed that the highest inhibition (96.91%) was obtained with the 1–3 kDa fraction, which was statistically different ($p < 0.05$) with the others.

12.5.7 Antithrombotic Activity

Thrombotic events due to blood clotting represent a serious problem in cardiovascular diseases (Ishihara et al. 2014; Goto and Tomita 2013; Lindholm and Mendis 2007). Although heparin has been used for this purpose in the last 50 years, its use results in the development of thrombocytopenia and immune response when used for a long time (Kim et al. 2013). In addition, other anticoagulant drugs such as aspirin and clopidogrel also produce serious secondary effects (Eikelboom et al. 2012), so the discovery of new BAPs with antithrombotic activity would be an attractive strategy against thrombosis.

In a preliminary study, Valicenti et al. (2019) reported that BAPs obtained from defatted expeller of chia seeds with commercial papain showed antithrombotic activity against intrinsic and extrinsic pathways. For the aPTT assay, the 3–10 kDa fraction (80 mg/mL) produced the complete inhibition, while an IC50 value of 50 mg/mL was found for the prothrombin time. These results encourage to deepen on the biological potential of BAPs from chia as antithrombotic agents of natural origin.

12.6 HYDROLYSATES FROM BY-PRODUCTS OF CHIA SEED

Agrifood industries are a relevant sector of the economy, frequently associated with the production of significant volumes of bio-waste that impact the sustainability of the planet. In recent years, the search for alternatives for the recycling of proteins with high biological and nutritional value from by-products of the oil industry has increased as a sustainable alternative to minimize the final disposal of waste, adding value to a market with a high impact on nutrition and simultaneous beneficial effects on health (Faustino et al. 2019; Torres-León et al. 2018; Helkar et al. 2016). In particular, the expeller from chia seeds is currently used in the manufacture of defatted flour and animal feed. Recently, the production of hydrolysates from chia expeller with antioxidant activity has been reported (Cotabarren et al. 2019; López García et al. 2019). For this reason, BAPs from by-products of the chia oil seed processing have become particularly interesting in a world with increased awareness of food waste and its use, thus contributing to the circular economy.

12.7 PROSPECTS

Nutrition plays an important role in human health and in the prevention of many chronic diseases. The consumer trend toward vegetarianism in recent years has meant that plant sources have gained more importance; plant by-products are a good alternative to animal sources in bioactive peptide production due to the high quantity and low unit costs. BAPs obtained from chia could be used in the development of functional food (Madruga et al. 2020; Segura-Campos et al. 2013b). Functional properties and biological activities of bioactive peptides depend on some factors including the amino acid type, sequence, and molecular weight. Therefore, studies are still going on to examine the changes in the final product quality caused by amino acid sequence in the bioactive peptide chains, and the interactions among them and between the food matrices. In this context, more studies should be done on novel food sources such as endemic, medicinal, and aromatic plants, as well as the production, purification, and characterization of bioactive peptides derived from these sources. In addition to these new sources, plant waste and by-products could be recovered to produce value-added commodities. Furthermore, scale-up studies will also be necessary to produce peptides for the food and pharmaceutical industries.

To date, studies related to sequential knowledge of proteins from chia sources are scarce. Twenty proteins from chia seeds have been identified in the literature based on their amino acids sequences (Grancieri et al. 2019b), but none of them corresponds to the globulins, albumins, prolamins and glutelins. Twelve of the identified proteins are responsible for the metabolic functions needed for the existence of the seed while eight proteins are related to lipid production and storage. *In silico* analysis using the BIOPEP database showed many bioactive peptides encrypted within the sequence of the parent protein, mainly comprising hypoglycemic, hypotensive, DPP-IV, and ACE-I inhibitory activity. Nevertheless, these sequences represent potential bioactive peptides, and further study should be performed to verify the predicted biological activities. San Pablo-Osorio et al. (2019) identified five peptides from the chia seed hydrolysates. Molecular docking showed that peptide sequences could block ACE by interacting with its catalytic site, encouraging the development of future studies on the antihypertensive potential of the identified peptides. Grancieri et al. 2019b also observed by molecular docking, a direct interaction of two peptides produced from digested albumin with LOX-1 and COX-2, and one peptide obtained from the digestion of the total protein with p65-NF-κB. Finally, Aguilar-Toalá et al. (2020) identified 16 peptides from the fraction lower than 3 kDa of the hydrolysate, all cationic and hydrophobic in nature, providing specificity in binding to negatively charged bacterial membranes through electrostatic interaction, thus promoting their penetration into the hydrophobic core of the bacterial membrane to destabilize the bilayer and/or promoting the cell depolarization.

12.8 CONCLUSIONS

Peptides identified in the protein hydrolysates from chia seeds have potentially auspicious biological activities, mainly antioxidant, antihypertensive, and hypoglycemic properties. Among other bioactive compounds that exert biological functions on health, BAPs may be responsible for the positive effects reported in research in humans, who consume the whole chia seed, although many results are still inconclusive. Further studies will be needed to determine the precise mechanism of action of each identified peptide that contributes to the observed health benefits, to clarify the relationship with structure-activity of the peptide, and to validate the biological activities in food and *in vivo* models. Finally, further studies should be conducted regarding the bioactivity associated with each individual peptide derived from key proteins of the chia seed, with potential benefits for human health.

ACKNOWLEDGMENTS

This work was supported by Universidad Nacional de Luján (Departamento de Ciencias Básicas) and UAV 2017–2018 Projects (Res.SPU N° 5157/17 y 190/18) grants to M.G.P, and Universidad Nacional de La Plata (projects PPID X/014 and PPID X/038) and PICT-2018–03271 (ANPCyT) grants to W.D.O.

REFERENCES

Aguilar-Toalá, José, Amanda Deering, and Andrea Liceaga. 2020. "New Insights into the Antimicrobial Properties of Hydrolysates and Peptide Fractions Derived from Chia Seed (*Salvia hispanica* L.)." *Probiotics and Antimicrobial Proteins* 12: 1571–1581. https://doi.org/10.1007/s12602-020-09653-8.

Ali, Norlaily Mohd, Yeap Swee Keong, Ho Wan Yong, Beh Boon Kee, Tan Sheau Wei, and Tan Soon Guan. 2012. "The Promising Future of Chia, *Salvia hispanica* L." *BioMed Research International*: 1–9. https://doi.org/10.1155/2012/171956.

Arango, Duque Guillermo, and Albert Descoteaux. 2014. "Macrophage Cytokines: Involvement in Immunity and Infectious Diseases." *Frontiers in Immunology* 5 (491): 1–12. https://doi.org/10.3389/fimmu.2014.00491.

Ayerza, Ricardo, and Wayne Coates. 2011. "Protein Content, Oil Content and Fatty Acid Profiles as Potential Criteria to Determine the Origin of Commercially Grown Chia (*Salvia hispanica* L.)." *Industrial Crops and Products* 24: 1366–1371. https://doi.org/10.1016/j.indcrop.2010.12.007.

Bethapudi, Prathyusha, Anila Kumari, Jessie Suneetha, and Naga Sai Srujana. 2019. "Chia Seeds for Nutritional Security." *Journal of Pharmacognosy and Phytochemistry* 8 (3): 2702–2707. http://doi.org/10.22271/phyto.2019.v8.i3e.702.

Bochicchio, Rocco, Tim Philips, Stella Lovelli, Rosanna Labella, Fernanda Galgano, Antonio Di Marisco, Michele Perniola, and Mariana Amato. 2015. "Innovative Crop Productions for Healthy Food: The Case of Chia (*Salvia hispanica* L.)." *The Sustainability of Agro-Food and Natural Resource Systems in the Mediterranean Basin* 29–45. https://doi.org/10.1007/978-3-319-16357-4_3.

Bougatef, Ali, Rafik Balti, Naima Nedjar-Arroume, Adje Estelle Ravallec Rozenn, Lassoued Imen Souissi Nabil, Didier Guillochon, and Moncef Nasri. 2010. "Evaluation of Angiotensin I-Converting Enzyme (ACE) Inhibitory Activities of Smooth Hound (*Mustelus mustelus*) Muscle Protein Hydrolysates Generated by Gastrointestinal Proteases: Identification of the Most Potent Active Peptide."

European Food Research Technology 231: 127–135. https://doi.org/10.1007/s00217-010-1260-4.

Butterfield, Allan, Hafiz Mohmmad Abdul, Wycliffe Opii, Shelley Newman, Gururaj Joshi, Mubeen Ahmad Ansari, and Rukhsana Sultana. 2006. "Pin1 in Alzheimer's Disease." *Journal of Neurochemistry* 98 (6): 1697–1706. https://doi.org/10.1111/j.1471-4159.2006.03995.x.

Cahill, Joseph. 2004. "Genetic Diversity Among Varieties of Chia (*Salvia hispanica* L.)." *Genetic Resources and Crop Evolution* 51: 773–781. https://doi.org/10.1023/B:GRES.0000034583.20407.80.

Chakrabarti, Subhadeep, Forough Jahandideh, and Jianping Wu. 2014. "Food-Derived Bioactive Peptides on Inflammation and Oxidative Stress." *BioMed Research International*: 1–12. https://doi.org/10.1155/2014/608979.

Chan-Zapata, Ivan, Víctor Emilio Arana-Argáez, Julio Cesar Torres-Romero, and Maira Rubí Segura-Campos. 2019. "Anti-inflammatory Effects of the Protein Hydrolysate and Peptide Fractions Isolated from *Salvia hispanica* L. Seeds." *Food and Agricultural Immunology* 30 (1): 796–803. https://doi.org/10.1080/09540105.2019.1632804.

Chim-Chi, Yasser, Santiago Gallegos-Tintoré, Cristian Jiménez-Martínez, Gloria Dávila-Ortiz, and Luis Chel-Guerrero. 2017. "Antioxidant Capacity of Mexican Chia (*Salvia hispanica* L.) Protein Hydrolyzates." *Journal of Food Measurement and Characterization* 12 (1): 323–331. https://doi.org/10.1007/s11694-017-9644-9.

Cicero, Arrigo, Federica Fogacci, and Alessandro Colletti. 2017. "Potential Role of Bioactive Peptides in Prevention and Treatment of Chronic Diseases: A Narrative Review." *British Journal of Pharmacology* 174 (11): 1378–1394. https://doi.org/10.1111/bph.13608.

Cotabarren, Juliana, Adriana Mabel Rosso, Mariana Edith Tellechea, Javier García-Pardo, Julia Lorenzo-Rivera, Walter David Obregón, and Mónica Graciela Parisi. 2019. "Adding Value to the Chia (*Salvia hispanica* L.) Expeller: Production of Bioactive Peptides with Antioxidant Properties by Enzymatic Hydrolysis with Papain." *Food Chemistry* 274: 848–856. https://doi.org/10.1016/j.foodchem.2018.09.061.

Eikelboom, John, Jack Hirsh, Frederick Spencer, Trevor Baglin, and Jeffrey Weitz. 2012. "Antiplatelet Drugs: Antithrombotic Therapy and Prevention of Thrombosis." *9th Ed: American College of Chest Physicians evidence-based Clinical Practice Guidelines* 141 (2): e89S–e119S. https://doi.org/10.1378/chest.11-2293.

Faustino, Margarida, Mariana Veiga, Pedro Sousa, Eduardo Costa, Sara Silva, and Manuela Pintado. 2019. "Agro-Food Byproducts as a New Source of Natural Food Additives." *Molecules* 24: 1056. https://doi.org/10.3390/molecules24061056.

Gautam, Raju, and Sanjay Jachak. 2009. "Recent Developments in Anti-inflammatory Natural Products." *Medicinal Research Reviews* 29 (5): 767–820. https://doi.org/10.1002/med.20156.

Girgih, Abraham, Chibuike Udenigwe, Huan Li, Abayomi Peter Adebiyi, and Rotimi Aluko. 2011. "Kinetics of Enzyme Inhibition and Antihypertensive Effects of Hemp Seed (*Cannabis sativa* L.) Protein Hydrolysates." *Journal of the American Oil Chemists' Society* 88: 1767–1774. https://doi.org/10.1007/s11746-011-1841-9.

Grancieri, Mariana, Hercia Stampini Duarte Martino, and Elvira Gonzalez de Mejia. 2019a. "Digested Total Protein and Protein Fractions from Chia Seed (*Salvia hispanica* L.) Had High Scavenging Capacity and Inhibited 5-LOX, COX-1-2, and iNOS Enzymes." *Food Chemistry* 289: 204–214. https://doi.org/10.1016/j.foodchem.2019.03.036.

Grancieri, Mariana, Hercia Stampini Duarte Martino, and Elvira Gonzalez de Mejia. 2019b. "Chia Seed (*Salvia hispanica* L.) as a Source of Proteins and Bioactive Peptides with Health Benefits: A Review." *Comprehensive Reviews in Food Science and Food Safety* 18: 480–499. https://doi.org/10.1111/1541-4337.12423.

Grundy, Scott, and Gloria Veja. 1988. "Plasma Cholesterol Responsiveness to Saturated Fatty Acids." *American Journal of Clinical Nutrition* 47: 822–824. https://doi.org/10.1093/ajcn/47.5.822.

Goto, Shinya, and Aiko Tomita. 2013. "Antithrombotic Therapy for Prevention of Various Thrombotic Diseases." *Drug Development Research* 74: 568–574. https://doi.org/10.1002/ddr.21116.

Helkar, Prathamesh Bharat, Aksasha Kumar Sahoo, and Namita Patil. 2016. "Review: Food Industry By-Products used as a Functional Food Ingredients." *International Journal of Waste Resources* 6: 248. https://doi.org/10.4172/2252-5211.1000248.

Hernández-Pérez, Talia, María Elena Valverde, Domancar Orona-Tamayo, and Octavio Paredes-López. 2020. "Chia (*Salvia hispanica*): Nutraceutical Properties and Therapeutic Applications." *Proceedings* 53 (1): 17. https://doi.org/10.3390/proceedings2020053017.

Holman, Rury, Sanjoy Paul, Angelyn Bethel, David Matthews, and Andrew Neil. 2008. "10-Year Follow-Up of Intensive Glucose Control in Type 2 Diabetes." *New England Journal of Medicine* 359: 1577–1589. https://doi.org/10.1056/NEJMoa0806470.

Hrnčič, Maša Knez, Maja Ivanovski, Darija Cör, and Željko Knez. 2020. "Chia Seeds (*Salvia Hispanica* L.): An Overview—Phytochemical Profile, Isolation Methods, and Application." *Molecules* 25: 11. https://doi.org/10.3390/molecules25010011.

Ishihara, Tsukasa, Yuji Koga, Kenichi Mori, Keizo Sugasawa, Yoshiyuki Iwatsuki, and Fukushi Hirayama. 2014. "Novel Strategy to Boost Oral Anticoagulant Activity of Blood Coagulation Enzyme Inhibitors Based on Biotransformation into Hydrophilic Conjugates." *Bioorganic and Medicinal Chemistry* 22: 6324–6332. https://doi.org/10.1016/j.bmc.2014.09.059.

Ixtaina, Vanesa Y., Susana M. Nolasco, and Mabel C. Tomás. 2008. "Physical Properties of Chia (*Salvia hispanica* L.) Seeds." *Industrial Crops and Products* 28 (3): 286–293. https://doi.org/10.1016/j.indcrop.2008.03.009.

Ixtaina, Vanesa Y., Marcela L. Martínez, Viviana G. Spotorno, Carmen M. Mateo, Damián M. Maestri, Bernd W. K. Diehl, Susana M. Nolasco, and Mabel C. Tomás. 2011. "Characterization of Chia Seed Oils Obtained by Pressing and Solvent Extraction." *Journal of Food Composition and Analysis* 24 (2): 166–174. https://doi.org/10.1016/j.jfca.2010.08.006.

Jamshidi, Amir, Mariana Amato, Ali Ahmadi, Rocco Bochicchio, and Roberta Rossi. 2019. "Chia (*Salvia hispanica* L.) as a Novel Forage and Feed Source: A Review." *Italian Journal of Agronomy* 14: 1297. https://doi.org/10.4081/ija.2019.1297.

Jemil, Ines, Leticia Mora, Rim Nasri, Ola Abdelhedi, María Concepción Aristoy, Mohamed Nasri Hajji, and Fidel Toldrá. 2016. "A Peptidomic Approach for the

Identification of Antioxidant and ACE-Inhibitory Peptides in Sardinelle Protein Hydrolysates Fermented by *Bacillus subtilis* A26 and *Bacillus amyloliquefaciens* An6." *Food Research International* 89 (1): 347–358. https://doi.org/10.1016/j.foodres.2016.08.020.

Johnson, Michelle, Lucius Anita, Meyer Tessa, and de Mejia Elvira González. 2011. "Cultivar Evaluation and Effect of Fermentation on Antioxidant Capacity and *in Vitro* Inhibition of α-Amylase and α-Glucosidase by Highbush Blueberry (*Vaccinium corombosum*)." *Journal of Agricultural and Food Chemistry* 59: 8923–8930. https://doi.org/10.1021/jf201720z.

Kačmárová, Kvetoslava, Blazena Lavová, Peter Socha, and Dana Urminská. 2016. "Characterization of Protein Fractions and Antioxidant Activity of Chia Seeds (*Salvia hispanica* L.)." *Slovak Journal of Food Sciences* 10 (1): 78–82. https://doi.org/10.5219/563.

Karaś, Monika. 2019. "Influence of Physiological and Chemical Factors on the Absorption of Bioactive Peptides." *International Journal of Food Science and Technology* 54: 1486–1496. https://doi.org/10.1111/ijfs.14054.

Khodabandehloo, Hadi, Satar Gorgani-Firuzjaee, Ghodratollah Panahi, and Reza Meshkani. 2016. "Molecular and Cellular Mechanisms Linking Inflammation to Insulin Resistance and β-Cell Dysfunction." *Translational Research* 167 (1): 228–256. https://doi.org/10.1016/j.trsl.2015.08.011.

Kim, Jin-Young, Ramamourthy Gopal, Sang Young Kim, Chang Ho Seo, Hyang Burm Lee, Hyeonsook Cheong, and Yoonkyung Park. 2013. "PG-2, a Potent AMP Against Pathogenic Microbial Strains, from Potato (*Solanum tuberosum* L. cv. Gogu Valley) Tubers Not Cytotoxic against Human Cells." *International Journal of Molecular Sciences* 14: 4349–4360. https://doi.org/10.3390/ijms14024349.

Kulczyński, Bartosz, Joanna Kobus-Cisowska, Maciej Taczanowski, Dominik Kmiecik, and Anna Gramza-Michałowska. 2019. "The Chemical Composition and Nutritional Value of Chia Seeds—Current State of Knowledge." *Nutrients* 11: 1242. https://doi.org/10.3390/nu11061242.

Lawrence, Toby, and Carol Fong. 2010. "The Resolution of Inflammation: Anti-Inflammatory Roles for NF-κB." *The International Journal of Biochemistry and Cell Biology* 42 (4): 519–523. https://doi.org/10.1016/j.biocel.2009.12.016.

Lindholm, Lars, and Shanthi Mendis. 2007. "Prevention of Cardiovascular Disease in Developing Countries." *Lancet*: 720–722. https://doi.org/10.1016/S0140-6736(07)61356-7.

López Débora, Natalia, Romina Ingrassia, Pablo Busti, Julia Bonino, Juan Francisco Delgado, Jorge Wagner, Valeria Boeris, and Dario Spelzini. 2018. "Structural Characterization of Protein Isolates Obtained from Chia (*Salvia hispanica* L.) Seeds." *LWT-Food Science and Technology* 90: 396–402. https://doi.org/10.1016/j.lwt.2017.12.060.

López-García, Sonia, Gerónimo Arámbula-Villa, Juan Torruco-Uco, Adriana Contreras-Oliva, Francisco Hernández-Rosas, Mirna López-Espíndola, and José Herrera-Corredor. 2019. "Antioxidant Activity of Chia (*Salvia hispanica* L.) Protein Fraction Hydrolyzed with Alcalase and Flavourzyme." *Agrociencias* 53 (4): 505–520.

Madruga, Karina, Meritaine Rocha, Sibele Santos Fernandes, and Myriam de las Mercedes Salas-Mellado. 2020. "Properties of Heat and Rice Breads Added with

Chia (*Salvia hispanica* L.) Protein Hydrolysate." *Food Science and Technology* 40 (3): 596–603. https://doi.org/10.1590/fst.12119.

Miranda-Ramos, Karla, María Carmen Millán-Linares, and Claudia Monika Haros. 2020. "Effect of Chia as Breadmaking Ingredient on Nutritional Quality, Mineral Availability and Glycemic Index of Bread." *Foods* 9: 663. https://doi.org/103390/foods9050663.

Mittal, Manish, Mohammad Rizwan Siddiqui, Khiem Tran, Sekhar Reddy, and Asrar Malik. 2014. "Reactive Oxygen Species in Inflammation and Tissue Injury." *Antioxidants and Redox Signaling* 20 (7): 1126–1167. https://doi.org/10.1089/ars.2012.5149.

Montesano, Domenico, Mónica Gallo, Francesca Blasi, and Lina Cossignani. 2020. "Biopeptides from Vegetable Proteins: New Scientific Evidences." *Current Opinion in Food Science* 31: 31–37. https://doi.org/10.1016/j.cofs.2019.10.008.

Morettini, Micaela, Fabio Alejandro Storm, Massimo Sacchetti, Aurelio Cappozzo, and Claudia Mazzà. 2015. "Effects of Walking on Low-Grade Inflammation and Their Implications for Type 2 Diabetes." *Preventive Medicine Reports* 2: 538–547. https://doi.org/10.1016/j.pmedr.2015.06.012.

Muñoz, Loreto, Ángel Cobos, Olga Díaz, and José Miguel Aguilera. 2013. "Chia Seed (*Salvia hispanica*): An Ancient Grain and a New Functional Food." *Food Reviews International* 29 (4): 394–408. http://doi.org/10.1080/87559129.2013.818014.

Nathan, Carl, and Aihao Ding. 2010. "Nonresolving Inflammation." *Cell* 140 (6): 871–882. https://doi.org/10.1016/j.cell.2010.02.029.

Ngo, Dai-Hung, Thanh-Sang Vo, Dai-Nghiep Ngo, Isuru Wijesekara, and Se-Kwon Kim. 2012. "Biological Activities and Potential Health Benefits of Bioactive Peptides Derived from Marine Organisms." *International Journal of Biological Macromolecules* 51 (4): 378–383. https://doi.org/10.1016/j.ijbiomac.2012.06.001.

Ngo, Dai-Hung, Isuru Wijesekara, Thanh-Sang Vo, Ta Quang Van, and Se-Kwon Kim. 2011. "Marine Food-Derived Functional Ingredients as Potential Antioxidants in the Food Industry: An Overview." *Food Research International* 44 (2): 523–529. https://doi.org/10.1016/j.foodres.2010.12.030.

Nongonierma, Alice, and Richard FitzGerald. 2015. "Investigation of the Potential of Hemp, Pea, Rice and Soy Protein Hydrolysates as a Source of Dipeptidyl Peptidase IV (DPP-IV) Inhibitory Peptides." *Food Digestion: Research and Current Opinion* 6: 19–29. https://doi.org/10.1007/s13228-015-0039-2.

Okin, Daniel, and Ruslan Medzhitov. 2012. "Evolution of Inflammatory Diseases." *Current Biology* 22 (17): R733–R740. https://doi.org/10.1016/j.cub.2012.07.029.

Orona-Tamayo, Domancar, María Elena Valverde, Blanca Nieto-Rendón, and Octavio Paredes-López. 2015. "Inhibitory Activity of Chia (*Salvia hispanica* L.) Protein Fractions Against Angiotensin I-Converting Enzyme and Antioxidant Capacity." *LWT-Food Science and Technology* 64: 236–242. http://doi.org/10.1016/j.lwt.2015.05.033.

Orona-Tamayo, Domancar, María Elena Valverde, and Octavio Paredes-López. 2017. "Chia—The New Golden Seed for the 21st Century: Nutraceutical Properties and Technological Uses." *Sustainable Protein Sources*: 265–281. http://doi.org/10.1016/B978-0-12-802778-3.00017-2.

Osman, Ali, Hanan Goda, Mahmoud Abdel-Hamid, Sanaa Badran, and Jeanette Otte. 2016. "Antibacterial Peptides Generated by Alcalase Hydrolysis of Goat Whey."

LWT—Food Science and Technology 65: 480–486. https://doi.org/10.1016/j.lwt.2015.08.043.

Ottesen, Martin, and Ib Svendsen. 1970. "The Subtilisins." *Methods in Enzymology* 19: 199–215. https://doi.org/10.1016/0076-6879(70)19014-8.

Park In, Yup, Ju Hyun Cho, Key Sun Kim, Yun-Bae Kim, Mi Sun Kim, and Sun Chang Kim. 2004. "Helix Stability Confers Salt Resistance upon Helical Antimicrobial Peptides." *Journal of Biological Chemistry* 279 (14): 13896–13901. https://doi.org/10.1074/jbc.M311418200.

Peiretti, Pier Giorgio. 2010. "Ensilability Characteristics of Chia (*Salvia hispanica* L.) During Its Growth Cycle and Fermentation Pattern of Its Silages Affected by Wilting Degrees." *Cuban Journal of Agricultural Science* 44 (1): 33–36.

Peiretti, Pier Giorgio, and Francesco Gai. 2009. "Fatty Acid and Nutritive Quality of Chia (*Salvia hispanica*) Seeds and Plant During Growth." *Animal Feed Science and Technology* 148 (2): 267–275. https://doi.org/10.1016/j.anifeedsci.2008.04.006.

Peláez, P., D. Orona-Tamayo, S. Montes-Hernández, M. E. Valverde, O. Paredes-López, and A. Cibrián-Jaramillo. 2019. "Comparative Transcriptome Analysis of Cultivated and Wild Seeds of *Salvia hispanica* (chia)." *Scientific Reports* 9: 9761. https://doi.org/10.1038/s41598-019-45895-5.

Rader, Daniel. 2003. "Regulation of Reverse Cholesterol Transport and Clinical Implications." *The American Journal of Cardiology* 92: 42–49. https://doi.org/10.1016/S0002-9149(03)00615-5.

Ramalingam, Latha, Kalhara Menikdiwela, Monique Le Mieux, Jannette Dufour, Gurvinder Kaur, Nishan Kalupahana, and Naima Moustaid-Moussa. 2017. "The Renin Angiotensin System, Oxidative Stress and Mitochondrial Function in Obesity and Insulin Resistance." *Biochimica et Biophysica Acta (BBA)—Molecular Basis of Disease* 1863 (5): 1106–1114. https://doi.org/10.1016/j.bbadis.2016.07.019.

Reyes-Caudillo, E., Alberto Tecante, and María Angeles Valdivia-Lopez. 2008. "Dietary Fibre Content and Antioxidant Activity of Phenolic Compounds Present in Mexican Chia (*Salvia hispanica* L.) Seeds." *Food Chemistry* 107 (2): 656–663. https://doi.org/10.1016/j.foodchem.2007.08.062.

Rydlo, Tali, Joseph Miltz, and Amram Mor. 2006. "Eukaryotic Antimicrobial Peptides: Promises and Premises in Food Safety." *Journal of Food Science* 71: 125–135. https://doi.org/10.1111/j.1750-841.2006.00175.x.

Salazar-Vega, Irene, Patricia Quintana Owen, and Maira Segura-Campos. 2020. "Physicochemical, Thermal, Mechanical, Optical and Barrier Characterization of Chia (*Salvia hispanica* L.) Mucilage-Protein Concentrate Biodegradable Films." *Journal of Food Science* 85 (4): 892–902. http://doi.org/10.1111/1750-3841.14962.

Sanchez, Del Angel Sandra, Ernesto Moreno-Martínez, and María Angeles Valdivia-López. 2003. "Study of Denaturation of Corn Proteins During Storage Using Differential Scanning Calorimetry." *Food Chemistry* 83 (4): 531–540. https://doi.org/10.1016/S0308-8146(03)00149-3.

Sandoval-Oliveros, María, and Octavio Paredes-López. 2013. "Isolation and Characterization of Proteins from Chia Seeds (*Salvia hispanica* L.)." *Journal of Agricultural and Food Chemistry* 61 (1): 193–201. http://doi.org/10.1021/jf3034978.

San Pablo-Osorio, Brenda, Luis Mojica, and Esmeralda Urias-Silvas. 2019. "Chia Seed (*Salvia hispanica* L.) Pepsin Hydrolysates Inhibit Angiotensin-Converting Enzyme by Interacting with Its Catalytic Site." *Journal of Food Science* 84 (5): 1170–1179. http://doi.org/10.1111/1750-3841.14503.

Schmidt, Stacy, and Matthew Hickey. 2009. "Regulation of Insulin Action by Diet and Exercise." *Journal of. Equine Veterinary Science* 29: 274–284. https://doi.org/10.1016/j.jevs.2009.04.185.

Segura-Campos, Maira Rubi, Fanny Peralta González, Luis Chel Guerrero, and David Betancur Ancona. 2013a. "Angiotensin I-Converting Enzyme Inhibitory Peptides of Chia (*Salvia hispanica*) Produced by Enzymatic Hydrolysis." *International Journal of Food Science*: 1–8. http://doi.org/10.1155/2013/158482.

Segura-Campos, Maira Rubi, Ine M. Salazar-Vega, Luis A. Chel-Guerrero, and David A. Betancur-Ancona. 2013b. "Biological Potential of Chia (*Salvia hispanica* L.) Protein Hydrolysates and Their Incorporation into Functional Foods." *LWT-Food Science and Technology* 50: 723–731. http://doi.org/10.1016/j.lwt.2012.07.017.

Silveira, Coelho Michele, Rosana Aparecida Manólio Soares-Freitas, José Alfredo Gomes Areas, Eliezer Avila Gandra, and Myriam de las Mercedes Salas-Mellado. 2018. "Peptides from Chia Present Antibacterial Activity and Inhibit Cholesterol Synthesis." *Plant Foods for Human Nutrition* 73 (2): 101–107. https://doi.org/10.1007/s11130-018-0668-z.

Silveira, Coelho Michele, Sabrine Araujo Aquino, Juliana Machado Latorres, and Myriam de las Mercedes Salas-Mellado. 2019. "*In Vitro* and *in Vivo* Antioxidant Capacity of Chia Protein Hydrolysates and Peptides." *Food Hydrocolloids* 91: 19–25. https://doi.org/10.1016/j.foodhyd.2019.01.018.

Skov, Annebeth, Soren Toubro, Birgitte Ronn, Lotte Holm, and Arne Astrup. 1999. "Randomized Trial on Protein vs Carbohydrate in *Ad Libitum* Fat Reduced Diet for the Treatment of Obesity." *International Journal of Obesity and Related Metabolic Disorders* 23 (5): 528–536. https://doi.org/10.1038/sj.ijo.0800867.

Sortino, María Angela, Tiziana Sinagra, and Pier Luigi Canonico. 2013. "Linagliptin: A Thorough Characterization Beyond Its Clinical Efficacy." *Frontiers in Endocrinology* 4: 1–9. https://doi.org/10.3389/fendo.2013.00016.

Sosa-Crespo, Irving, Hugo Laviada Molina, Luis Chel-Guerrero, Rolffy Ortiz-Andrade, and David Betancur-Ancona. 2018. "Efecto Inhibitorio de Fracciones Peptídicas Derivadas de la Hidrólisis de Semillas de Chía (*Salvia hispanica*) sobre las Enzimas α-amilasa y α-Glucosidasa." *Nutrición Hospitalaria* 35 (4): 928–935. http://doi.org/10.20960/nh.1713.

Timilsena, Yakindra Prasad, Jitraporn Vongsvivut, Raju Adhikari, and Benu Adhikari. 2017. "Physicochemical and Thermal Characteristics of Australian Chia Seed Oil." *Food Chemistry* 228: 394–402. https://doi.org/10.1016/j.foodchem.2017.02.021.

Torres-León, Cristian, Nathiely Ramírez-Guzmán, Liliana Londoño-Hernández, Gloria Martinez-Medina, Rene Díaz-Herrera, Víctor Navarro-Macias, Olga Alvarez-Perez, Brian Picazo, María Villarreal-Vázquez, Juan Ascacio-Valdes, and Cristóbal Aguilar. 2018. "Food Waste and Byproducts: An Opportunity to Minimize Malnutrition and Hunger in Developing Countries." *Frontiers in Sustainable Food Systems* 2 (52): 1–17. https://doi.org/10.3389/fsufs.2018.00052.

Udenigwe, Chibuike and Rotimi Aluko. 2012. "Food Protein-Derived Bioactive Peptides: Production, Processing, and Potential Health Benefits." *Journal of Food Science* 77 (1): R11–R24. https://doi.org/10.1111/j.1750-3841.2011.02455.x.

Ullah, Rahman, Muhammad Nadeem, Abdul Khalique, Muhammad Imran, Shahid Mehmood, Arshad Javid, and Hafiz Hassain. 2015. "Nutritional and Therapeutic Perspectives of Chia (Salvia hispanica L.): A Review." *Journal of Food Science and Technology* 53: 1750–1758. https://doi.org/10.1007/s13197-015-1967-0.

Urbizo-Reyes, Uriel, Fernanda San Martín-González, José García-Bravo, Vigil Aurelio López Malo, and Andrea Liceaga. 2019. "Physicochemical Characteristics of Chia Seed (*Salvia hispanica*) Protein Hydrolysates Produced Using Ultrasonication Followed by Microwave-Assisted Hydrolysis." *Food Hydrocolloids* 97: 105187. https://doi.org/10.1016/j.foodhyd.2019.105187.

USDA. 2011. "National Nutrient Database for Standard Reference, Release 24." Nutrient Data Laboratory Home Page, Ed. U.S. Department of Agriculture, Agricultural Research Service. https://data.nal.usda.gov/dataset/usda-national-nutrient-database-standard-reference-legacy-release.

Valicenti Tania, Juliana Cotabarren, Jorgelina A. Rodríguez Gastón, Walter D. Obregón, Adriana M. Rosso, and Mónica G. Parisi. 2019. "Evaluación Preliminar de la Actividad Antitrombótica de Hidrolizados de Expeller de Chía." Proceedings XXI CYTAL-ALACCTA. ISBN 978-987-22165-9-7.

Velnar, Tomaz, Tracey Bailey, and Vladimir Smrkolj. 2009. "The Wound Healing Process: An Overview of the Cellular and Molecular Mechanisms." *Journal of International Medical Research* 37 (5): 1528–1542. https://doi.org/10.1177/147323000903700531.

Yedida, Hima, Venkata S.P. Bitra, Sunita Venkata Seshamamba Burla, Veeranianevulu Gudala, Sharmila Kondeti, Rajendra Kumar Vuppula, and Samuel Jaddu. 2020. "Hydration Behavior of Chia Seed and Spray Drying of Chia Mucilage." *Journal of Food Processing and Preservation* 44 (6): e14456. https://doi.org/10.1111/jfpp.14456.

Zhang, Jun-Ming, and Jianxiong An. 2007. "Cytokines, Inflammation and Pain." *International Anesthesiology Clinics* 45 (2): 27–37. https://doi.org/10.1097/AIA.0b013e318034194e.

Chapter 13

Effects of Phytochemicals in Native Berries on the Reduction of Risk Factors of Age-Related Diseases

Mariane Lutz and Marcelo Arancibia

CONTENTS

13.1 INTRODUCTION

Native berries that grow in the southern area of Chile, including the Patagonia, are recognized as foods that exert beneficial effects on health (Schreckinger et al. 2010), mainly due to their high content of antioxidant compounds (Seeram 2008; Gironés-Vilaplana et al. 2014; Ramírez et al. 2015; Rothwell et al. 2016). Chile is a very long and narrow land in which the wild edible berry species are found mainly in the geographical macrozone ranging from 36°S to 56°S, which comprise evergreen forests with an oceanic climate that gets colder toward the southern territory. The majority of Chilean berry fruits (mostly shrub

DOI: 10.1201/9781003087618-13

TABLE 13.1 MAIN BERRY FRUITS FOUND IN CHILE

Common name	Botanical name	Family
Maqui	*Aristotelia chilensis* (Mol.) Stuntz	Elaeocarpaceae
Calafate	*Berberis microphylla* G. Forst	Berberidaceae
Murtilla or murta	*Ugni molinae* Turcz	Myrtaceae
Chaura	*Gaultheria pumilla*	Ericaceae
Chilean strawberry	*Fragaria chiloensis* L. Duch.	Rosaceae
	fma. *Patagonica* Staudt (red)	
	fma. *Chiloensis* Staudt (white)	

forest species) belong to the Rosaceae, Ericaceae, Myrtaceae, Berberidaceae, Elaeocarpaceae, and Grossulariaceae families (Table 13.1), many of which have been traditionally used by the locals as food and herbal medicines.

Berries are dietary sources of bioactive molecules such as phenolic compounds (PC), carotenoids, and dietary fiber, which exert a variety of beneficial effects. These fruits have been described as the richest dietary source of PC (Nowak et al. 2016). In this chapter, the study of the physiological effects of the intake of Chilean native berries evaluated using *in vitro* assays and *in vivo* animal model studies is described. Since the scientific substantiation of the beneficial effects of the intake of bioactive compounds requires dietary intervention studies (Lutz 2019), the results of some randomized clinical trials are also described. The chapter focuses on the PC that are abundant in Chilean native berries that exert major antioxidant and anti-inflammatory effects, and the possible mechanisms of action that may contribute to the reduction of risk factors for age related diseases such as cardiovascular diseases (CVD), diabetes mellitus, and cancer, as well as their potential effects on mental health.

13.1.1 Characteristics of Chilean Berry Fruits

Although many of the small, highly colored fruits considered in this chapter are not real berries, the term "berry fruits" is widely used both in the scientific literature and in the commercial trade of these goods, recognized for their high content of PC (Seeram 2008). Among these, anthocyanins are the most prevailing molecules, including cyanidin, delphinidin, pelargonidin, peonidin, malvidin, and petunidin (Castañeda-Ovando et al. 2009; López et al. 2020).

Maqui: *Aristotelia chilensis* (Mol.) Stuntz is an edible black-colored berry produced by a shrub that thrives in the temperate forests of central to southern Chile and western Argentina (Rodríguez et al. 1983; Figure 13.1a). It plays important roles in the *Mapuche* culture, symbolizing good intentions. The leaves and fruits have been used in folk medicine to treat a variety of ailments including sore throat, kidney pains, ulcers, fever, inflammation, and diarrhea (Hoffmann 1991; Molares and Ladio 2009; Schmeda-Hirschmann et al. 2019). Maqui has a very high content of anthocyanins, responsible for the intense dark color and the high antioxidant (AOX) capacity of this fruit, one of the highest among berries (Miranda-Rottmann et al. 2002; Rothwell et al. 2016; Li et al. 2017a). PC in maqui are highly dependent on the genotypes, and the anthocyanins increase with the maturation of the fruit, along with the change to darker colors (Fredes et al. 2014). One of the main medical uses of maqui is as an anti-inflammatory agent for kidney associated pain, gastric ulcers, fever, and wound healing (Céspedes et al. 2010b, 2017). Numerous potential beneficial effects of the intake of maqui berries and leaves have been described, although not all of them are based on scientific evidence (Romanucci et al. 2016). Recognized as a rich natural source of delphinidins, an extract has been developed: Delphinol, standardized to 25% delphinidin, which is commercially available (Watson and Schönlau 2015). The development of food products containing maqui or its constituents represent an interesting potential of beneficial effects on health (Peña Araos 2015), and currently some food products such as the dehydrated berries and juices are available on the international market.

Calafate: *Berberis buxifolia* is an evergreen bush that grows in the Patagonia zone, both in Chile and Argentina (Figure 13.1b). It produces dark purple-black or bluish berries. The roots have been used to control fever and inflammation, abdominal pain, indigestion, and the fruits are used to obtain their natural pigments (Montenegro 2002; Freile et al. 2003). It is widely used in jams, juices, and alcoholic beverages. The fruit contains a great variety of anthocyanidins, including 3-O-monoglycosylated and 3,5-O-diglycosylated delphinidins, cyanidins, petunidins, peonidins, and malvidins. The concentrations of anthocyanins (mean of 17.81 μmol/g f.w.) and hydroxycinnamic acid derivatives (2.62 μmol/g f.w.) in calafate are very high (Ruiz et al. 2010, 2014; Ramírez et al. 2015; Fredes et al. 2020; Sánchez Gutiérrez and Guzmán-Pincheira 2020). The content of the alkaloid berberine does not represent a risk, and the levels found in seeds reaches nearly 0.001% (Freile et al. 2003). Moreover, calafate berries have been recorded as the highest source of phenolic antioxidants (Speisky et al. 2012; Mariangel et al. 2013; Brito et al. 2014). In a pharmacokinetic study using gerbils, Bustamante et al. (2018) measured a series of phenolic catabolites in plasma, predominating phenylacetic acid derivatives, with the main increase 2 hours post-intake, while 3-hydroxyphenylacetic and phenylacetic acids increased at 4–8 hours post-intake. The levels of the catabolites found in gerbil plasma were between 0.1 and 1 μM, and no original berry anthocyanins were detected.

Murtilla, murta, or **uñi** (*Ugni molinae* Turcz) is a wild perennial shrub growing in the south of Chile, mainly in the coastal and the Andean mountains (Landrum 1988). Different plant ecotypes develop a variety of fruit colors including soft green, yellow, fuchsia (purplish), and light and dark red (Seguel et al. 2000). It is known as *murta* in the zones of Valdivia and Chiloé, and *murtilla* toward the northern zone (Biobío), while the *Mapuche* call it *uñi* (Urban 1934). (Figure 13.1c). *Mapuche* people used a "sweet wine" based on murta as an appetite stimulant (Fredes et al. 2020), leaf infusions have been used to treat urinary and throat infections, and the fruits are used for their astringent properties (Muñoz et al. 2001; Montenegro 2002). Currently the fruit is mainly used in food preparations such as jams, marmalades, and jellies, in which the berry pectins act as a texturizing agent. The fruity, sweet, and floral aroma of murta is due to a variety of volatile compounds including ethyl hexanoate and 4-meth oxy-2,5-dimethyl-furan-3-one (Scheuermann et al. 2008).

Chaura or **Mutilla** (*Gaultheria pumila*) is a native Chilean low bush that produces fleshy, flavored, and aromatic fruits. It inhabits the Andes Mountains, mainly in the southern area of the country. The berries exhibit morphometric differences depending on color (white, red, purple, and pink), shape (pepper, full cone, and round), and size (Rubilar et al. 2006) and in their phytochemical composition (Villagra et al. 2014; Figure 13.1d). The berry fruit was used by the people from West Fuego–Patagonia natives in ceremonies, as food, and for medicinal purposes as an analgesic and digestive (Domínguez 2010; Teillier and Escobar 2013).

Chilean strawberry (*Fragaria chiloensis* L. Duch. ssp. *chiloensis* fma. *patagonica* Staudt and fma. *chiloensis* Staudt) are perennial herbaceous plants that grow in the southern area of Chile. There are two main varieties: *patagonica* (red fruits) and *chiloensis* (white fruits; Figure 13.1e). The thalamus constitutes 98% of the fresh fruit, while the achenes reach 1.1%. The PC in these parts reach 59.4% and 40.6%, respectively. The achenes concentrate 85% of the total anthocyanins in the fruit (Cheel et al. 2007). The fruit contains a variety of PC (Morales-Quintana et al. 2019).

Figure 13.1 (a) Maqui: *Aristotelia chilensis* (Mol.) Stuntz. (b) Calafate: *Berberis microphylla* G. Forst. (c) Murtilla: *Ugni molinae* Turcz. (d) Chaura: *Gaultheria pumilla*. (e) Chilean strawberry: *Fragaria chiloensis* L. Duch.

No traditional medicinal uses have been described, and the fruits are commonly used as fresh or preserved foods. Retamales et al. (2005) have proposed a strategy to preserve this native fruit, since it has been increasingly displaced by commercial varieties such as *Fragaria ananassa*, which is bigger and redder, however it does not present the flavor and aroma of the native white fruit.

13.1.2 Chemical Composition of Berry Fruits

López et al. (2020) described the nutritional composition of the Chilean native berries. In general, berry fruits are rich in PC, fiber, and other bioactive compounds (Escribano-Bailón et al. 2006; Ruiz et al. 2010; USDA 2011; Nile and Park 2014). PC are secondary metabolites that participate in the defense of the plant against environmental stressors such as microorganisms, fungi, insects, and UV radiation; they also chelate toxic heavy metals and exert AOX protection from reactive oxygen species (ROS) generated during the photosynthetic process (Gould and Lister 2006). The main PC are flavonoids (flavonols, flavanols), phenolic acids, anthocyanins and proanthocyanidins, and condensed and hydrolysable tannins, among others (Seeram 2008), and their content in berry fruits depends on a series of botanical and environmental factors (Wang and Lin 2000). Most PC are found as glycosylated forms or as esters or polymers.

The main PC found in native Chilean berries are anthocyanins, water-soluble flavonoids (Clifford 2000) responsible for the pink, red, blue, and purple color, depending on pH (Lohachoompol et al. 2014). The degree of ripeness of the fruits contributes to variation in total PC content, and in some fruits, anthocyanins accumulate during maturation (Arena and Curvetto 2008). Anthocyanins vary in the number and position of hydroxyl and methoxyl groups attached to the core anthocyanidin structure. Although there are fewer than 20 naturally occurring anthocyanidins, there are many hundreds of different anthocyanins, including cyanidin, delphinidin, petunidin, peonidin, pelargonidin, and malvidin (Prior and Wu 2006; Yoshida et al. 2009), and their amount varies across species and depends on environmental factors (e.g., solar radiation, temperature, soil composition; Alfaro et al. 2013). The aglycon (anthocyanin) may be linked to one or more glycosidic units: glucose and galactose (hexoses), and rhamnose, arabinose, and xylose (pentoses), attached at various positions (Clifford 2000; Yoshida et al. 2009). The amount of anthocyanins is especially high in maqui and calafate berries, along with a variety of PC identified in this berry (Escribano-Bailón et al. 2006; Céspedes et al. 2010b, 2017; Ruiz et al. 2010; 2013; Tanaka et al. 2013; Mariangel et al. 2013; Fredes et al. 2014; Genskowsky et al. 2016; Brauch et al. 2016; Ulloa-Inostroza et al. 2017; Fredes et al. 2020). In fact, Escribano-Bailón et al. (2006) described the presence of 73% of delphinidin derivatives in maqui, prevailing over the cyanidin derivatives (37%). The main species described was delphinidin-3-sambubioside-5-glucoside (34% total anthocyanins). The mean content of anthocyanins was 137.6 delphinidin-3-glucoside equivalents/100 g f.w. In calafate, the mean amount of anthocyanins was 761.30 mg cyanidin-3-glucoside/100 g f.w. (Arena and Curvetto 2008); while Chilean strawberries and chaura do not exhibit such high amounts of anthocyanins, they contain a varied profile of PC species (Simirgiotis et al. 2009; Giampieri et al. 2012; Ulloa-Inostroza et al. 2017; Oyarzún et al. 2020). A variety of anthocyanin pigments have been described in strawberries of different varieties and selections, and pelargonidin-3-glucoside has been identified as the major molecule, independent of genetic and environmental factors (Lopes da Silva et al. 2007).

Other important flavonoids in native berries are elagitannins and flavonol derivatives, such as quercetin-3-O-glucoside and kaempferol, with quercetin derivatives being the most abundant in strawberries (Cheel et al. 2007; Giampieri et al. 2012), flavanols (epicathequin) and flavonols (myricetin, quercetin), and hydroxycinnamic acid derivatives in maqui (Rubilar et al. 2006; Peña-Neira et al. 2007; Avello 2009; Ruiz et al. 2010). The main flavonol glucoside in Chilean strawberries is quercetin-3-glucuronide, and anthocyanins include cyanidin-malonyl-glucoside and pelargonidin-malonyl-glucoside, with a total amount of anthocyanins of 2.20 mg/100 g (Fredes et al. 2009). Villagra et al. (2014) described a wide variation in the anthocyanins content in chaura (*Gaultheria pumila*), in association with color: the content of monomeric anthocyanins (mg/100 g) was 5942 ± 422 in red fruits, 3854 ± 192 in pink fruits, and 626.2 ± 41 in white fruits. To date, the phytochemical profile of each of the berries described in this chapter is still not complete. A description of the PC that have been identified so far in the Chilean berry fruits is shown in Table 13.2.

Although various PC have been identified in these native berry extracts, their quantification has more scarcely been described. Moreover, data on the content of these compounds in berry fruits are hard to compare due to the different methodologies used to obtain the extracts, the analytical procedures, and the dissimilar expression of the results (units). Table 13.3 shows the amounts of various anthocyanins in maqui (*Aristotelia chilensis* (Mol.) Stuntz), the most extensively native berry studied, which are considered as responsible to a great extent for the beneficial effects of maqui intake on health.

13.1.3 Antioxidant Capacity

Chilean native berries exhibit an elevated total content of PC that confers on them a high AOX capacity (Ruiz et al. 2010; Speisky et al. 2012; Brito et al. 2014; Fredes et al. 2014b; Chamorro et al. 2019). This property is usually assessed through the classical *in vitro* chemical methods DPPH, ORAC, TEAC, ABTS, and FRAP, among others based on the use of an arbitrarily selected oxidant

TABLE 13.2 MAIN PHENOLIC COMPOUNDS FOUND IN CHILEAN BERRIES

Berry	Phenolic and hydroxy-cinnamic acids	Flavonoids	Anthocyanidins	Tannins	References
Maqui	Gallic, protocatechuic, p-coumaric	Myricetin, myricetin-3-glc, quercetin, quercetin-3-glc, 3,5-dimethoxy-quercetin, quercetin-3-gal, quercetin-galloil-hexoside, rutin, catechin	Delphinidin-3-glc, delphinidin-3,5-diglc, delphinidin-3-sam, delphinidin-3-sam-5-glc, cyanidin-3-glc, cyanidin-3-sam, cyanidin-3,5-diglc, cyaninin-3-sam-5-glc	Ellagic acid derivatives	Escribano-Bailón et al. (2006); Schreckinger et al. (2010); Céspedes et al. (2010a); Tanaka et al. (2013); Gironés-Vilaplana et al. (2014); Fredes et al. (2014a); Genskowsky et al. (2016); Brauch et al. (2016); Ulloa-Inostroza et al. (2017); Li et al. (2017a)

(Continued)

TABLE 13.2 (CONTINUED)

Berry	Phenolic and hydroxy-cinnamic acids	Flavonoids	Anthocyanidins	Tannins	References
Calafate	Gallic, ferulic, chlorogenic, neochlorogenic, caffeic, coumaric, feruloyl-derivate, cafeoylquinic	Rutin, myricetin, myricetin-3-glc, myricetin-3-rut-glc, myricetin-3-rut, myricetin-3-rham, myricetin-3-gal, quercetin, isoquercitrin, isorhamnetin, kaempferol, quercetin-3-rut, quercetin-3-glc, quercetin-3-gal, quercetin-hexoside derivate, quercetin-3-rham, isorham-3-glc, isorham-3-rut, isorham-3-rut-7-glc, isorham-3-gal, isorham-3-malonylgal, isorham-3-malonylglc, hyperoside	Malvidin-3-glc, malvidin-3-gal, malvidin-3-ara, malvidin-3-rut, malvidin-3-rut-5-glc, petunidin-3-glc, petunidin-3-gal, petunidin-3-rut, petunidin-3-rut-5-glc, petunidin-3,5-dihexoside, peonidin-3-glc, peonidin-3-rut, peonidin-3,5-dihexoside, delphinidin-3-glc, delphinidin-3-ara, delphinidin-3-gal, delphinidin-3-rut-5-glc, pelargonidin-3-glc, cyanidin-3-glc, cyanidin-3-rut, cyanidin-3-gal, cyanidin-3-ara, cyanidin-3,5-dihexoside	Ellagic acid derivatives	Ruiz et al. (2010, 2014); Mariangel et al. (2013); Ramírez et al. (2015); Ulloa-Inostroza et al. (2017); Fredes et al. (2020); Sánchez-Gutiérrez and Guzmán-Pincheira (2020)

Murtilla	Gallic, chlorogenic, neochlorogenic, protocatechuic, p-coumaric, feruloyl-quinic	Kaempferol, catechin, isoquercitrin, myricetin, myricetin-glc, quercetin, quercetin-glc, quercetin-dirham, rutin, vanillic-4-hydroxy-3-mehoxybenzoic acid, hyperoside	Delphinidin-3-glc, cyanidin-3-glc, peonidin-3-glc, peonidin-3-ara, peonidin-3-dihexoside, petunidin-3-glc, petunidin-3-gal, cyanidin-3-rut	Ellagic acid derivatives	Rubilar et al. (2006); Peña-Neira et al. (2007); Avello (2009); Schreckinger et al. (2010); Junqueira-Gonçalves et al. (2015); Ramírez et al. (2015); Ulloa-Inostroza et al. (2017); Fredes et al. (2020)
Chaura	Gallic, protocatechuic, ρ-coumaric, cinnamic, caffeic	Kaempferol, kaempferol-3-glc, myricetin, luteolin, luteolin-3-glc, quercetin, quercetin 3-glc, quercetin-3-rut, myricetin, catechin	Delphinidin-3-glc, Malvidin-3-glc, Peonidin-3-ara, Peonidin-3-glc	Ellagic acid derivatives	Liu et al. (2013); Villagra et al. (2014); Oyarzún et al. (2020)
Chilean strawberry	Ferulic, 4-Coumaric	Cinnamoyl-rham-pyranoside, cinnamoyl-xylofuranosyl-(1–6)-glc-pyranose, cinnamoyl-xylopyranoside, kaempferol, quercetin, quercetin-3-rut, kaempferol-7-glc, catechin, daidzein, genistein	Procyanidin B, pelargonidin-3-glc, pelargonidin-3-rut, cyanidin, cyanidin-3-glc, cyanidin-malonyl-glc, cyanidin-3-glc-pyranoside, pelargonidin-malonyl-glc	Ellagic acid derivatives	Lopes da Silva et al. (2007); Cheel et al. (2007); Fredes et al. (2014b); Simirgiotis et al. (2009); Schreckinger et al. (2010); Giampieri et al. (2012); Ulloa-Inostroza et al. (2017); López et al. (2018)

Abbreviations: glc, glucoside; rut, rutinoside; gal, galactoside; ara, arabinoside; rham, rhamnoside; sam, sambubioside.

TABLE 13.3 ANTHOCYANINS CONTENT IN MAQUI (*ARISTOTELIA CHILENSIS* (MOL.) STUNTZ)

Anthocyanin	mg/100 g f.w.	PAE (%)	ANC (%)	mg	mg/100 g d.b.	mg/ 100 g d.b.	mg/ 100 g d.b.	g/kg d.b.	% ext	g/kg d.b. juice	g/kg d.b. dried
Delphinidin-3-glc	17.1 ± 0.2	17.5	22.3	32.53	210.90 ± 1.84	110.69 ± 3.41	167.55 ± 12.58	9.48 ± 0.25	1.6	2.6 ± 0.0	3.8 ± 0.3
Delphinidin-3,5-diglc	23.7 ± 0.2	7.5	11.0	49.80	240.35 ± 8.45	251.45 ± 19.3	114.38 ± 1.34	7.23 ± 0.04	14.3	2.5 ± 0.0	15.9 ± 1.4
Delphinidin-3-sam	–	9.3	12.9	30.51	63.22 ± 0.37	18.63 ± 1.87	72.45 ± 6.22	7.06 ± 0.15	6.1	3.1 ± 0.0	1.2 ± 0.1
Delphinidin-3-sam-5-glc	46.4 ± 0.1	8.4	11.9	101.05	150.25 ± 11.44	125.21 ± 24.65	354.51 ± 27.49	4.36 ± 0.01	7.1	4.3 ± 0.1	8.1 ± 0.9
Cyanidin-3-glc	8.6 ± 0.05	9.3	12.5	17.20	–	–	–	1.24 ± 0.02	0.9	–	–
Cyanidin-3-sam	8.9 ± 0.04	0.3	0.4	17.37	82.21 ± 0.48	23.44 ± 2.71	56.29 ± 4.61	0.73 ± 0.11	0.6	5.8 ± 0.5	1.9 ± 0.1
Cyanidin-3,5-diglc	–	–	–	18.71	134.65 ± 3.28	77.07 ± 3.60	11.47 ± 0.90	5.36 ± 0.05	2.1	–	–
Cyaninidin-3-sam-5-glc	18.7 ± 0.02	6.1	8.7	20.73	–	–	–	6.89 ± 0.06	2.7	7.8 ± 0.6	6.0 ± 0.5
Cyanidin-3 glc-5-rham	–	–	–	–	2.54 ± 1.87	7.59 ± 1.87	1.96 ± 0.75	–	–	–	–
TOTAL	137.6 ± 0.4	58.4	79.7	289.9	984.12 ± 7.32	614.08 ± 47.44	881.84 ± 46.07	42.35 ± 0.08	–	26.1 ± 1.1	36.9 ± 3.2
References	Escribano-Bailón et al. (2006)	Schreckinger et al. (2010)		Céspedes et al. (2010a)	Gironés-Vilaplana et al. (2014)			Genskowsky et al. (2016)	Tanaka et al. (2013)	Brauch et al. (2016)	

Abbreviations: glc, glucoside; rut, rutinoside; gal, galactoside; ara, arabinoside; rham, rhamnoside; sam, sambubioside; f.w., fresh weight; d.b., dry basis; Ext, extract; PAE, post amberlite extract; ANC, anthocyanin-enriched fraction.

generator that does not resemble the physiological medium, and when applied to biological samples, it is removed from its context, which is characterized by enzymatic maintenance of a steady-state (Prior et al. 2005; Sies 2007). The AOX capacity measured by the *in vitro* ORAC assay (Prior et al. 2005) in calafate is the highest described in berries, followed by maqui and murtilla (Table 13.4). The values are significantly higher from those exhibited by other berries, such as raspberry, blueberry, and blackberry species (Speisky et al. 2012), and correlate well with their total PC content. Mariangel et al. (2013) observed an association between the geographic zone and the AOX activity: using the DPPH assay, these authors report higher AOX activity in calafate grown in the southern, coldest region (9.4 TE/g d.b.) compared to the berry grown in a more template zone (3.3 TE/g d.b.). These values are associated with total PC content, which ranged from 16.1 to 34.6 mg GAE/g d.b. in the more temperate and coldest regions, respectively. A high AOX capacity of calafate was also described by Brito et al. (2014), Ruiz et al. (2010), and Brauch et al. (2016). The values are significantly higher than those of well-recognized fruits as good AOX sources (e.g., pomegranate, cranberry). Ruiz et al. (2010) reported a total PC content (μmol GAE/g f.w.) of 97 in maqui, 87 in calafate, and 32 in murtilla. The anthocyanin content was higher in calafate, followed by maqui, and then murtilla, which is consistent with the weaker color of the pink fruit. Brauch et al. (2016) described the effect of juice processing on the PC content in maqui, determined by the Folin Ciocalteu assay (Singleton and Rossi 1965): in fresh, juice, and dried maqui the PC amounts were 19.7 ± 0.9, 7.3 ± 0.2, and 32.0 ± 2.1 g/kg, respectively (Table 13.4).

The effect of thermal processing on native berries has also been described: Rodríguez et al. (2016) reported that drying maqui in a range of temperatures of 40–80 °C did not affect the AOX activity, while Quispe-Fuentes et al. (2020) observed that various methods of drying (freezing, convective, sun, infrared, vacuum) did not reduce significantly the PC content in this berry. Moreover, Ah-Hen et al. (2017) described a good retention of total PC in murta berries submitted to a chemical extraction in a steam juicer for different extraction times: fresh fruits contained 591 ± 22 mg GAE/100 mL, and the juice extract could retain around 80% of the PC and anthocyanins after 21 days of storage at 5 °C; and in microencapsulated maqui, Bastías-Montes et al. (2019) reported changes in the total PC, total anthocyanins, and DPPH antioxidant assay values by spray-drying in a range of 120–170 °C. A comparison of the total PC and anthocyanins contents and antioxidant activity of native berries is shown in Table 13.4.

Although the *in vitro* AOX tests are valuable to compare and rank berries according to their ability to reduce compounds or capture free radicals and other ROS, the results do not reflect what happens inside the body. Moreover, when acting as quenchers of free radicals, PC convert themselves into free

TABLE 13.4 ANTIOXIDANT CAPACITY, TOTAL PHENOLIC COMPOUNDS, AND TOTAL ANTHOCYANINS IN CHILEAN BERRIES

BERRY	ORAC	DPPH	TEAC	TPC	Tantho	Reference
Maqui	19850 ± 966 μmol TE/100 g	–	–	1664 ± 83 mg GAE/100 g	–	Speisky et al. (2012)
Calafate	25622 ± 3322 μmol TE/100 g	–	–	1201 ± 34 mg GAE/100 g	–	Speisky et al. (2012)
Murtilla	10770 ± 453 μmol TE/100 g	–	–	863 ± 30 mg GAE/100 g	–	Speisky et al. (2012)
Maqui	–	–	88.1 ± 21.5 μmol TE/g	97 ± 10 μmol GAE/g	17.88 ± 1.15 μmol/g	Ruiz et al. (2010)
Calafate	–	–	74.5 ± 5.9 μmol TE/g	87 ± 9 μmol GAE/g	17.81 ± 0.98 μmol/g	Ruiz et al. (2010)
Murtilla	–	–	11.7 ± 2.3 μmol TE/g	32 ± 4 μmol GAE/g	0.21 ± 1.28 μmol/g	Ruiz et al. (2010)
Calafate	–	2.33 ± 0.21 μg/mL	–	65.53 ± 1.35 μmol TE/g d.b.	51.62 ± 1.78 mg cyan-3-glc/g d.b.	Brito et al. (2014)
Murtilla	–	10.94 ± 0.32 μg/mL	–	9.24 ± 0.28 μmol TE/g d.b.	6.85 ± 0.10 mg cyan-3-glc/g d.b.	Brito et al. (2014)
Calafate_C1	–	5.2 ± 0.02 mg TE/g d.b.	–	22.2 ± 0.08 mg GAE/g d.b.	0.24 ± 0.00 mg yaniding/g d.b.	Mariangel et al. (2013)
Calafate_C2	–	9.4 ± 0.05 mg TE/g d.b.	–	34.9 ± 0.05 mg GAE/g d.b.	0.12 ± 0.00 mg yaniding/g d.b.	Mariangel et al. (2013)
Maqui fresh	–	–	73.6 ± 2.1 mmol TE/L	19.7 ± 0.9 g/kg	26.1 ± 1.1 g/kg	Brauch et al. (2016)

Maqui juice	–	–	63.7 ± 1.6 mmol TE/L	7.3 ± 0.2 g/L	11.6 ± 0.1 g/L	Brauch et al. (2016)
Maqui dried	–	–	154.6 ± 8.8mmol TE/kg	32.0 ± 2.1 g/kg	36.9 ± 3.2 g/kg	Brauch et al. (2016)
Chilean strawb.	–	–	–	106.3 mg GAE/100 g	2.3 mg cyan-3-glc/100 g	Cheel et al. (2007)
Murta juice 1	–	328 ± 49 µg TE/mL	–	769.6 ± 7.2 mg GAE/100 g	2.1 ± 0.1 mg/L	Ah-Hen et al. (2018)
Murta juice 2	–	331061 µg TE/mL	–	2573.3 ± 1.9 mg GAE/100 g	4.4 ± 0.2 mg/L	Ah-Hen et al. (2018)
Maqui Micro-encaps_1	–	93.38 ± 0.93% inhibition	–	35.66 ± 13.71 mg GAE/g d.b.	30.53 ± 0.66 mg cyan-3-glc/g d.b.	Bastías et al. (2019)
Maqui Micro-encaps_2	–	85.76 ± 2.22 % inhibition	–	23.05 ± 0.64 mg GAE/g d.b.	21.46 ± 0.56 cyan-3-glc/g d.b.	Bastías et al. (2019)

Antioxidant capacity assays: ORAC (oxygen-radical absorbing capacity), DPPH (2,2-diphenyl-1-picryl-hydrazyl hydrate), TEAC (Trolox equivalent antioxidant capacity); TPC (total phenolic compounds); Tantho (total anthocyanins). Results are expressed by fresh weight unless otherwise specified.

Abbreviations: d.b., dry basis; calafate_C1, grown in template zone; calafate_C2, grown in cold zone; Murta juice 1, fresh; Murta juice 2, after 15 minutes' extraction in steam juicer; Maqui Microencaps_1, microencapsulated powder; Maqui Microencaps_2, microencapsulated powder dried at 170 °C.

radicals, acting as pro-oxidants (Stevenson and Hurst 2007; Halliwell 2008). PC act as relevant AOX *in vivo*, inducing cellular endogenous enzymic protective mechanisms by modulating the Nrf2-ARE (antioxidant response element) signaling pathway to increase Nrf2 transcription (Kropat et al. 2013). Circulating PC upregulate endogenous AOX enzymes (catalase, superoxide dismutase, glutathione peroxidase) at physiological levels, providing a relevant mechanism of protection against the oxidative stress damage involved in the development of age-related diseases such as CVD, cancer, and neurodegeneration. In fact, a better measure of the AOX impact of dietary PC *in vivo* are assays such as the lag time to low density lipoprotein (LDL) oxidation (Heinonen et al. 1998) or the determination of specific markers of peroxidative damage to biomolecules, as in the case of F2-isoprostanes (Milne et al. 2007; Croft 2015).

13.1.4 Absorption and Bioavailability of PC

After ingestion, most PC molecules are hydrolyzed to release the aglycones that can be absorbed. Some PC can be absorbed as glycosides (Passamonti et al. 2002) and are detected as such in blood (D'Archivio et al. 2007; Kay et al. 2017). After absorption, PC are metabolized by phase I and II enzymes mainly into sulfated, glucuronidated, and methylated forms. Since anthocyanins are not extensively metabolized (Cermak et al. 2009; Krga et al. 2016), their bioavailability is lower than that of other flavonoids (Yang et al. 2010), and for the most part reach the colon and are extensively metabolized by the gut microbiota, generating metabolites that may be absorbed (Hidalgo et al. 2012). These metabolites, generated by endogenous intestinal enzymes, and those of the gut microbiota may be different from the parent compounds found in the berry ingested, and may or may not exert biological actions (Denev et al. 2012). De Ferrars et al. (2014) identified a series of anthocyanin metabolites in human plasma, urine, and feces, of which the main were small molecules such as hippuric acid, vanillic acid, vanillic acid sulphate, and 4-hydroxybenzaldehyde. In a systematic review of clinical dietary intervention studies, Chandra et al. (2019) observed that most metabolites derive from flavonoids and hydroxycinnamic acids, highlighting the fact that metabolites change significantly when 24 hours of urine is used as a biological matrix, reflecting that it is a more stable and concentrated metabolite pool than blood (Curtis et al. 2019). A different approach was used by Rothwell et al. (2016), who analyzed the polyphenol metabolome using the Phenol-Explorer database (http://phenol-explorer.eu/), reporting nearly 400 metabolites, of which ~75% were the original molecules found in foods. The large variability observed in the response to the intake of anthocyanins may be due to inter-individual variations in the absorption and metabolism of these compounds and/or a variability on their intrinsic effect to target genes or proteins (e.g., genetic variation for enzymes involved in the absorption and metabolism that produce differences in the expression of functional enzymes). The variations are also dependent on the food matrix and food processing, both affecting the content and bioaccessibility of anthocyanins, as well as the biomass and/or biodiversity of the intestinal microbiota (Eker et al. 2020). These effects were shown by Burgos-Edwards et al. (2020) in Chilean currant berries (*Ribes* spp.), although the study, as well as other with similar results, were performed using *in vitro* assays of simulated human digestion, which do not reflect the physiology in human subjects (Lutz 2019). The low bioavailability of anthocyanins has augmented the interest in the development of a series of formulations of micro-/nano-encapsulation systems (Sharif et al. 2020), although the beneficial effects of anthocyanins are probably due to the synergy effect between the parent compounds in food, the metabolites generated into the body, the conjugated products, and those products generated by the action of the intestinal microbiota (Cavalcante et al. 2018; Tena et al. 2020).

13.2 BERRIES AND RISK OF AGE-RELATED DISEASES

The progressive accumulation of oxidative damage to macromolecules and mitochondria leads to pathophysiological alterations, functional decline, and accelerated ageing (Luo et al. 2020; Vega-Gálvez et al. 2021). The roles of dietary PC on health have been a major subject of study (Shahidi and Ambigaipalan 2015). A wide number of tissue and cell culture, and animal models are used to study the effects and possible mechanisms of action of PC, establishing the adequate endpoint to be measured. They are based mainly upon the possible biochemical/molecular mechanisms involved in the protective roles of PC on the deleterious effects of oxidative stress and inflammation, including the modulation of cell signaling pathways, nuclear transcription factors, and the synthesis of inflammatory mediators, among others, that may explain how they act at very low levels.

Sirtuins (SIRT) 1–7 are histone deacetylase enzymes playing important roles in controlling gene expression, DNA repair, metabolism, oxidative stress response, mitochondrial function, and biogenesis and, if deregulated, may contribute to the development of several age-related pathologies, including cancer, diabetes mellitus, and CVD (O'Callaghan and Vassilopoulos 2017). They regulate the expression of the endogenous AOX enzymes superoxide dismutase and glutathione peroxidase and also regulate inflammatory response (Iside et al. 2020). Among the PC that act on SIRTs is the flavonoid quercetin, which is present in Chilean berries. Quercetin exerts antiproliferative, chemopreventive, and anticarcinogenic activities *in vitro* by modulating gene expression. As an AOX, quercetin increases the defense mechanisms and reduces free radicals, factors involved in pathologies such as diabetes mellitus, atherosclerosis, hypertension, neurodegenerative diseases, inflammation, and cancer (Oboh et al. 2016).

13.2.1 Inflammation

Chronic inflammation caused by oxidative damage increases the risk of many chronic disorders, including heart, CVD, and neurodegenerative diseases, obesity, insulin resistance, and diabetes mellitus (Geto et al. 2020). Many age-related diseases exhibit systemic, low-grade inflammation that compromises the metabolic functions, and plays a pivotal role in the pathogenesis of obesity. Examples of downregulation of inflammation pathways include transcription factors for gene expression of compounds such as interleukin receptors and Toll-like receptor TLR-4, NF-κB, AP-1, and JNK, or acting as a ligand for PPAR-γ (Vauzour et al. 2010; Joseph et al. 2016). For example, a widely used type of cell culture to assess anti-inflammatory effects is human PBMC (peripheral blood mononuclear cells), on which mitogens stimulate the response. Thus, the production of inflammatory markers or the breakdown of tryptophan can be used to indicate an immunomodulatory potential (Becker et al. 2013). However, it should be kept in mind that cell culture models do not represent what happens under physiological conditions if the compounds are consumed in the diet. Additionally, numerous animal models are used to evidence an anti-inflammatory effect. For instance, Quispe-Fuentes et al. (2020) used an induced mice ear edema model to study the preventive effect of maqui intake on inflammation. The authors applied fresh and dried maqui extracts to the inner and outer surfaces of the right ear and observed that fresh extract produced a moderate anti-inflammatory activity in induced mice ear skin edema associated to protein kinase C activation with the subsequent NFκB activation, while both freezing and vacuum drying processes increased the anti-inflammatory effect. The consumption of various phytochemicals, such as flavonoids, has been inversely correlated with systemic inflammatory markers in human populations (Corley et al. 2015). However, cancer is very complex and several conditions affect the development of the disease acting as risk modulators, such as oxidative stress, chronic inflammation, obesity, metabolic syndrome, angiogenesis, apoptosis, autophagy, and proliferation, among others (Kristo et al. 2016).

13.2.2 Diabetes Mellitus

The intake of a diet high in PC is associated to a lower incidence of type 2 diabetes mellitus (Costacou and Mayer-Davis 2003). The antidiabetic activity of PC has been attributed to a variety of actions, including the regulation of carbohydrate metabolism, improvement of glucose uptake, protection of pancreatic β-cells, enhancement of insulin action, and regulation of crucial signaling pathways to cell homeostasis (Dias et al. 2017). One of the mechanisms by which berries may contribute to reduce risk factors of this disease is retarding the absorption of glucose through the reduction of starch hydrolysis by inhibiting pancreatic α-amylase and intestinal glucose absorption by α-glucosidase enzymes, as has been demonstrated *in vitro* for some PC (Da Silva Pinto et al. 2010), such as proanthocyanidins, which exhibit an inhibitory effect on α-amylase and α-glycosidase (Grussu et al. 2011; Chamorro et al. 2019). Moreover, anthocyanidins such as pelargonidin may stimulate insulin secretion from rodent pancreatic β-cells, a strong action of the 3-monoglucosides of cyanidin and delphinidin (Jayaprakasam et al. 2005), actions that are added to the AOX activity that protect pancreatic α-cells from glucose-induced oxidative stress.

13.2.3 Cardiovascular Diseases

Various PC may exert beneficial effects mainly by reducing risk factors of the onset of CVD (Rangel-Huerta et al. 2015), an age-related disease that involves "inflamm-aging" since an upregulation of inflammatory responses induces senescence and inflammatory changes (Ricordi et al. 2015). Most studies on the effects of PC have been realized on *in vitro* systems, demonstrating their ability to affect endothelial damage and platelets reactivity, among other processes that are relevant in the development of CVD (Lutz et al. 2019a). Additionally, PC may exert beneficial effects on the vascular system through the induction of AOX defenses, lowering blood pressure, and reducing inflammatory responses (Scalbert et al. 2005).

The importance of diet on the maintenance of vascular function has been known for over a decade, with polyphenols playing major roles (Vita 2005; Varadharaj et al. 2017). Calfio and Huidobro-Toro (2019) assayed the vasodilator activity of hydroalcoholic or acetone extracts of calafate in an animal rat model of arterial mesenteric bed, observing that both extracts exerted endothelium-dependent vasodilator responses via a NO-dependent mechanism. The authors observed that the AOX capacity bioassay of glycosylated anthocyanins is lower than in the *in vitro* chemically based assay, suggesting that the glycosylated anthocyanins are not major contributors to the *in vivo* AOX potential. The vasorelaxation elicited by conjugated anthocyanins is endothelium-dependent, revealing that the anthocyanins may target eNOS activity to promote NO production. Besides, flavonoids are able to prevent endothelial cell apoptosis and to modulate various signaling pathways leading to inflammation (Dayoub et al. 2013). In a meta-analysis of randomized control trials with sequential analysis to estimate the effect of berries consumption on CVD risk factor, Huang et al. (2016) observed that berries consumption significantly lowered low density lipoprotein (LDL)-cholesterol, systolic blood pressure, fasting glucose, body mass index, hemoglobin A1c (HbA1c), and TNF-α. In a similar meta-analysis,

Ângelo et al. (2018) concluded that the intake of berries reduces biomarkers of risk factors of CVD such as blood total cholesterol, (LDL)-cholesterol, and triglycerides, while increasing the level of high-density lipoprotein (HDL)-cholesterol and reducing blood pressure. These clinical studies substantiate the beneficial effect of berries intake on the reduction of risk factors of CVD.

13.2.4 Cancer

Numerous *in vitro, in vivo* and human studies have suggested that diet is critical for reducing the risk factors of cancer, mainly through the consumption of certain foods (Stewart and Wild 2014). The antiproliferative activity of berries has been assessed mainly using *in vitro* models, using a number of cell culture models, including MCF-7 (breast), NCI-ADR (breast, multidrug-resistant phenotype), UACC-62 (melanoma), and NCI 460 (lung, non-small cell). Wang et al. (2007) report that strawberry fruit extracts (including *F. chiloensis*) inhibit the growth of human lung epithelial cancer A549 cells to a different degree, according to their genotypes. The effect has been attributed mainly to the potential anticancer effects of flavonols and anthocyanins (Hou et al. 2003), taking into account that the effect may be the result of the synergistic combination of PC. The consumption of berry fruits may exert numerous effects on the reduction of risk factors for cancer, including the AOX effect of PC, reduction in DNA oxidative damage, and improvement of the DNA repair capacity (Meyers et al. 2003; Katsube et al. 2003; Olsson et al. 2004), as well as antigenotoxic, antimutagenic, and anti-invasive activities through different mechanisms (Forbes-Hernandez et al. 2016). Various berry fruits extracts inhibit the growth and stimulate apoptosis of tumor cell lines (e.g., black raspberry and strawberry extracts), and exhibit strong pro-apoptotic effects against HT-29 cell line (Seeram et al. 2006). The reduction of the carcinogenic process has been attributed to various mechanisms, including the modulation of different cellular processes, such as arrest of cell cycle by increasing levels of cyclin-dependent kinase inhibitor proteins and inhibition of cyclins, induction of apoptosis via cytochrome c release, activation of caspases and regulation of Bcl-2 family members, inhibition of signaling pathways (Akt, MAPK, NF-κB, Wnt/β-catenin) and inflammation (COX-2, TNF, IL secretion), as well as suppression of key proteins involved in angiogenesis and metastasis and other crucial cellular processes (Kristo et al. 2016; Li et al. 2017b).

13.2.5 Mental Health

Mental health considers complex theoretical concepts, such as the symptom and the mental disorder, which involve a multicausality. However, in recent years research has conferred an important role on nutrition. The generation of ROS induced by oxidative stress plays major roles in the development and pathogenesis of various neurodegenerative diseases, including Alzheimer's disease and Parkinson's disease (Ullah et al. 2019). As mentioned above, berries are the richest source of PC, especially anthocyanins, which can interact with brain function in various ways: as AOX, anti-neuroinflammation, and anti-apoptotic agents (Salehi et al. 2020). These PC also act by modifying molecular transport at the blood-brain barrier level, improving cerebral blood flow, and modulating neuron and glial cell function (Schaffer and Halliwell 2012; Godos et al. 2020), through action on specific receptors, neuronal signaling pathways and the promotion of proteins involved in neuronal plasticity (Rendeiro et al. 2015). Besides, anthocyanins modulate neuronal cell death signaling pathways, mitochondrial function, inhibit protein aggregation, and potentiate autophagy, while they also prevent excitotoxicity-induced neuronal cell death by maintaining calcium homeostasis (Henriques et al. 2020).

The most direct effects of PC on brain metabolism could be mediated by low-molecular-weight metabolites capable of diffusing through the blood-brain barrier (Figueira et al. 2017; Carecho et al. 2020). A systematic review conducted by Głąbska et al. (2020) found that the consumption of fruits and vegetables, of which berries were considered a subgroup, improved levels of optimism, self-efficacy, psychological distress and depressive symptoms, but these results correspond to aggregate data from highly heterogeneous studies. Nonetheless, there is preclinical and clinical evidence of the positive effect of berries on cognitive functioning. The MIND study (Mediterranean-DASH Diet Intervention for Neurodegenerative Delay) analyzed the effects of the Mediterranean diet, which considers foods rich in polyphenols, including berries, on five cognitive domains (episodic memory, working memory, semantic memory, visuospatial ability, perceptual speed). Using a mixed linear model with random effects adjusted according to clinical and biodemographic variables, it was shown that diet substantially decreased the cognitive decline associated with aging, verifying that diet would also be associated with a reduction in the incidence of Alzheimer's disease (Morris et al. 2015). Similar effects on cognitive function were reported by Philip et al. (2019), who in a crossover clinical trial found that supplementation with an extract rich in polyphenols, obtained from blueberries and grapes, significantly improved cognitive performance, particularly in working memory and attention. In another randomized clinical trial, Whyte et al. (2019) studied the cognitive effect of ingesting 400 mL of a berry smoothie containing blueberry, strawberry, raspberry, and blackberry. This intervention was associated with sustained accuracy in responses to cognitive tests up to 6 hours after ingestion and a shorter latency time in responses to cognitive assessment. The authors emphasized the role that flavonoid-rich berries have in improving cognitive functioning.

Current evidence has pointed to the effects of berries on cognitive outcomes. However, due to the implication of these functions in the processes that regulate emotions and behavior, it would be expected to find effects on other spheres of mental health, such as affective disorders. Likewise, it could be pointed out that, due to the functions of PC on neurotransmission, there are direct effects on domains other than cognitive.

13.3 CONCLUSION

It is well established that berries affect a variety of risk factors for the most frequent age-related diseases, mainly associated with oxidative stress and inflammation. The highest level of evidence to substantiate the effects of a food constituent into the human body is obtained through randomized clinical dietary intervention trials, often analyzed through systematic reviews and meta-analysis (Lutz et al. 2019b). These studies are just above the epidemiological evidence and may prove a causality for the effects of the intake of berries on surrogate biomarkers such as blood lipids, blood pressure, endothelial function, and arterial stiffness related to aging.

As described in this chapter, the chemistry of each of the native Chilean berries is still under analysis, while there are myriad preclinical tests demonstrating potential effects on the development the non-communicable diseases that are prevalent in aging, while randomized clinical dietary intervention studies are currently increasing to satisfy the need to substantiate the promising healthy properties. The evidence available so far has led to the common recognition of these native berries as "superfruits."

REFERENCES

Ah-Hen, Kong S., Karen Mathias-Rettig, Luis Salvador Gómez-Pérez, Gabriela Riquelme-Asenjo, Roberto Lemus-Mondaca, and Ociel Muñoz-Fariña. 2018. "Bioaccessibility of Bioactive Compounds and Antioxidant Activity in Murta (*Ugni molinae* T.) Berries Juices." *Food Measure* 12: 602–615. https://doi.org/10.1007/s11694-017-9673-4.

Alfaro, Susana, Ana Mutis, Rubén Palma, Andrés Quiroz, Ivette Seguel, and Erick Scheuermann. 2013. "Influence of Genotype and Harvest Year on Polyphenol Content and Antioxidant Activity in Murtilla (*Ugni molinae* Turcz) Fruit." *Journal of Soil Science and Plant Nutrition* 13 (1): 67–78. http://doi.org/10.4067/S0718-95162013005000000.

Ângelo, Luis, Fernanda Domingues, and Luisa Pereira. 2018. "Association Between Berries Intake and Cardiovascular Diseases Risk Factors: A Systematic Review with Meta-Analysis and Trial Sequential Analysis of Randomised Controlled Trials." *Food and Function* 9 (2): 740–757. https://doi.org/10.1039/c7fo01551h.

Arena, Miriam E., and Nestor Curvetto. 2008. "*Berberis buxifolia* Fruiting: Kinetic Growth Behavior and Evolution of Chemical Properties During the Fruiting Period and Different Growing Seasons." *Scientia Horticulturae* 118 (2): 120–127. https://doi.org/10.1016/j.scienta.2008.05.039.

Avello, Marcia, Rogelio Valdivia, Ruth Sanzana, María A. Mondaca, Sigrid Mennickent, Valeska Aeschlimann, Magalis Bittner, and José Becerra. 2009. "Extractos Antioxidantes y Antimicrobianos de *Aristotelia chilensis* y *Ugni molinae* y sus aplicaciones como preservantes en productos cosméticos." *Boletín Latinoamericano y del Caribe de Plantas Medicinales y Aromáticas* 8 (6): 479–486. ISSN:07177917.

Bastías-Montes, José Miguel, Mónica Consuelo Choque-Chávez, Julio Alarcón-Enos, Roberto Quevedo-León, Ociel Muñoz-Fariña, and Carla Vidal-San-Martín. 2019. "Effect of Spray Drying at 150, 160, and 170°C on the Physical and Chemical Properties of Maqui Extract (*Aristotelia chilensis* (Molina) Stuntz)." *Chilean Journal of Agricultural Research* 79 (1). http://doi.org/10.4067/S0718-58392019000100144.

Becker, Kathrin, Sebastian Schroecksnadel, Johanna Gostner, Cathrine Zacknun, Harald Schennah, Florian Uberall, and Dietmar Fuchs. 2013. "Comparison of in Vitro Tests for Antioxidant and Immunomodulatory Capacities of Compounds." *Phytomedicine: International Journal of Phytotherapy and Phytopharmacology* 21 (2): 164–171. https://doi.org/10.1016/j.phymed.2013.08.008.

Brauch, Johanna E., Maria Buchweitz, Ralf Martin Schweiggert, and Reinhold Carle. 2016. "Detailed Analysis of Fresh and Dried Maqui (*Aristotelia chilensis* (Mol.) Stuntz) Berries and Juice." *Food Chemistry* 190 (1): 308–316. https://doi.org/10.1016/j.foodchem.2015.05.097.

Brito, Anghel, Carlos Areche, Beatriz Sepúlveda, Edward J. Kennelly, and Mario Simirgiotis. 2014. "Anthocyanin Characterization, Total Phenolic Quantification and Antioxidant Features of Some Chilean Edible Berry Extracts." *Molecules* 19 (8): 10936–1055. https://doi.org/10.3390/molecules190810936.

Burgos-Edwards, Alberto, Felipe Jiménez-Aspee, Cristina Theoduloz, Guillermo Schmeda-Hirschmenn. (2020). "Effects of Gastrointestinal Digested Polyphenolic Enriched Extracts of Chilean Currants (*Ribes magellanicum* and *Ribes punctatum*) on in vitro Fecal Microbiota." *Food Chemistry* 258 (3):144–155. https://doi.org/ 10.1016/j.foodchem.2018.03.053.

Bustamante, Luis, Edgar Pastene, Daniel Duran-Sandoval, Carola Vergara, Dietrich von Baer, and Claudia Mardones. 2018. "Pharmacokinetics of Low Molecular Weight Phenolic Compounds in Gerbil Plasma After the Consumption of Calafate Berry (*Berberis microphylla*) Extract." *Food Chemistry* 268: 347–354. https://doi.org/10.1016/j.foodchem.2018.06.048.

Calfio, Camila, and Juan Pablo Huidobro-Toro. 2019. "Potent Vasodilator and Cellular Antioxidant Activity of Endemic Patagonian Calafate Berries (*Berberis microphylla*) with Nutraceutical Potential." *Molecules* 24 (15): 2700–2718. https://doi.org/10.3390/molecules24152700.

Carecho, Rafael, Diego Carregosa, and Claudia Nunes dos Santos. 2020. "Low Molecular Weight (Poly)Phenol Metabolites Across the Blood-Brain Barrier: The

Underexplored Journey." *Brain Plasticity (Prepress)*: 1–22. https://doi.org/10.3233/BPL-200099.

Castañeda-Ovando, Arceli, María de Lourdes Pacheco-Hernández, María Páez-Hernández, José A. Rodríguez, and Carlos A. Galán-Vidal. 2009. "Chemical Studies of Anthocyanins: A Review." *Food Chemistry* 113 (4): 859–871. https://doi.org/10.1016/j.foodchem.2008.09.001.

Cavalcante, Braga, Anna Rafaela, Daniella C. Murador, Leonardo Mendes de Souza Mesquita, and Veridiana Vera de Rosso. 2018. "Bioavailability of Anthocyanins: Gaps in Knowledge, Challenges and Future Research." *Journal of Food Composition and Analysis* 68: 31–40. https://doi.org/10.1016/j.jfca.2017.07.031.

Cermak, Rainer, Alessandra Durazzo, Giuseppe Maiani, Volker Bohm, Dietmar R. Kammerer, Reinhold Carle, Wieslaw Wiczkowski, Mariusz K. Piskula, and Rudolf Galensa. 2009. "The Influence of Postharvest Processing and Storage of Foodstuffs on the Bioavailability of Flavonoids and Phenolic Acids." *Molecular Nutrition and Food Research* 53 (52): 184–193. https://doi.org/10.1002/mnfr.200700444.

Céspedes, Carlos L., Maribel Valdez-Morales, José G. Ávila, Mohammed El-Hafidi, Julio Alarcón, and Octavio Paredes-López. 2010a. "Phytochemical Profile and Thee Antioxidant Capacity of Chilean Wild Black-Berry Fruits *Aristotelia chilensis* (Mol.) Stuntz (Elaeocarpaceae)." *Food Chemistry* 119 (3): 886–895. https://doi.org/10.1016/j.foodchem.2009.07.045.

Céspedes, Carlos L., Julio Alarcón, José G. Ávila, and Antonio Nieto. 2010b. "Anti-inflammatory activity of *Aristotelia chilensis* Mol. (Stuntz) (Elaeocarpaceae)." *Boletín Latinoamericano y del Caribe de Plantas Medicinales y Aromáticas* 9 (2): 127–135.

Céspedes, Carlos L., Natalia Pavón, Mariana Domínguez, Julio Alarcón, Cristián Balbontín, Isao Kubo, Mohammed El-Hafidi, and Jose G. Ávila. 2017. "The Chilean Superfruit Black-Berry *Aristotelia chilensis* (Elaeocarpaceae), Maqui as Mediator in Inflammation-Associated Disorders." *Food and Chemical Toxicology* 108 (Part B): 438–450. https://doi.org/10.1016/j.fct.2016.12.036.

Chamorro, Melina F., Gabriela Reiner, Cristina Theoduloz, Ana Ladio, Guillermo Schmeda-Hirschmann, Sergio Gómez-Alonso, and Felipe Jiménez-Aspee. 2019. "Polyphenol Composition and (bio)activity of *Berberis* Species and Wild Strawberry from the Argentinean Patagonia." *Molecules* 24 (18): 3331. https://doi.org/10.3390/molecules24183331.

Chandra, Preeti, Atul S. Rathore, Kristine L Kay, Jessica L. Everhart, Peter Curtis, Britt Burton-Freeman, Aedin Cassidy, and Colin D. Kay. 2019. "Contribution of Berry Polyphenols to the Human Metabolome." *Molecules* 24 (23): 4220. https://doi.org/10.3390/molecules24234220.

Cheel, José, Cristina Theoduloz, Jaime Rodríguez, Peter Caligari, and Guillermo Schmeda-Hirschmann. 2007. "Free Radical Scavenging Activity and Phenolic Content in Achenes and Thalamus from *Fragaria chiloensis* ssp. *Chiloensis*, *F. vesca* and *F.* x *ananassa* cv. Chandler." *Food Chemistry* 102 (1): 36–44. https://doi.org/10.1016/j.foodchem.2006.04.036.

Clifford, Michael N. 2000. "Anthocyanins-Nature, Occurrence and Dietary Burden." *Journal of the Science of Food and Agriculture* 80 (7): 1063–1072. https://doi.org/10.1002/(sici)1097-0010(20000515)80:7 < 1063::aid-jsfa605 > 3.0.co;2-q.

Corley, Janie, Janet A. Kyle, John M. Starr, Geraldine McNeill, and Ian J. Deary. 2015. "Dietary Factors and Biomarkers of Systemic Inflammation in Older People: The Lothian Birth Cohort 1936." *British Journal of Nutrition* 114 (7): 1088–1098. https://doi.org/10.1017/S000711451500210X.

Costacou, Tina, and Elizabeth J. Mayer-Davis. 2003. "Nutrition and Prevention of Type 2 Diabetes." *Annual Review of Nutrition* 23 (1): 147–170. https://doi.org/10.1146/annurev.nutr.23.011702.073027.

Croft, Kevin D. 2015. "Dietary Polyphenols: Antioxidants or Not?" *Archives of Biochemistry and Biophysics* 595 (Special Issue): 120–124. https://doi.org/10.1016/j.abb.2015.11.014.

Curtis, Peter J., Vera van der Velpen, Lindsey Berends, Amy Jennings, Martin Feelisch, A. Margot Umpleby, Mark Evans, Bernadette O. Fernandez, Mia S. Meiss, Magdalena Minion, John Potter, Anne Marie Minihane, Colin K. Day, Eric B. Rimm, and Aedin Cassidy. 2019. "Blueberries Improve Biomarkers of Cardiometabolic Function in Participants with Metabolic Syndrome-Results from a 6-Month, Double-Blind, Randomized Controlled Trial." *American Journal of Clinical Nutrition* 109 (6): 1535–1545. https://doi.org/10.1093/ajcn/nqy380.

D'Archivio, Massimo, Carmela Filesi, Roberta Di Benedetto, Raffaella Gargiulo, Claudio Giovannini, and Roberta Masella. 2007. "Polyphenols, Dietary Sources and Bioavailability." *Annali dell'Istituto Superiore di Sanità* 43 (4): 348–361. PMID:18209268.

Da Silva Pinto, Marcia, Joao E. de Carvalho, Franco M. Lajolo, María Inés Genovese, and Kalidas Shetty. 2010. "Evaluation of Antiproliferative, Anti—Type 2 Diabetes, and Antihypertension Potentials of Ellagitannins from Strawberries *(Fragaria ananassa* Duch.) Using in Vitro Models." *Journal of Medicinal Food* 13 (5): 1027–1035. https://doi.org/10.1089/jmf.2009.0257.

Dayoub, Ousama, Ramaroson Andriantsitohaina, and Nicolas Clere. 2013. "Pleiotropic Beneficial Effects of Epigallocatechin Gallate, Quercetin and Delphinidin on Cardiovascular Diseases Associated with Endothelial Dysfunction." *Cardiovascular and Hematological Agents in Medicinal Chemistry* 11 (4): 249–264. https://doi.org/10.2174/1871525712666140309233048.

De Ferrars, Rachel M., Charls Czank, Qingzhi Zhang, Nigel P. Botting, Paul A. Kroon, Aniel Cassidy, and Colin D. Kay. 2014. "The Pharmacokinetics of Anthocyanins and Their Metabolites in Humans." *British Journal of Pharmacology* 171 (13): 3268–3282. https://doi.org/10.1111/bph.12676.

Denev, Petko N., Christo G. Kratchanov, Milan Ciz, Antonin Lojek, and Maria G. Kratchanova. 2012. "Bioavailability and Antioxidant Activity of Black Chokeberry (*Aronia melanocarpa*) Polyphenols: In Vitro and in Vivo Evidences and Possible Mechanisms of Action: A Review." *Comprehensive Reviews in Food Science and Food Safety* 11 (5): 471–489. https://doi.org/10.1111/j.1541-4337.2012.00198.x.

Dias, Tania R., Marco G. Alves, Susana Casal, Pedro F. Oliveira, and Branca M. Silva. 2017. "Promising Potential of Dietary (Poly)phenolic Compounds in the Prevention and Treatment of Diabetes Mellitus." *Current Medicinal Chemistry* 24 (4): 334–354. https://doi.org/10.2174/0929867323666160905150419.

Domínguez, Erwin. 2010. "Flora de interés etnobotánico usada por los pueblos originarios: Aónikenk, Selk'nam, Kawésqar, Yagan y Haush en la Patagonia Austral." *Dominguezia* 26 (2): 19–29. www.dominguezia.org/volumen/articulos/2622.pdf.

Eker, Merve Eda, Kjersti Aaby, Irena Budic-Leto, Suzana Rimac Brnčić, Sedef Nehir El, Sibel Karakaya, Sebnem Simsek, Claudine Manach, Wieslaw Wiczkowski, and Sonia de Pascual-Teresa. 2020. "A Review of Factors Affecting Anthocyanin Bioavailability: Possible Implications for the Inter-Individual Variability." *Foods 9* (1): 2. https://doi.org/10.3390/foods9010002.

Escribano-Bailón, María T., Cristina Alcalde-Eon, Orlando Muñoz, Julián Rivas-Gonzalo, and Celestino Santos-Buelga. 2006. "Anthocyanins in Berries of Maqui (*Aristotelia chilensis* (Mol) Stuntz)." *Phytochemical Analysis* 17 (1): 8–14. https://doi.org/10.1002/pca.872.

Figueira, Inés, Goncalo Garcia, Rui Carlos Pimpão, A. Filipa Almeida, María Alexandra Brito, and Catarina Brito. 2017. "Polyphenols Journey Through Blood-Brain Barrier Towards Neuronal Protection." *Scientific Reports* 7: 11456. https://doi.org/10.1038/s41598-017-11512-6.

Forbes-Hernandez, Tamara Y., Massimiliano Gasparrini, Sadia Afrin, Stefano Bompadre, Bruno Mezzetti, Jose L. Quiles, Francesca Giampieri, and Maurizio Battino. 2016. "The Healthy Effects of Strawberry Polyphenols: Which Strategy Behind Antioxidant Capacity?" *Critical Reviews in Food Science and Nutrition* 56 (Suppl 1): S46–S59. https://doi.org/10.1080/10408398.2015.1051919.

Fredes, Carolina, Gad G. Yousef, Paz Robert, Mary H. Grace, Mary Ann Lila, Miguel Gómez, Marlene Gebauer, and Gloria Montenegro. 2014a. "Anthocyanin Profiling of Wild Maqui Berries (*Aristotelia chilensis* [Mol.] Stuntz) from Different Geographical Regions in Chile." *Journal of the Science and Food and Agriculture* 94 (13): 2639–2648. https://doi.org/10.1002/jsfa.6602.

Fredes, Carolina, Gloria Montenegro, Juan P. Zoffoli, Francisca Santander, and Paz Robert. 2014b. "Comparison of the Total Phenolic Content, Total Anthocyanin Content and Antioxidant Activity of Polyphenol-Rich Fruits Grown in Chile." *Ciencia e Investigación Agrar*ia 41 (1): 49–60. https://doi.org/10.4067/S0718-16202014000100005.

Fredes, Carolina, Alejandra Parada, Jaime Salinas, and Paz Robert. 2020. "Phytochemicals and Traditional Use of Two Southernmost Chilean Berry Fruits: Murta (*Ugni molinae* Turcz) and Calafate (*Berberis buxifolia* Lam.)." *Foods* 9 (1): 54. https://doi.org/10.3390/foods9010054.

Freile, Mónica L., Franca Giannini, Graciela Pucci, Andrés Sturniolo, Laura L. Rodero, Vilma Balzareti, and Ricardo D. Enriz. 2003. "Antimicrobial Activity of Aqueous Extracts and of Berberine Isolated from *Berberis heterophylla*." *Fitoterapia* 74 (7–8): 702–705. https://doi.org/10.1016/s0367-326x(03)00156-4.

Genskowsky, Estefanía, Luis A. Puente, José A. Pérez-Álvarez, Juana Fernández-López, Loreto A. Muñoz, and Manuel Viuda-Martos. 2016. "Determination of Polyphenolic Profile, Antioxidant Activity and Antibacterial Properties of Maqui [*Aristotelia chilensis* (Molina) Stuntz] a Chilean Blackberry." *Journal of the Science of Food and Agriculture* 96 (1): 4235–4242. https://doi.org/10.1002/jsfa.7628.

Geto, Zeleke, Meseret Derbeu Molla, Feyissa Challa, Yohannes Belay, and Tigist Getahun. 2020. "Mitochondrial Dynamic Dysfunction as a Main Triggering Factor for Inflammation Associated Chronic Non-Communicable Diseases." *Journal of Inflammation Research* 13: 97–107. https://doi.org/10.2147/JIR.S232009.

Giampieri, Francesca, Sara Tulipani, José M. Alvarez-Suarez, José L. Quiles, Bruno Mezzetti, and Maurizio Battino. 2012. "The Strawberry: Composition, Nutritional

Quality, and Impact on Human Health." *Nutrition* 28 (1): 9–19. https://doi.org/10.1016/j.nut.2011.08.009.

Gironés-Vilaplana, Amadeo, Nieves Baenas, Débora Villaño, Hernán Speisky, Cristina García Viguera, and Diego A. Moreno. 2014. "Evaluation of Latin-American Fruits Rich in Phytochemicals with Biological Effects." *Journal of Functional Foods* 7 (3): 599–608. https://doi.org/10.1016/j.jff.2013.12.025.

Głąbska, Dominika, Barbara Groele, Dominika Guzek, and Krystyna Gutkowska. 2020. "Fruit and Vegetable Intake and Mental Health in Adults: A Systematic Review." *Nutrients* 12 (1): 115. https://doi.org/10.3390/nu12010115.

Godos, Justyna, Walter Currenti, Donato Angelino, Pedro Mena, Sabrina Castellano, Filippo Caraci, Fabio Galvano, Daniele Del Rio, Raffaele Ferri, and Giuseppe Grosso. 2020. "Diet and Mental Health: Review of the Recent Updates on Molecular Mechanisms." *Antioxidants* 9 (4): 346. https://doi.org/10.3390/antiox9040346.

Gould, Kevin S., and Carolyn Lister. 2006. "Flavonoid Functions in Plants." In O. M. Andersen and K. R. Markham (Eds.), *Flavonoids, Chemistry, Biochemistry and Applications*. Boca Raton, FL: CRC Press.

Grussu, Dominic, Derek Stewart, and Gordon J. McDougall. 2011. "Berry Polyphenols Inhibit α-Amylase in Vitro: Identifying Active Components in Rowanberry and Raspberry." *Journal of the Agriculture and Food Chemistry* 59 (6): 2324–2331. https://doi.org/10.1021/jf1045359.

Halliwell, Barry. 2008. "Are Polyphenols Antioxidants or Pro-Oxidants? What Do We Learn from Cell Culture and in Vivo Studies?" *Archives of Biochemistry and Biophysics* 476 (2): 107–112. https://doi.org/10.1016/j.abb.2008.01.028.

Henriques, Joana F., Diana Serra, Teresa C. P. Dinis, and Leonor M. Almeida. 2020. "The Anti-Neuroinflammatory Role of Anthocyanins and Their Metabolites for the Prevention and Treatment of Brain Disorders." *International Journal of Molecular Sciences* 21 (22): 8653. https://doi.org/10.3390/ijms21228653.

Heinonen, Marina, Anne Strunge Meyer, and Edwin N. Frankel. 1998. "Antioxidant Activity of Berry Phenolics on Human Low-Density Lipoprotein and Liposome Oxidation." *Journal of the Agriculture and Food Chemistry* 46 (10): 4107–4112. https://doi.org/10.1021/jf980181c.

Hidalgo, María, M. José Oruna-Concha, Sofía Kolida, Gemma E. Walton, Stamatina Kallithraka, Jeremy P. E. Spencer, and Sonia de Pascual-Teresa. 2012. "Metabolism of Anthocyanins by Human Gut Microflora and Their Influence on Gut Bacterial Growth." *Journal of the Agriculture and Food Chemistry* 60 (15): 3882–3890. https://doi.org/10.1021/jf3002153.

Hoffmann, Adriana. 1991. *Flora Silvestre de Chile, Zona Araucana*. Santiago, Chile: Fundación Claudio Gay, 2da edición, 94.

Hou, De-Xing, Takashi Ose, Shigang Lin, Kazuhiro Harazoro, Izumi Imamura, Mayumi Kubo, Takuhiro Uto, Norihiko Terahara, Makoto Yoshimoto, and Makoto Fujii. 2003. "Anthocyanidins Induce Apoptosis in Human Promyelocytic Leukemia Cells: Structure—Activity Relationship and Mechanisms Involved." *International Journal of Oncol*ogy 23 (3): 705–712. PMID:12888907.

Huang, Haohai, Guangzhao Chen, Dan Liao, Yongkun Zhu, and Xiaoyan Xue. 2016. "Effects of Berries Consumption on Cardiovascular Risk Factors: A Meta-Analysis with Trial Sequential Analysis of Randomized Controlled Trials." *Scientific Reports* 6: 23625. https://doi.org/10.1038/srep23625.

Iside, Concetta, Marika Scafuro, Angela Nebbioso, and Luia Altucci. 2020. "SIRT1 Activation by Natural Phytochemicals: An Overview." *Frontiers in Pharmacology* 11: Art.1225. https://doi.org/10.3389/fphar.2020.01225.

Jayaprakasam, Bolleddula, Shaiju K. Vareed, Karl Olson, and Muraleedharan G. Nair. 2005. "Insulin Secretion by Bioactive Anthocyanins and Anthocyanidins Present in Fruits." *Journal of the Agriculture and Food Chemistry* 53 (1): 28–31. https://doi.org/10.1021/jf049018+.

Joseph, Shama V., Indika Edirisinghe, and Britt M. Burton-Freeman. 2016. "Fruit Polyphenols: A Review of Anti-Inflammatory Effects in Humans." *Critical Reviews in Food Science and Nutrition* 56 (3): 419–444. https://doi.org/10.1080/10408398.2013.767221.

Junqueira-Gonçalves, Maria Paula, Lina Yáñez, Carolina Morales, Muriel Navarro, Rodrigo A. Contreras, and Gustavo E. Zúñiga. 2015. "Isolation and Characterization of Phenolic Compounds and Anthocyanins from Murta (*Ugni molinae* Turcz.) Fruit: Assessment of Antioxidant and Antibacterial Activity." *Molecules* 20 (4): 5698–5713. https://doi.org/10.3390/molecules20045698.

Katsube, Naomi, Keiko Iwashita, Tojiro Tsushida, Koji Yamaki, and Masuko Kobori. 2003. "Induction of Apoptosis in Cancer Cells by Bilberry (*Vaccinium myrtillus*) and the Anthocyanins." *Journal of the Agriculture and Food Chemistry* 51 (1): 68–75. https://doi.org/10.1021/jf025781x.

Kay, Colin D., Gema Pereira-Caro, Iziar A. Ludwig, Alan Crozier, and Michael N. Clifford. 2017. "Anthocyanins and Flavanones Are More Bioavailable Than Previously Perceived: A Review of Recent Evidence." *Annual Review of Food Science and Technology* 8: 155–180. https://doi.org/10.1146/annurev-food-030216-025636.

Krga, Irena, Laurent Emmanuelle Monfoulet, Aleksandra Konic-Ristic, Sylvie Mercier, Maria Glibetic, Christine Morand, and Dragan Milenkovic. 2016. "Anthocyanins and Their Gut Metabolites Reduce the Adhesion of Monocyte to TNF Alpha-Activated Endothelial Cells at Physiologically Relevant Concentrations." *Archives of Biochemistry and Biophysics* 599: 51–59. https://doi.org/10.1016/j.abb.2016.02.006.

Kristo, Alekandra S., Dorothy Klimis-Zacas, and Angelos K. Sikalidis. 2016. "Protective Role of Dietary Berries in Cancer." *Antioxidants* 5 (4): 37. https://doi.org/10.3390/antiox5040037.

Kropat, Christopher, Dolores Mueller, Ute Boettler, Kristin Zimmermann, Elke H. Heiss, Verena M. Dirsch, Dorothee Rogoll, Ralph Melcher, Elke Richling, and Doris Marko. 2013. "Modulation of Nrf2-Dependent Gene Transcription by Bilberry Anthocyanins in Vivo." *Molecular Nutrition and Food Research* 57 (3): 545–550. https://doi.org/10.1002/mnfr.201200504.

Landrum, Leslie R. 1988. "The Myrtle Family (*Myrtaceae*) in Chile." *Proceedings of the California Academy of Sciences (USA)* 45 (12): 277–317.

Li, Jie, ChunhuaYuan, Li Pan, Annécie Benathrehina, Heebyung Chai, William J. Keller, C. Benhamin Naman, and A. Douglas Kinghorn. 2017a. "Bioassay-Guided Isolation of Antioxidant and Cytoprotective Constituents from a Maqui Berry (*Aristotelia chilensis*) Dietary Supplement Ingredient as Markers for Qualitative and Quantitative Analysis." *Journal of the Agriculture and Food Chemistry* 65 (39): 8634–8642. https://doi.org/10.1021/acs.jafc.7b03261.

Li, Daotong, Pengpu Wang, Yinghua Luo, Mengyao Zhao, and Fang Chen. 2017b. "Health Benefits of Anthocyanins and Molecular Mechanisms: Update from Recent Decade." *Critical Reviews in Food Science and Nutrition* 57 (8): 1729–1741. https://doi.org/10.1080/10408398.2015.1030064.

Liu, Wei-Rui, Wen-Lin Qiao, Zi-Zhen Liu, Xiao-Hong Wang, Rui Jiang, Shu-Yi Li, Ren-Bing Shi, and Gai-Mei She. 2013. "*Gaultheria*: Phytochemical and Pharmacological Characteristics." *Molecules* 18: 12071–12108. https://doi.org/10.3390/molecules181012071.

Lohachoompol, Virachnee, George Srzednicki, and John Craske. 2004. "The Change of Total Anthocyanins in Blueberries and Their Antioxidant Effect After Drying and Freezing." *Journal of Biomedicine and Biotechnology* 5: 248–252. https://doi.org/10.1155/S1110724304406123.

Lopes da Silva, Fátima, M. Teresa Escribano-Bailón, José J. Pérez Alonso, Julián Rivas-Gonzalo, and Celestino Santos-Buelga. 2007. "Anthocyanin Pigments in Strawberry." *LWT Food Science and Technol*ogy 40 (2): 374–382. https://doi.org/10.1016/j.lwt.2005.09.018.

López, Jessica, Carlos Vera, Rubén Bustos, and Jennyfer Florez-Mendez. 2020. "Native Berries of Chile: A Comprehensive Review on Nutritional Aspects, Functional Properties, and Potential Health Benefits." *Journal of Food Measurement and Characterization*, October 24. https://doi.org/10.1007/s11694-020-00699-4.

López, María Dolores, Nieves Baenas Jorge Retamal-Salgado Nelson Zapata, and Diego A. Moreno. 2018. "Underutilized Native Biobío Berries: Opportunities for Foods and Trade." *Natural Products Communications* 13 (12): 1681–1684. https://doi.org/10.1177/1934578X1801301226.

Luo, Jiao, Kevin Mills, Saskia le Cessie, Raymond Noordam, and Diana van Heemst. 2020. "Ageing, Age-Related Diseases and Oxidative Stress: What to Do Next?" *Ageing Research Reviews* 57: 100982. https://doi.org/10.1016/j.arr.2019.100982.

Lutz, Mariane. 2019. "Science Behind the Substantiation of Health Claims in Functional Foods: Current Regulations." In K. Shetty and D. Sarkar (Eds.), *Functional Foods and Biotechnology: Sources of Functional Foods and Ingredients*. Boca Raton: CRC Press, Taylor & Francis Group, Chapter 2.

Lutz, Mariane, Marcelo Arancibia, and Jana Stojanova. 2019b. "Using Systematic Reviews in the Scientific Substantiation of Health Properties of Foods and Food Constituents." *Medwave* 19 (6): e7664. https://doi.org/10.5867/medwave.2019.06.7664.

Lutz, Mariane, Eduardo Fuentes, Felipe Ávila, Marcelo Alarcón, and Iván Palomo. 2019a. "Roles of Phenolic Compounds in the Reduction of Risk Factors of Cardiovascular Diseases." *Molecules* 24 (2): 366. https://doi.org/10.3390/molecules24020366.

Mariangel, Emilia, Marjorie Reyes-Diaz, Walter Lobos, Emma Bensch, Heidi Schalchli, and Pamla Ibarra. 2013. "The Antioxidant Properties of Calafate (*Berberis microphylla*) Fruits from Four Different Locations in Southern Chile." *Ciencia e Investigación Agrícola* 40 (1): 161–170. http://doi.org/10.4067/S0718-16202013000100014.

Meyers, Katherine J., Christopher B. Watkins, Marvin P. Pritts, and Rui Hai Liu. 2003. "Antioxidant and Antiproliferative Activities of Strawberries." *Journal of the Agriculture and Food Chemistry* 51 (23): 6887–6892. https://doi.org/10.1021/jf034506n.

Milne, Ginger L., Stephanie C. Sanchez, Erik S. Musiek, and Jason D. Morrow. 2007. "Quantification of F2-Isoprostanes as a Biomarker of Oxidative Stress." *Nature Protocols* 2 (1): 221–226. https://doi.org/10.1038/nprot.2006.375.

Miranda-Rottmann, Soledad, Augusto A. Aspillaga, Druso D. Pérez, Luis Vásquez, Álvaro L. F. Martínez, and Federico Leighton. 2002. "Juice and Phenolic Fractions of the Berry *Aristotelia chilensis* Inhibit LDL Oxidation in Vitro and Protect Human Endothelial Cells Against Oxidative Stress." *Journal of the Agriculture and Food Chemistry* 50 (26): 7542–7547. https://doi.org/10.1021/jf025797n.

Molares, Soledad, and Ana Ladio. 2009. "Ethnobotanical Review of the Mapuche Medicinal Flora: Use Patterns on a Regional Scale." *Journal of Ethnopharmacology* 122 (2): 251–260. https://doi.org/10.1016/j.jep.2009.01.003.

Montenegro, Gloria. 2002. *Chile, nuestra flora útil. Guía de plantas de uso apícola alimentario, medicinal folclórico, artesanal y ornamental.* Santiago, Chile: Universidad Católica de Chile, 267.

Morales-Quintana, Luis, and Patricio Ramos. 2019. "Chilean Strawberry (*Fragaria chiloensis*): An Integrative and Comprehensive Review." *Food Research International* 119 (5): 769–776. https://doi.org/10.1016/j.foodres.2018.10.059.

Morris, Martha Clare, Christy C. Tangney, Yamin Wang, Frank M. Sacks, Lisa L. Barnes, David A. Bennett, and Neelum T. Aggarwal. 2015. "MIND Diet Slows Cognitive Decline with Aging." *Alzheimer's Dementia* 11 (9): 1015–1022. https://doi.org/10.1016/j.jalz.2015.04.011.

Muñoz, Orlando, Marco Montes, and Tatiana Wilkomirsky. 2001. *Plantas medicinales de uso en Chile*. Santiago, Chile: Química y Farmacología. Editorial Universitaria, 330.

Nile, Shivraj, and Se Won Park. 2014. "Edible Berries: Bioactive Components and Their Effect on Human Health." *Nutrition* 30 (2): 134–144. https://doi.org/10.1016/j.nut.2013.04.007.

Nowak, Dariusz, Michal Gośliński, and Elizbieta Wojtowicz. 2016. "Comparative Analysis of the Antioxidant Capacity of Selected Fruit Juices and Nectars: Chokeberry Juice as a Rich Source of Polyphenols." *International Journal of Food Properties* 19 (6): 1317–1324. https://doi.org/10.1080/10942912.2015.1063068.

Oboh, Ganiyu, Ayokunle O. Ademosun, and Opeyemi B. Ogunsuyi. 2016. "Quercetin and Its Role in Chronic Diseases." *Advances in Experimental Medicine and Bio*logy 929: 377–387. https://doi.org/10.1007/978-3-319-41342-6_17.

O'Callaghan, Carl, and Athanassios Vassilopoulos. 2017. "Sirtuins at the Crossroads of Stemness, Aging, and Cancer." *Aging Cell* 16 (6): 1208–1218. https://doi.org/10.1111/acel.12685.

Olsson, Marie E., Karl-Erik Gustavsson, Staffan Andersson, Ake Nilsson, and Rui-Dong Duan. 2004. "Inhibition of Cancer Cell Proliferation in Vitro by Fruit and Berry Extracts and Correlations with Antioxidant Levels." *Journal of the Agriculture and Food Chemistry* 52 (24): 7264–7271. https://doi.org/10.1021/jf030479p.

Oyarzún, Paulina, Pablo Cornejo, Sergio Gómez-Alonso, and Antonieta Ruiz. 2020. "Influence of Profiles and Concentrations of Phenolic Compounds in the Coloration and Antioxidant Properties of *Gaultheria poeppigii* Fruits from Southern Chile." *Plant Foods in Human Nutrition* 75 (4): 532–539. https://doi.org/10.1007/s11130-020-00843-x.

Passamonti, Sabina, Urska Vrhovsekb, and Fulvio Mattivi. 2002. "The Interaction of Anthocyanins with Bilitranslocase." *Biochemical and Biophysical Research Communications* 296 (3): 631–636. https://doi.org/10.1016/S0006-291X(02)00927-0.

Peña Araos, Jorge. 2015. "*Aristotelia chilensis*: A Possible Nutraceutical or Functional Food." *Medicinal Chemistry* 5 (8): 378–382. https://doi.org/10.4172/2161-0444.1000289.

Peña-Neira, Alvaro, Carolina Fredes, María L. Hurtado, Celestino Santos-Buelga, and Javier Pérez-Alonso. 2007. "Low Molecular Weight Phenolic and Anthocyanin Composition of the 'Murta' (*Ugni molinae* Turcz.), a Chilean Native Berry." Berry Health, Berry Symposium, Corvallis, June 11–12.

Philip, Pierre, Patricia Sagaspe, Jacques Taillard, Claire Mandon, Joël Constans, Line Pourtau, Camille Pouchieu, Donato Angelino, Pedro Mena, Daniela Martini, Daniele Del Rio, and David Vauzour. 2019. "Acute Intake of a Grape and Blueberry Polyphenol-Rich Extract Ameliorates Cognitive Performance in Healthy Young Adults During a Sustained Cognitive Effort." *Antioxidants* 8 (12): 650. https://doi.org/10.3390/antiox8120650.

Prior, Ronald L., and Xianli Wu. 2006. "Anthocyanins: Structural Characteristics That Results in Unique Metabolic Patterns and Biological Activities." *Free Radicals Research* 40 (10): 1014–1028. https://doi.org/10.1080/10715760600758522.

Prior, Ronald L., Xianli Wu, and Karen Schaich. 2005. "Standardized Methods for the Determination of Antioxidant Capacity and Phenolics in Foods and Dietary Supplements." *Journal of the Agriculture and Food Chemistry* 53 (10): 4290–4302. https://doi.org/10.1021/jf0502698.

Quispe-Fuentes, Issis, Antonio Vega-Gálvez, Mario Aranda, Jacqueline Poblete, Alexis Pasten, Cristina Bilbao-Sainz, Delilah Wood, Tara McHugh, and Carla Delporte. 2020. "Effects of Drying Processes on Composition, Microstructure and Health Aspects from Maqui Berries." *Journal of Food Science and Technology* 57: 2241–2250. https://doi.org/10.1007/s13197-020-04260-5.

Ramírez, Javier, Ricardo Zambrano, Beatriz Sepúlveda, Edward J. Kennelly, and Mario Simirgiotis. 2015. "Anthocyanins and Antioxidant Capacities of Six Chilean Berries by HPLC-HR-ESI-ToF-MS." *Food Chemistry* 176 (6): 106–114. https://doi.org/10.1016/j.foodchem.2014.12.039.

Rangel-Huerta, Oscar D., Belén Pastor-Villaescusa, Concepción Aguilera, and Ángel Gil. 2015. "A Systematic Review of the Efficacy of Bioactive Compounds in Cardiovascular Disease: Phenolic Compounds." *Nutrients* 7 (7): 5177–5216. https://doi.org/10.3390/nu7075177.

Rendeiro, Catarina, Justin S. Rhodes, and Jeremy P. E. Spencer. 2015. "The Mechanisms of Action of Flavonoids in the Brain: Direct Versus Indirect Effects." *Neurochemistry International* 89 (10): 126–139. https://doi.org/10.1016/j.neuint.2015.08.002.

Retamales, Jorge B., Peter D. S. Caligari, Basilio Carrasco, and Guillermo Saud. 2005. "Current Status of the Chilean Native Strawberry (*Fragaria chiloensis* L. Duch.) and the Research Needs to Convert the Species into a Commercial Crop." *Horticultural Science* 40 (6): 1633–1634. https://doi.org/10.21273/HORTSCI.40.6.1633.

Ricordi, Camillo, Marta Garcia-Contreras, and Sara Farnetti. 2015. "Diet and Inflammation: Possible Effects on Immunity, Chronic Diseases, and Life Span." *Journal of the American College of Nutrition* 34 (Suppl 1): 10–13. https://doi.org/10.1080/07315724.2015.1080101.

Rodríguez, Katia, Kong Ah-Hen, Antonio Vega-Gálvez, Valeria Vásquez, Issis Quispe-Fuentes, Pilar Rojas, and Roberto Lemus-Mondaca. 2016. "Changes in Bioactive Components and Antioxidant Activity of Maqui, *Aristotelia chilensis* (Mol.) Stuntz, Berries During Drying." *LWT (Lebensmittel-Wissenschaft-Technologie)—Food Science and Technology* 65 (1): 537–542. https://doi.org/10.1111/ijfs.12392.

Rodríguez, Roberto R., Oscar Matthei, and Max M. Quezada. 1983. *Flora arbórea de Chile*. Chile: Editorial Universitaria de Concepción, 408.

Romanucci, Valeria, Daniele D'Alonzo, Annalisa Guaragna, Cinzia Di Marino, Sergio Davinelli, Giovanni Scapagnini, Giovanni Di Fabio, and Armando Zarrelli. 2016. "Bioactive Compounds of *Aristotelia chilensis* Stuntz and Their Pharmacological Effects." *Current Pharmaceutical Biotechnology* 17 (6): 513–523. https://doi.org/10.2174/1389201017666160114095246.

Rothwell, Joseph A., Mireia Urpi Sarda, Maria Boto-Ordoñez, Rafael Llorach, Andreu Farran-Codina, Dinesh Kumar Barupal, Vanessa Neveu, Claudine Manach, Cristina Andres-Lacueva, and Augustin Scalbert. 2016. "Systematic Analysis of the Polyphenol Metabolome Using the Phenol Explorer Database." *Molecular Nutrition and Food Research* 60 (1): 203–211. https://doi.org/10.1002/mnfr.201500435.

Rubilar, Mónica, Manuel Pinelo, Mónica Ihl, Erick Scheuermann, Jorge Sineiro, and María J. Nuñez. 2006. "Murta Leaves (*Ugni molinae* Turcz) as a Source of Antioxidant Polyphenols." *Journal of the Agriculture and Food Chemistry* 54 (1): 59–64. https://doi.org/10.1021/jf051571j.

Ruiz, Antonieta, Isidro Hermosín-Gutiérrez, Claudia Mardones, Carola Vergara, Erika Herlitz, Mario Vega, Carolin Dorau, Peter Winterhalter, and Dietrich von Baer. 2010. "Polyphenols and Antioxidant Activity of Calafate (*Berberis microphylla*) Fruits and Other Native Berries from Southern Chile." *Journal of the Agriculture and Food Chemistry* 58 (10): 6081–6089. https://doi.org/10.1021/jf100173x.

Ruiz, Antonieta, Isidro Hermosín-Gutiérrez, Carola Vergara, Dietrich von Baer, Moisés Zapata, Antonieta Hitschfeld, Luis Obando, and Claudia Mardones. 2013. "Anthocyanin Profiles in South Patagonian Wild Berries by HPLC-DAD-ESI-MS/MS." *Food Research International* 51 (2): 706–713. http://doi.org/10.1016/j.foodres.2013.01.043.

Ruiz, Antonieta, Moisés Zapata, and Constanza Sabando. 2014. "Flavonols, Alkaloids, and Antioxidant Capacity of Edible Wild Berberis Species from Patagonia." *Journal of the Agriculture and Food Chemistry* 62 (51): 12407–12417. https://doi.org/10.1021/jf502929z.

Salehi, Bahare, Jafad Sharifi-Rad, Francesca Cappellini, Javad Sharifi-Rad, Francesca Cappellini, Željko Reiner, Debora Zorzan, Muhammad Imran, Bilge Sener, Mehtap Kilic, Mohamed El-Shazly, Nouran M. Fahmy, Eman Al-Sayed, Miquel Martorell, Chiara Tonelli, Katia Petroni, Anca Oana Docea, Daniela Calina, and Alfred Maroyi. 2020. "The Therapeutic Potential of Anthocyanins: Current Approaches Based on Their Molecular Mechanism of Action." *Frontiers in Pharmacology* 11: Art. 1300. https://doi.org/10.3389/fphar.2020.01300.

Sánchez Gutiérrez, Raúl, and Carla Guzmán Pincheira. 2020. "Description of the Antioxidant Capacity of Calafate Berries (*Berberis microphylla*) Collected in Southern Chile." *Food Science and Technology*, September 28. http://doi.org/10.1590/fst.25820.

Scalbert, Augustin, Claudine Manach, Christine Morand, Christian Rémésy, and Liliana Jiménez. 2005. "Dietary Polyphenols and the Prevention of Diseases." *Critical Reviews in Food Sciences and Nutrition* 45 (4): 287–306. https://doi.org/10.1080/1040869059096.

Schaffer, Sebastian, and Barry Halliwell. 2012. "Do Polyphenols Enter the Brain and Does It Matter? Some Theoretical and Practical Considerations." *Genes and Nutrition* 7 (2): 99–109. https://doi.org/10.1007/s12263-011-0255-5.

Scheuermann, Erick, Ivette Seguel, Adolfo Montenegro, Rubén Bustos, Emilio Hormazábal, and Andrés Quiroz. 2008. "Evolution of Aroma Compounds of Murtilla Fruits (*Ugni molinae* Turcz) during Storage." *Journal of the Science of Food and Agriculture* 88: 485–492. https://doi.org/10.1002/jsfa.3111.

Schmeda-Hirschmann, Guillermo, Felipe Jiménez-Aspee, Cristina Theoduloz, Ana Ladio. 2019. "Patagonian Berries as Native Food and Medicine." *Journal of Ethnopharmacology* 241: 111979. https://doi.org/10.1016/j.jep.2019.111979.

Schreckinger, María E., Jennifer Lotton, Mary Ann Lila, and Elvira González de Mejía. 2010. "Berries from South America: A Comprehensive Review on Chemistry, Health Potential, and Commercialization." *Journal of Medicinal Food* 13 (2): 233–246. https://doi.org/10.1089/jmf.2009.0233.

Seeram, Navindra P. 2008. "Berry Fruits: Compositional Elements, Biochemical Activities, and the Impact of Their Intake on Human Health, Performance, and Disease." *Journal of the Agriculture and Food Chemistry* 56 (3): 627–629. https://doi.org/10.1021/jf071988k.

Seeram Navindra P., Lynn S. Adams, Yanjun Zhang, Rupo Lee, Daniel Sand, Henry S. Scheuller, and David Heber. 2006. "Blackberry, Black Raspberry, Blueberry, Cranberry, Red Raspberry, and Strawberry Extracts Inhibit Growth and Stimulate Apoptosis of Human Cancer Cells in Vitro." *Journal of the Agriculture and Food Chemistry* 54 (25): 9329–9339. https://org.do/10.1021/jf061750g. PMID: 17147415.

Seguel, Ivette, Enrique Peñaloza, Nelba Gaete, Adolfo Montenegro, and Andrea Torres. 2000. "Colecta y caracterización molecular de germoplasma de murta (*Ugni molinae* Turcz.) en Chile." *Agro Sur* 28 (2): 32–41. https://doi.org/10.4206/agrosur.2000.v28n2-05.

Shahidi, Fereidoon, and Priyatharini Ambigaipalan. 2015. "Phenolics and Polyphenolics in Foods, Beverages and Spices: Antioxidant Activity and Health Effects—a Review." *Journal of Functional Foods* 18 (Part B): 820–897. https://doi.org/10.1016/j.jff.2015.06.018.

Sharif, Niloufar, Sara Khoshnoudi-Nia, and Seid Mahdi Jafari. 2020. "Nano/Microencapsulation of Anthocyanins; a Systematic Review and Meta-Analysis." *Food Research International* 132 (6): 109077. https://doi.org/10.1016/j.foodres.2020.109077.

Sies, Helmut. 2007. "Total Antioxidant Capacity: Appraisal of a Concept." *Journal of Nutrition* 137 (6): 1493–1495. https://doi.org/10.1093/jn/137.6.1493.

Simirgiotis, Mario J., Cristina Theoduloz, Peter D. Caligari, and Guillermo Schmeda-Hirschmann. 2009. "Comparison of Phenolic Composition and Antioxidant Properties of Two Native Chilean and One Domestic Strawberry Genotypes." *Food Chemistry* 113 (2): 377–385. https://doi.org/10.1016/j.foodchem.2008.07.043.

Singleton, Vernon L., and Joseph A. Rossi, Jr. 1965. "Colorimetry of Total Phenolics with Phosphomolybdic-Phosphotungstic Acid Reagents." *American Journal of Enology and Viticulture* 16 (3): 144–158.

Speisky, Hernán, Camilo López-Alarcón, Maritza Gómez, Jocelyn Fuentes, and Cristina Sandoval-Acuña. 2012. "First Web-Based Database on Total Phenolics and Oxygen Radical Absorbance Capacity (ORAC) of Fruits Produced and Consumed Within the South Andes Region of South America." *Journal of the Agriculture and Food Chemistry* 60 (36): 8851–8859. https://doi.org/10.1021/jf205167k.

Stevenson, David E., and Roger D. Hurst. 2007. "Polyphenolic Phytochemicals—Just Antioxidants or Much More?" *Cellular and Molecular Life Sciences* 64 (11): 2900–2916. https://doi.org/10.1007/s00018-007-7237-1.

Stewart, Bernard W., and Christopher P. Wild. 2014. *World Cancer Report: World Health Organization*. Geneva, Switzerland and Lyon, France: IARC Publication.

Tanaka, Junji, Takashi Kadekaru, Kenjirou Ogawa, Shoketsu Hitoe, Hiroshi Shimoda, and Hideaki Hara. 2013. "Maqui Berry (*Aristotelia chilensis*) and the Constituent Delphinidin Glycoside Inhibit Photoreceptor Cell Death Induced by Visible Light." *Food Chemistry* 139 (1–4): 129–137. https://doi.org/10.1016/j.foodchem.2013.01.036.

Teillier, Sebastián, and Felipe Escobar. 2013. "Revision of the Genus of *Gaultheria* L. (Ericaceae) in Chile." *Gayana Botanica* 70: 136–153. https://doi.org/10.4067/S0717-66432013000100014.

Tena, Noelia, Julia Martín, and Agustín G. Asuero. 2020. "State of the Art of Anthocyanins: Antioxidant Activity, Sources, Bioavailability, and Therapeutic Effect in Human Health." *Antioxidants* 9 (5): 451. https://doi.org/10.3390/antiox9050451.

Ullah, Rahat, Mehtab Khan, Shahid Ali Shah, Kamran Saeed, and Myeong O. Kim. 2019. "Natural Antioxidant Anthocyanins—A Hidden Therapeutic Candidate in Metabolic Disorders with Major Focus in Neurodegeneration." *Nutrients* 11 (6): 1195. https://doi.org/10.3390/nu11061195.

Ulloa-Inostroza, Elizabeth M., Eric G. Ulloa-Inostroza, Miren Alberdi, Daniela Peña-Sanhueza, Jorge González-Villagra, Laura Jaakola, and Marjorie Reyes-Díaz. 2017. "Native Chilean Fruits and the Effects of Their Functional Compounds on Human Health." In V. Waisundara and N. Shiomi (Eds.), *Superfood and Functional Food: An Overview of Their Processing and Utilization*. London: Intech Open. https://doi.org/10.5772/67067.

Urban, Otto. 1934. *Botánica de las plantas endémicas de Chile*. Chile: Soc Imp Lit Concepción, 291.

USDA. 2011. *United States Department of Agriculture: Database for the Flavonoid Content of Selected Foods (Release 3.0)*. Beltsville, MD: Agricultural Research Service Press. www.ars.usda.gov/ARSUserFiles/80400525/Data/Flav/Flav_R03-1.pdf.

Varadharaj, Saradhadevi, Owen J. Kelly, Rami N. Khayat, Purnima S. Kumar, Nasser Ahmed, and Jay L. Zweier. 2017. "Role of Dietary Antioxidants in the Preservation of Vascular Function and the Modulation of Health and Disease." *Frontiers in Cardiovascular Medicine* 4 (11): 64. https://doi.org/10.3389/fcvm.2017.00064.

Vauzour, David, Ana Rodriguez-Mateos, Giulia Corona, María J. Oruna-Concha, and Jeremy P. Spencer. 2010. "Polyphenols and Human Health: Prevention of Disease and Mechanisms of Action." *Nutrients* 2 (11): 1106–1131. https://doi.org/10.3390/nu2111106.

Vega-Gálvez, Antonio, Ángela Rodríguez, and Karina Stucken. 2021. "Antioxidant, Functional Properties and health-Promoting Potential of Native South American Berries: A Review." *Journal of the Science of Food and Agriculture.* 101 (2): 364–378. http://doi.org/10.1002/jsfa.10621.

Villagra, Evelyn, Carola Campos-Hernández, Pablo Cáceres, Gustavo Cabrera, Yamilé Bernardo, Ariel Arencibia, Basilio Carrasco, Peter D. S. Caligari, José Pico, and Rolando García-Gonzales. 2014. "Morphometric and Phytochemical Characterization of Chaura Fruits (*Gaultheria pumila*): A Native Chilean Berry with Commercial Potential." *Biological Research* 47: Art. 26. https://doi.org/10.1186/0717-6287-47-26.

Vita, Joseph A. 2005. "Polyphenols and Cardiovascular Disease: Effects on Endothelial and Platelet Function." *American Journal of Clinical Nutrition* 81 (1): 292S–297S. https://doi.org/10.1093/ajcn/81.1.292S.

Wang, Shiow Y., Kim S. Lewers, Linda Bowman, and Min Ding. 2007. "Antioxidant Activities and Anticancer Cell Proliferation Properties of Wild Strawberries." *Journal of the American Society for Horticultural Science* 132 (5): 647–658. https://doi.org/10.21273/JASHS.132.5.647.

Wang, Shiow Y., and Hai S. Lin. 2000. "Antioxidant Activity in Fruits and Leaves of Blackberry, Raspberry, and Strawberry Varies with Cultivar and Developmental Stage." *Journal of the Agriculture and Food Chemistry* 48 (2): 140–146. https://doi.org/10.1021/jf9908345.

Watson, Ronald R., and Frank Schönlau. 2015. "Nutraceutical and Antioxidant Effects of a Delphinidin-Rich Maqui Berry Extract Delphinol®: A Review." *Minerva Cardioangiology* 63 (2 Suppl 1): 1–12. PMID:25892567.

Whyte, Adrian R., Nancy Cheng, Laurie T., Butler, Daniel J. Lamport, and Claire M. Williams. 2019. "Flavonoid-Rich Mixed Berries Maintain and Improve Cognitive Function Over a 6h Period in Young Healthy Adults." *Nutrients* 11 (11): 685. https://doi.org/10.3390/nu11112685.

Yang, Meng, Sung I. Koo, Wong O. Song, and Chun Ock K. 2010. "Food Matrix Affecting Anthocyanin Bioavailability: Review." *Current Medicinal Chemistry* 18 (2): 291–300. https://doi.org/10.2174/092986711794088380.

Yoshida, Kumi, Mihoko Mori, and Tadao Kondo. 2009. "Blue Color Development by Anthocyanins: From Chemical Structure to Cell Physiology." *Natural Products Rep*orts 26 (7): 884–915. https://doi.org/10.1039/b800165k.

Index

www.ingramcontent.com/pod-product-compliance
Lightning Source LLC
Chambersburg PA
CBHW060752220326
41598CB00022B/2407

* 9 7 8 1 0 3 2 1 8 7 7 7 8 *